Lecture Notes in Physics

Springer-Verlag Berlin Heidelberg GmbH

The Editorial Policy for Proceedings

The series Lecture Notes in Physics reports new developments in physical research and teaching – quickly, informally, and at a high level. The proceedings to be considered for publication in this series should be limited to only a few areas of research, and these should be closely related to each other. The contributions should be of a high standard and should avoid lengthy redraftings of papers already published or about to be published elsewhere. As a whole, the proceedings should aim for a balanced presentation of the theme of the conference including a description of the techniques used and enough motivation for a broad readership. It should not be assumed that the published proceedings must reflect the conference in its entirety. (A listing or abstracts of papers presented at the meeting but not included in the proceedings could be added as an appendix.)

When applying for publication in the series Lecture Notes in Physics the volume's editor(s) should submit sufficient material to enable the series editors and their referees to make a fairly accurate evaluation (e.g. a complete list of speakers and titles of papers to be presented and abstracts). If, based on this information, the proceedings are (tentatively) accepted, the volume's editor(s), whose name(s) will appear on the title pages, should select the papers suitable for publication and have them refereed (as for a journal) when appropriate. As a rule discussions will not be accepted. The series editors and Springer-Verlag will normally not interfere with the detailed editing except in fairly obvious cases or on technical matters.

Final acceptance is expressed by the series editor in charge, in consultation with Springer-Verlag only after receiving the complete manuscript. It might help to send a copy of the authors' manuscripts in advance to the editor in charge to discuss possible revisions with him. As a general rule, the series editor will confirm his tentative acceptance if the final manuscript corresponds to the original concept discussed, if the quality of the contribution meets the requirements of the series, and if the final size of the manuscript does not greatly exceed the number of pages originally agreed upon. The manuscript should be forwarded to Springer-Verlag shortly after the meeting. In cases of extreme delay (more than six months after the conference) the series editors will check once more the timeliness of the papers. Therefore, the volume's editor(s) should establish strict deadlines, or collect the articles during the conference and have them revised on the spot. If a delay is unavoidable, one should encourage the authors to update their contributions if appropriate. The editors of proceedings are strongly advised to inform contributors about these points at an early stage.

The final manuscript should contain a table of contents and an informative introduction accessible also to readers not particularly familiar with the topic of the conference. The contributions should be in English. The volume's editor(s) should check the contributions for the correct use of language. At Springer-Verlag only the prefaces will be checked by a copy-editor for language and style. Grave linguistic or technical shortcomings may lead to the rejection of contributions by the series editors. A conference report should not exceed a total of 500 pages. Keeping the size within this bound should be achieved by a stricter selection of articles and not by imposing an upper limit to the length of the individual papers. Editors receive jointly 30 complimentary copies of their book. They are entitled to purchase further copies of their book at a reduced rate. As a rule no reprints of individual contributions can be supplied. No royalty is paid on Lecture Notes in Physics volumes. Commitment to publish is made by letter of interest rather than by signing a formal contract. Springer-Verlag secures the copyright for each volume.

The Production Process

The books are hardbound, and the publisher will select quality paper appropriate to the needs of the author(s). Publication time is about ten weeks. More than twenty years of experience guarantee authors the best possible service. To reach the goal of rapid publication at a low price the technique of photographic reproduction from a camera-ready manuscript was chosen. This process shifts the main responsibility for the technical quality considerably from the publisher to the authors. We therefore urge all authors and editors of proceedings to observe very carefully the essentials for the preparation of camera-ready manuscripts, which we will supply on request. This applies especially to the quality of figures and halftones submitted for publication. In addition, it might be useful to look at some of the volumes already published. As a special service, we offer free of charge LaTeX and TeX macro packages to format the text according to Springer-Verlag's quality requirements. We strongly recommend that you make use of this offer, since the result will be a book of considerably improved technical quality. To avoid mistakes and time-consuming correspondence during the production period the conference editors should request special instructions from the publisher well before the beginning of the conference. Manuscripts not meeting the technical standard of the series will have to be returned for improvement.

For further information please contact Springer-Verlag, Physics Editorial Department II, Tiergartenstrasse 17, D-69121 Heidelberg, Germany

G. M. Simnett   C. E. Alissandrakis
L. Vlahos (Eds.)

Solar and Heliospheric Plasma Physics

Proceedings of the
8th European Meeting on Solar Physics
Held at Halkidiki, Greece,
13–18 May 1996

 Springer

Editors

George M. Simnett
Astrophysics and Space Research Group
School of Physics and Space Research
The University of Birmingham
Egbaston, Birmingham B15 2TT, United Kingdom

Constantine E. Alissandrakis
Department of Physics
University of Ioannina
GR-45110 Ioannina, Greece

Loukas Vlahos
Department of Physics
Aristotle University of Thessaloniki
GR-54006 Thessaloniki, Greece

Cataloging-in-Publication Data applied for.
Die Deutsche Bibliothek - CIP-Einheitsaufnahme

Solar and heliospheric plasma physics : proceedings of the 8th
European Meeting on Solar Physics, held at Halkidiki, near
Thessaloniki, Greece, 13 - 18 May 1996 / G. M. Simnett ... (ed.).
(Lecture notes in physics ; Vol. 489)
ISBN 978-3-662-14144-1   ISBN 978-3-540-69124-2 (eBook)
DOI 10.1007/978-3-540-69124-2
ISSN 0075-8450
ISBN 978-3-662-14144-1

© Springer-Verlag Berlin Heidelberg 1997
Originally published by Springer-Verlag Berlin Heidelberg New York in 1997
Softcover reprint of the hardcover 1st edition 1997

Typesetting: Camera-ready by the authors/editors
Cover design: *design & production* GmbH, Heidelberg
SPIN: 10550811   55/3144-543210 - Printed on acid-free paper

Preface

Every three years the Solar Physics Section of the European Physical Society holds a major scientific meeting. The eighth one in the series took place at Halkidiki, near Thessaloniki, Greece during May 13-18, 1996. Previous meetings in this series were in Florence, Toulouse, Oxford, Noordwijkerhout, Debrecen, Titisee, and Catania. The meeting was given the title "Solar and Heliospheric Plasma Physics". This was carefully chosen to bring together theoretical ideas on the plasma physics of both hot and dense plasmas in the solar atmosphere, similar physics but applied to the tenuous and cooler plasmas found in the heliosphere, together with results from Ulysses and SOHO.

The scientific organising committee consisted of G.M. Simnett (UK; Chairman), C.E. Alissandrakis (Greece), G. Belvedere (Italy), P. Heinzel (Czech Republic), M. Huber (ESA/ESTEC), E. Landi Degl'Innocenti (Italy), J. Staude (Germany), N. Vilmer (France), M. Vazquez (Spain), L. Vlahos (Greece), J.-C. Vial (France), V. Zaitsev (Russia) and Yu.D. Zhugzhda (Russia). The local organising committee was led by C.E. Alissandrakis and L. Vlahos. The meeting attracted 170 participants from 24 countries. Generous support was given by the following: the Directorate General for Science Research and Development of the European Commission; the Greek Ministry of Education; the Greek General Secretariat of Science and Technology; the Greek National Committee for Astronomy; the Aristotle University of Thessaloniki; and the University of Ioannina.

The meeting was organised into seven different sessions, each with two invited reviews plus several oral contributions selected by the scientific organising committee from the papers submitted for the meeting. The remainder of the submitted papers were presented as posters. Wherever possible the two invited reviews for a given session were chosen to give the theoretical and observational positions of the primary session topic. It has become the custom over recent years for the Solar Physics Section to award a prize for the best poster given by a young (<35 years) participant. This time the Kluwer prize (two personally selected books from the Kluwer publication list) was presented to Panagiota Petkaki.

The meeting was held in the splendid facilities of the Athos Palace Hotel in Halkidiki. There was a lively atmosphere, due in part to the presence of many young scientists. The proximity of the beach and the beautiful spring weather did not affect the attendance at the scientific sessions. The participants were very grateful to Loukas Vlahos for selecting such an attractive location. There was a boat trip around the Athos peninsula on the Wednesday, which was an interesting diversion, especially for the majority who had not experienced it before. The excellent Conference dinner on Thursday evening was accompanied with traditional Greek dancers, and they were followed by not-so-traditional EPS dancers! The success of the meeting owed much to the tireless efforts of the local organising committee. Special thanks go to Despoina Papadiki and the students of the Astrophysics Group in the Physics Department of the University of Thessaloniki.

February 1997

George Simnett
Costas Alissandrakis
Loukas Vlahos

Contents

The Structure and Dynamics of the Sun from Helioseismology, and the Neutrino Problem

Lucio Paternò

Istituto di Astronomia, Università di Catania, Città Universitaria, 95125 Catania, Italy

Abstract. The inversion of thousands of p-modes, identified with great accuracy, has clearly demonstrated the substantial correctness of the standard solar models. The use of the new equation of state and opacity tables, together with the element diffusion towards the central regions of the Sun, has allowed the construction of solar models which are consistent with helioseismology within 1%. This result, combined with the neutrino fluxes measured by the four exsisting experiments and some nuclear reaction model-independent constraints, seems to exclude the possibility that the neutrino problem is astrophysical in origin, but it should rely upon non standard neutrino properties, as the MSW effect. The internal dynamics of the Sun are deduced from the inversion of p-mode rotational splittings which has revealed unexpected features in the distribution of the angular velocity in the convection zone and in the underlying radiative envelope from the base of the convection zone to a depth of $0.3\,R_\odot$. This poses severe constraints on the location of the solar cycle generation mechanism and the evolution of the solar angular momentum. Only recently, full operation of the helioseismic networks IRIS and BISON has allowed us to measure the lowest degree p-mode splittings and to infer the rotation of the central regions of the Sun below $0.3\,R_\odot$, though with contradictory results. While the p-mode analysis gives a reliable picture of the real Sun, except in the innermost regions where only average properties can be deduced, the detailed knowledge of the structural and dynamical properties of the solar core will only be possible by means of the detection and identification of g-modes, if they are really excited in the Sun, from the forthcoming data of the GOLF instrument on the SOHO satellite.

1 Introduction

The majority of solar physicists believe that the Sun is a rotating star with a core in which the energy is generated by the conversion of hydrogen into helium. This core is surrounded by a radiative envelope, a convection zone, which produces surface phenomena such as the granulation, supergranulation and large scale motions, and a radiating photosphere from which most of the energy produced in the interior escapes. Owing to the presence of magnetic fields a multitude of complex phenomena are observable in the photospheric, chromospheric and coronal layers, including sunspots, active regions, spicules, coronal arches and holes, and the solar wind. It is unimaginable that all the details of the phenomenology we observe in the external layers can be described by a global model of the Sun. The various phenomena are produced, on large scales, by the interaction of convection with rotation and magnetic fields and, on smaller scales, by magnetized plasma instabilities Therefore for theorists who study the interior of the Sun all the photospheric, chromospheric and coronal phenomena are summarized by two boundary conditions taken at the Sun's surface, namely its luminosity and effective temperature.

As we shall see, the interior of the Sun is described by the so-called standard solar model, which is a commonly accepted recipe for the construction of a solar model in agreement with the oberved luminosity, $L_\odot$, and effective temperature, T_e. Some 15-20 years ago these were the only constraints to which any solar model had to be subjected. Nowadays, with the advent of the helioseismology, which probes the solar structure, and the full operation of four neutrino experiments, which probe the nuclear processes occurring in the solar core, additional constraints are imposed by the observed oscillation spectrum and neutrino fluxes.

The continuous improvements in the description of physical processes which determine the energy generation, the interaction of radiation with matter, the state of the matter, the gravitational settling of the elements and the convective transport in the stellar interiors have produced solar models ever more consistent with helioseismic data, namely models whose internal stratification approaches closely, within 1%, the internal stratification as deduced from the inversion of the oscillation frequencies. The very significant result that the standard solar model is essentially equivalent to the so-called seismic Sun has been reached in recent times (Basu et al. 1996). However, while on the one hand the improvements in the description of physical processes occurring in the Sun's interior were leading the stratification of the solar models to coincide ever more with the stratification of the *true* or seismic Sun, on the other hand the same improvements were leading to models which predicted neutrino fluxes ever more diverging from those measured by the experiments.

At the present we face two different Suns, the seismic and neutrino Sun, with a little chance of overlapping them, unless the present ideas about stellar evolution, which are confirmed by the helioseismology, or the standard model of weak interactions are dramatically changed. However, since there is a consensus that the Sun is unique, a solution must exist which brings the seismic Sun in line with the neutrino Sun or viceversa. The first solution is known as the astrophysical solution to the neutrino problem, and it relies upon *ad hoc* changes to physical processes such as the opacity or nuclear cross-sections, while the second solution is known as the particle physics solution to the neutrino problem, and it relies upon non-standard properties of the neutrino.

The study of the internal dynamics of the Sun is accomplished by inverting the observed p-mode rotational splittings, thus permitting us to infer the angular velocity behaviour, as a function of radius and latitude, from the surface to the region near the Sun's core. While the p-mode analysis has established with a great degree of accuracy this behaviour down to $0.3\,R_\odot$, the observations concerning the lowest degree rotational splittings, obtained by the IRIS and BISON helioseismic networks, give conflicting results about the core rotation, which cannot be reconciled by the measurement error bars. While one set of observations is consistent with a solar core rotating slightly faster than the surface, the other one requires the core to rotate slower than the surface. This problem should be clarified by the oscillation observations from the SOHO satellite, especially if the GOLF instrument will be able to reveal the presence of g-modes in the Sun.

However, while I think that the new data from SOHO are extremely important for a better understanding of the core dynamics, I do not believe that they can alter in a dramatic way the stratification of the Sun as deduced from p-mode ground-based observations, which describe accurately the structure of the core.

2   The Standard Solar Model

The standard solar model is an astrophysical recipe, derived from stellar structure and evolution theory in main sequence, which is based on some sound physical assumptions and two parameters of ignorance: the initial helium abundance and the α parameter of the mixing-length theory of convection. These assumptions are the following: i) no rotation; ii) no magnetic field; iii) no mass loss; iv) energy production by the conversion of hydrogen into helium; v) convective transport described by the mixing-length theory. For a gaseous sphere, the first two assumptions mean that the Sun is spherically symmetric, namely that all quantities depend only on radial distance. In the case of a slowly rotating, weakly magnetic star with a negligible mass loss like the Sun, these assumptions are well justified. Additional hypotheses are continuity of mass, and mechanical and thermal equilibrium. Thus the basic equations which describe the the solar structure are the following:

$$\frac{\mathrm{d}M_r}{\mathrm{d}r} = 4\pi\varrho r^2 \tag{1}$$

$$\frac{\mathrm{d}P}{\mathrm{d}r} = -\varrho\frac{GM_r}{r^2} \tag{2}$$

$$\frac{\mathrm{d}L_r}{\mathrm{d}r} = 4\pi\varrho r^2\left(\varepsilon + T\frac{\partial S}{\partial t}\right) \tag{3}$$

$$\frac{\mathrm{d}T}{\mathrm{d}r} = -\frac{T}{H_P}[(1-\zeta)\nabla_R + \zeta\nabla_A] \tag{4}$$

where the variables are indicated by the usual symbols, $H_P = -\mathrm{d}r/\mathrm{d}\ln P$ is the pressure scale height, $0 \leq \zeta(\alpha,\kappa,\ldots) \leq 1$ is the efficiency of convection with α a free parameter of the mixing-length theory and κ the radiative opacity, and ∇_R and ∇_A are the radiative and adiabatic gradients, respectively. Equation (4) describes the energy transport which ranges from purely radiative ($\zeta = 0$) to fully convective or adiabatic ($\zeta = 1$), the border between purely radiative and convective transport depending on the Schwarzschild criterion for convective instability. The four independent boundary conditions are taken two at the centre ($L_r = 0$, $M_r = 0$) and two at the surface ($T = T_e$, $P = g_\odot/\kappa$), where $g_\odot$ is the surface gravity of the Sun. To be integrated Eqs.(1-4) need three additional relationships which describe the equation of state, the opacity and the rate of energy production:

$$P = P(\varrho, T, C_i) \tag{5}$$

$$\kappa = \kappa(\varrho, T, C_i) \tag{6}$$

$$\varepsilon = \varepsilon(\varrho, T, C_i) \tag{7}$$

where C_i stands for the chemical composition in relative mass abundances of hydrogen (X), helium (Y), and all the elements heavier than helium (Z), such that $X + Y + Z = 1$. In order to construct a solar model with the present age of the Sun, $\tau_\odot = 4.57 \pm 0.02 \times 10^9$ y (Bahcall et al. 1995), one needs to know the distribution of chemical composition from the Sun's centre to the surface. This is obviously impossible as the radiation we observe carries out information only from the outermost layers. Three supplementary hypotheses are therefore necessary to construct a model of the present Sun: i) the Sun was initially (before entering the Zero Age Main Sequence) chemically homogeneous with the same chemical composition we observe today at the surface; ii) the changes in the chemical composition are only due to the conversion of hydrogen into helium; iii) the age of the Sun is about $\tau_\odot$ as deduced from meteoritic datation of the Earth's age.

Therefore one can describe the Sun's evolution through equilibrium states, each of them differing from the previous one from a small quantity δt, until the present age of the Sun is reached:

$$x_i(t + \delta t) = x_i(t) + \frac{\partial x_i(t)}{\partial t}\delta t \tag{8}$$

Eq.(8) is in reality a system of n equations for the x_i's ($i = 1, 2, 3, \ldots n$), the abundances of the nuclear species which determine the burning reactions. Since the derivative which appears in the RHS member of Eq.(8) is known from the nuclear reaction rates which are in turn functions of ϱ, T and C_i, it is possible to determine the new chemical composition and integrate Eqs.(1-4) for a new equilibrium state. Two problems are in order now which are connected with the two ignorance parameters already mentioned. From spectroscopy, we know the ratio Z/X from meteoritic abundances [$Z/X = 0.0267$ (Anders and Grevesse (1989), and more recently $Z/X = 0.0245$ (Grevesse and Noels 1993)], but the determinations of Y from both helium lines and solar wind are very uncertain so that the initial helium abundance Y_i is used as a free parameter (1st ignorance parameter). The second problem is connected with the mixing-length parameter α which is the ratio of the mixing-length to the pressure scale height. In the fluid dynamics laboratory experiments α is of the order of unity, in agreement with the behaviour of convection

Table 1. Neutrinos produced in nuclear reactions, their type of spectrum and energy.

number	reaction	spectrum	energy (MeV)
1	$p(p,e^+\nu_e)D$	continuous	$0-0.42$
2	$p(p,e^-,\nu_e)D$	line	1.44
3	$^3He(p,e^+\nu_e)^4He$	continuous	$0-19.79$
4	$^7Be(e^-,\nu_e)^7Li$	lines	$0.38\,(10\%),0.86\,(90\%)$
5	$^8B(e^+\nu_e)^8Be^*$	continuous	$0-17.98$
6	$^{13}N(e^+\nu_e)^{13}C$	continuous	$0-1.20$
7	$^{15}O(e^+\nu_e)^{15}N$	continuous	$0-1.73$
8	$^{17}F(e^+\nu_e)^{17}O$	continuous	$0-1.73$

in viscous fluids. However the behaviour of convection in the inviscid stellar interiors may be largely different so that α remains undetermined and plays the role of the second ignorance parameter. In spite of their uncertainty these parameters are useful to construct solar models which are able to fit the two main constraints of reproducing the present solar luminosity and effective temperature after an evolution time equal to the present age of the Sun. In fact Y_i controls $L_\odot$ almost independently of α and this controls T_e almost independently of $L_\odot$. Therefore, by varying Y_i and α, it is possible to position correctly the present Sun in the H-R diagram.

The standard solar model predicts the emitted electron neutrino spectrum on the basis of the nuclear reactions of the p-p chain and CNO cycle (Bahcall 1989). From the spectrum it is possible to deduce the neutrino flux at the Earth (neutrinos cm^{-2} s^{-1}) one expects to measure with the detectors of the experiments based on the neutrino capture from ^{37}Cl (Homestake) and ^{71}Ga (GALLEX, SAGE) atoms, or on neutrino scattering by water electrons (Kamiokande), which measures only the neutrinos emitted by ^8B β decay. In the first two cases the detected flux is measured in SNU (Solar Neutrino Unit $\equiv 10^{-36}$ captures per second per detector-atom), in the third case in terms of the ratio of measured to predicted flux. The nuclear reactions which emit neutrinos occurring in the solar core, the type of neutrino spectrum and energy are listed in Table 1. In recent times, besides the improvements in describing the physical processes which lead to the formulation of the equation of state [Mihalas et al. 1988, Mihalas et al. 1990 (MHD), Rogers et al. 1996 (EOS/OPAL95)], opacity [Rogers and Iglesias 1992 (OPAL92), Iglesias and Rogers 1996 (OPAL95)] and nuclear reaction rates (Kamionkowski and Bahcall 1994a,b), the standard solar models have been improved by taking

Table 2. Main characteristics of solar models by Bahcall and Pinsonneault (1992) and Bahcall et al. (1995) without (no diff.) and with (diff.) elemental diffusion for different initial Z/X. The quantities are expressed in c.g.s. units, temperatures in Kelvin and the number in parenthesis indicates the exponent. Explanation for individual quantities is given in the text.

quantity	no diff. (92)	He diff. (92)	no diff. (95)	He+metal diff. (95)
$(Z/X)_i$	0.0267	0.0267	0.0245	0.0245
Y_i	0.2716	0.2727	0.2679	0.2775
X_s	0.7094	0.7328	0.7146	0.7351
Y_s	0.2716	0.2466	0.2679	0.2469
X_c	0.3541	0.3428	0.3613	0.3333
Y_c	0.6270	0.6376	0.6213	0.6456
T_c	1.559 (7)	1.569 (7)	1.556 (7)	1.584 (7)
ϱ_c	1.518 (2)	1.545 (2)	1.524 (2)	1.562 (2)
P_c	2.345 (17)	2.373 (17)	2.373 (17)	2.402 (17)
α	1.27	1.36	1.96	2.09
x_{cz}	0.721	0.707	0.726	0.712
T_{cz}	2.237 (6)	2.261 (6)	2.085 (6)	2.204 (6)

into account the gravitational settling of the elements heavier than hydrogen during solar evolution. Bahcall and Pinsonneault (1992) have considered only helium diffusion, while Bahcall et al. (1995) have included also the diffusion of elements heavier than helium. As we shall see, the inclusion of elemental diffusion produces better agreement with helioseismic data. The main characteristics of the above mentioned models are listed in Table 2, where the suffixes i, s and c indicate initial, surface and central quantities respectively, α is the mixing-length parameter, and x_{cz} and T_{cz} are the fractional radius and temperature of the convection zone base respectively. The predicted neutrino flux at the Earth in neutrinos cm^{-2} s^{-1} (ϕ_ν) for each reaction listed in Table 1 and the total flux $\sum \phi_\nu$ are given in Table 3 together with the total neutrino flux in SNU expected to be measured with the chlorine [$\Phi_\nu(Cl)$] and gallium [$\Phi_\nu(Ga)$] experiments. In reality, one does not need to know the details of the reactions for predicting the total neutrino flux at the Earth, because it only depends on the solar luminosity and on the hypothesis that the energy is produced by the conversion of hydrogen into helium through the following global reaction:

$$4\,^1\mathrm{H} \rightarrow\,^4\mathrm{He} + 2e^+ + 2\nu_e + \gamma \tag{9}$$

Table 3. Neutrino fluxes predicted by Bahcall and Pinsonneault (1992) and Bahcall et al. (1995) for solar models without (no diff.) and with (diff.) elemental diffusion (see Table 2). Explanation for individual quantities is given in the text.

neutrino flux	no diff. (92)	He diff. (92)	no diff. (95)	He+metal diff. (95)
$\phi_\nu(pp)$	6.04 (10)	6.00 (10)	6.01 (10)	5.91 (10)
$\phi_\nu(pep)$	1.43 (8)	1.43 (8)	1.44 (8)	1.40 (8)
$\phi_\nu(hep)$	1.25 (3)	1.23 (3)	1.27 (3)	1.21 (3)
$\phi_\nu(^7Be)$	4.61 (9)	4.89 (9)	4.53 (9)	5.15 (9)
$\phi_\nu(^8B)$	5.06 (6)	5.69 (6)	4.85 (6)	6.62 (6)
$\phi_\nu(^{13}N)$	4.35 (8)	4.92 (8)	4.07 (8)	6.18 (8)
$\phi_\nu(^{15}O)$	3.72 (8)	4.26 (8)	3.45 (8)	5.45 (8)
$\phi_\nu(^{17}F)$	4.67 (6)	5.39 (6)	4.02 (6)	6.48 (6)
$\sum \phi_\nu$	6.60 (10)	6.60 (10)	6.55 (10)	6.56 (10)
$\Phi_\nu(Cl)$	7.2	8.0	7.0	9.3
$\Phi_\nu(Ga)$	127	132	126	137

which produces the total luminosity of the Sun and delivers an energy $Q = 26.73\,\mathrm{MeV}$, a small part of which ($Q_\nu = 0.6\,\mathrm{MeV}$) is carried away by neutrinos. Since for each reaction two neutrinos are emitted, the total number of neutrinos arriving on the Earth's surface is:

$$n_\nu = \frac{2L_\odot}{Q - Q_\nu} \frac{1}{4\pi d^2} \simeq 6.59 \times 10^{10} \text{ neutrinos cm}^{-2}\,\mathrm{s}^{-1} \qquad (10)$$

where d is the mean Sun-Earth distance.

3   Helioseismology

Even though motions with period around 5 minutes were detected on the Sun's surface in the early 60's (Leighton 1961, Evans and Michard 1962), helioseismology was born 15 years later, when Deubner (1975) was able to produce the $k_h - \omega$ diagram of 5 minute solar oscillations, so giving the incontrovertible proof that the observed oscillations were standing acoustic waves trapped in resonant cavities below the Sun's surface. The normal mode nature of solar oscillations was predicted by Ulrich (1970) and independently by Leibacher and Stein (1971). These authors interpreted the oscillatory character of the observed motions in terms of acoustic waves reflected on the one hand at the surface, owing to the rapid variation of density, and on the other hand in regions below the surface, located at different depths depending on the wave characteristics, owing to the temperature increase towards the Sun's centre.

In agreement with these predictions, the $k_h - \omega$ diagram obtained by Deubner in 1975 showed the existence of well separated regions in which the oscillation power was concentrated, demonstrating that, for a given horizontal wavenumber k_h, not every angular frequency ω was admittable, but only a discrete number of frequencies. These correspond to the eigenfrequencies of the standing waves trapped in resonant cavities whose bottom wall is located at different depths in the internal regions of the Sun. Since each wave, characterized by a different k_h and ω, propagates in a different region, reflecting the physical properties of the crossed medium, it is then possible to deduce the internal stratification of the Sun from its vibrational properties.

Owing to rotation, the frequencies of the modes are split into multiplets, whose separation in frequency depends on the rotational strength. Since each mode splitting reflects the rotation at different depths and latitudes, it is possible to determine the behaviour of the Sun's angular velocity from the surface to the innermost layers.

Since Deubner's discovery up to now, more than 4000 acoustic modes, with related splittings, have been identified with an average accuracy in frequency of the order of 10^{-4}. This has given us a depth of knowledge of the internal structure and dynamics of the Sun, not imaginable 20 years ago.

The stably-stratified character of the radiative envelope below the convection zone would permit the existence of another class of oscillation modes, the internal gravity modes, not yet detected by means of ground based observations because their amplitude decays rapidly in the convection zone and it is very small at the surface. If the forthcoming space observations can identify these modes, then this would constitute a fundamental point for the detailed knowledge of the structure and dynamics of the Sun's core.

3.1 Physical Background

Small amplitude oscillations of a spherically symmetric star can be separated into normal modes whose temporal dependence is described by a sine-wave and spatial dependence by radial functions proportional to the spherical harmonics of degree ℓ and azimuthal order m. The displacement eigenvector $\boldsymbol{\xi}$ of such a wave is described, in spherical polar coordinates r, θ, ϕ, as:

$$\boldsymbol{\xi} = \Re \left\{ \left[\xi_{r,n\ell}(r), \xi_{h,n\ell}(r)\frac{\partial}{\partial\theta}, \xi_{h,n\ell}(r)\frac{\partial}{\sin\theta\partial\phi} \right] Y_\ell^m(\theta,\phi) \exp(i\omega_{n\ell}t) \right\} \quad (11)$$

where $\Re$ stands for the real part, $\xi_{r,n\ell}(r)$ and $\xi_{h,n\ell}(r)$ are the radial and horizontal eigenfunctions respectively, and $\omega_{n\ell}$ is the mode eigenfrequency. The normal modes are therefore identified by three quantum numbers, the radial order n $(n = 0, 1, 2, \ldots)$, which is the number of nodal surfaces in the radial direction, the spherical degree ℓ $(\ell = 0, 1, 2, \ldots)$, which is the number of total nodal lines on the surface, and the azimuthal order m $(m = -\ell, -\ell+1, \ldots, -1, 0, 1, \ldots, \ell-1, \ell)$ which is the number of nodal lines along the longitude on the surface. On spherical surfaces, the number of nodal lines ℓ is connected with the horizontal wavenumber k_h and wavelength λ_h through the following relationship:

$$ k_h = \frac{2\pi}{\lambda_h} = \frac{\sqrt{\ell(\ell+1)}}{r} $$

and, for each k_h, the radial wavenumber k_r must satisfy the stationary condition on the number of radial nodes along the path of the wave:

$$ \int_{r_1}^{r_2} k_r(\omega^2, r) dr = \pi(n + \phi) $$

where r_1 and r_2 are the turning points at the bottom and the top of the acoustic cavity, and ϕ is a phase constant which depends on the effective polytropic index in the upper layers.

In the case of spherical symmetry $(m = 0)$ there are no preferential directions in a sphere; therefore the modes show $(2\ell + 1)$-fold degeneracy in m and both the eigenfrequencies and eigenfunctions are independent of m. Rotation introduces a preferred direction and removes the degeneracy causing multiple frequency splittings, whose separation depends on the rotational strength. In the case of slow rotation, which applies to the Sun, perturbation theory indicates that the separation depends linearly on the angular velocity of rotation Ω as:

$$ \Delta\omega_{n\ell m} = \omega_{n\ell m} - \omega_{n\ell} = \pm |m| \, \Omega C_{n\ell} \tag{12} $$

where $\omega_{n\ell}$ and $C_{n\ell}$ are the eigenfrequency and structure parameter of a nonrotating spherically symmetric star, respectively.

The modes with $\ell = 0$ are called radial, and in such a case the star expands and contracts conserving its spherical shape, while those with $\ell > 0$ are called nonradial. In such a case the star shows some zones which expand adjacent to zones which simultaneously contract, giving rise to different configurations on the surface. The presence of millions of acoustic modes $(\simeq 10^7)$ simultaneously excited, gives rise to the 5 min acoustic spectrum extending from 1.8 mHz to 5.5 mHz, and renders the identification of the individual modes in ℓ and m difficult by data reduction analysis. The identification in n is possible only by a theoretical prediction, since the radial nodes below the surface are not observed.

Since the amplitude of oscillations in the Sun is small when compared with its radius $(\delta r/R < 10^{-4})$, the basic equations describing the dynamics of the oscillations (mass, momentum, energy conservation and Poisson equations) can be linearized. Moreover, since in the Sun the wave propagation time is much shorter than the thermal diffusion time, the energy equation can be replaced by the adiabatic relation. On taking into account the above simplifications and the fact that the horizontal spatial structure of the modes is described by spherical harmonics, the linear adiabatic oscillations of spherically symmetric stars are described by a system of four ordinary differential equations for the eigenfunctions $\xi_r(r)$, $\xi_h(r)$, and perturbed gravitational potential $\Phi'(r)$, whose coefficients are quantities derived from the unperturbed spherical model (Christensen-Dalsgaard and Berthomieu 1991). For a given ℓ, the system can be solved, with the appropriate boundary conditions, as an eigenvalue problem for the frequencies, which, in this case, are real. These eigenvalues obey a variational principle (Chandrasekhar 1964):

$$\omega^2 \xi = \mathcal{L}(\xi) \tag{13}$$

where $\mathcal{L}(\xi)$ is a linear operator which contains the mode eigenfunctions and quantities of the unperturbed structure with their derivatives.

In a star in which the main forces acting in the body are pressure and gravity, two kinds of oscillations can be maintained by these restoring forces, namely pressure (or acoustic) and internal gravity waves, which form the classes of p-modes and g-modes respectively. The former modes constitute the 5 minute acoustic mode spectrum and their excitation mechanism the action of the turbulent convection (Goldreich and Keeley 1977, Goldreich et al. 1994). The latter modes are still not detected with certainty and their excitation mechanism is not well-established (Dziembowski et al. 1985).

The solution of the equations indicates that the frequency of all the modes increases with ℓ, while, for a given ℓ, it increases with n for p-modes, producing a series of *overtones*, and decreases with n for g-modes, producing a series of *undertones*. A complete $\ell - \nu$ diagram, where $\nu = \omega/2\pi$ is the cyclical frequency of the modes, can be found in Christensen-Dalsgaard and Berthomieu (1991).

The dispersion relation, which links the wavenumber $k = \sqrt{k_r^2 + k_h^2}$ to the angular frequency ω, describes the acoustic and gravity wave propagation in a medium, whose vibrational properties are defined by the three characteristic frequencies S_ℓ (Lamb), N (Brunt-Väisälä), and ω_c (cut-off) (see Unno et al. 1989 for definitions):

$$c_s^2 k_r^2 + S_\ell^2 \left(1 - \frac{N^2}{\omega^2}\right) = \omega^2 - \omega_c^2 \tag{14}$$

where $c_s = \sqrt{\Gamma_1 P/\varrho}$ is the adiabatic sound speed, with Γ_1 the adiabatic exponent. The condition for propagation requires that the radial wavenumber must be real ($k_r^2 > 0$). Equation (14) indicates that this condition is fulfilled in two domains of ω, $\omega^2 > \omega_+^2$ and $\omega^2 < \omega_-^2$, where ω_+^2 and ω_-^2 are the greater and smaller roots of Eq.(14), respectively. Otherwise the waves are evanescent and do not show an oscillatory character in space, but decay exponentially. For $\omega^2 > \omega_+^2 \simeq \max[S_\ell, \omega_c]$ the waves propagate as p-modes, while for $\omega^2 < \omega_-^2 \simeq \min[S_\ell, N]$ they propagate as g-modes. In the Sun, for $\ell > 2$, the high frequency propagation region of p-modes is well separated from the low frequency region of g-modes, which are trapped below the Brunt-Väisälä frequency. For $\ell \leq 2$, modes can behave as p-modes in some regions and as g-modes in some other regions. A propagation diagram can be found in Christensen-Dalsgaard and Berthomieu (1991).

The eigenfunctions of p-modes show their largest amplitudes and radial variations in the surface layers, while those of g-modes are strongest near the centre. The lower the degree of the mode, the deeper are the regions of the Sun which p-modes penetrate, the highest degree modes being trapped only in the surface regions. The radial p-modes can penetrate to the Sun's centre if their frequency is larger than ω_c there ($\simeq 1\,\mathrm{mHz}$), but their eigenfunctions have small amplitudes and vary slowly, so reflecting only average properties of the medium in the central regions. Examples of the behaviour of some p and g-mode radial eigenfunctions are given in Christensen-Dalsgaard and Berthomieu (1991). Therefore p-modes are more suitable for investigating the detailed structure and dynamics in the convective and immediately underlying layers and the average structure and dynamics in the central regions, while g-modes are more suitable for studying the detailed structure and dynamics near the centre.

3.2   Forward and Inverse Approach

The structure and dynamics of the Sun's interior can be inferred from helioseismology through two different but complementary strategies. The first is the direct approach to the problem, namely to start from a given model and vary the relevant physics (equation of state, opacity, nuclear reaction rates) in order to reproduce the observed oscillation spectrum, or, in the case of rotation, to find rotational behaviours which are consistent with the observed splittings. The second is the inverse approach, namely to start from the observed mode frequencies and rotational splittings for deducing the internal structure and rotation of the Sun by means of data inversion.

One of the most widely used methods of data inversion is based on the variational principle, expressed in Eq.(13), which makes use of a reference standard solar model. It is then possible to formulate an integral equation which describes the changes in eigenfrequencies caused by virtual changes in the equilibrium structure of the reference model:

$$\delta\omega^2 \int_0^{M_\odot} \boldsymbol{\xi}^* \cdot \boldsymbol{\xi} \, dM_r = \int_0^{M_\odot} \boldsymbol{\xi}^* \cdot \delta\mathcal{L}(\boldsymbol{\xi}) \, dM_r \qquad (15)$$

In Eq.(15), the known functions are the differences between the observed eigenfrequencies, ω, and those of the reference model, ω_0, while the unknowns are the differences between the physical quantities of the real Sun, y, and those of the reference model, y_0. If we assume that the reference model is close to the real Sun, and this assumption seems to be plausible on the basis of the small differences between the observed and calculated eigenfrequencies, then the differences $\Delta\omega_{n\ell}^2 = \omega_{n\ell}^2 - \omega_{n\ell,0}^2$ are linear functions of the differences in physical quantities $\Delta y_i = y_i - y_{0i}$ as:

$$\frac{\Delta\omega_{n\ell}^2}{\omega_{n\ell,0}^2} = \int_0^{R_\odot} \left\{ \sum_i K_{n\ell,i}[Q_0(r), \xi_{r,n\ell}(r), \xi_{h,n\ell}(r), r] \frac{\Delta y_i(r)}{y_{0i}(r)} \right\} dr \qquad (16)$$

$K_{n\ell,i}$, the kernels of the integral, are sensitivity functions which depend on the quantities of the reference model, Q_0, and its eigenfunctions. Equation (16) shows that the eigenfrequencies are weighted averages of the equilibrium quantities with weighting functions which depend on the eigenfunctions. Each eigenfrequency, $\omega_{n,\ell}$, then reflects the structure of the Sun in the layers where its eigenfunction has the major weight. A set of Eqs.(16) for all the observed modes can be regarded as an integral equation which has to be solved in order to determine $\Delta y_i(r)$, namely the corrections which have to be imposed to the reference model in order to obtain the observed eigenfrequencies.

An equation analogous to Eq.(16) can be derived, in the case of slow rotation ($\Omega \ll \omega_{n\ell}$), for determining the behaviour of the angular velocity $\Omega(r, \theta)$ in the Sun's interior, where the differences $\Delta\omega_{n\ell}^2$ are replaced by the measured splittings $\Delta\omega_{n\ell m}$ and the quantities Δy by $\Omega(r, \theta)$ (Hansen et al. 1977):

$$\Delta\omega_{n\ell m} = \int_0^{R_\odot} \int_0^\pi K_{n\ell m, \Omega} \times$$
$$[m, Q_0(r), \xi_{r,n\ell}(r), \xi_{h,n\ell}(r), r, \cos\theta]\Omega(r, \theta)dr \sin\theta d\theta \qquad (17)$$

There are different methods for solving Eqs.(16) and (17) (Craig and Brown 1986). However the solution is far from being unique both because the observed frequencies constitute a finite set of data and the observations are not free of errors. Procedures are then used, which give a good compromise between amplitude and localization errors, as in the case of the Backus-Gilbert (1968) method widely used in helioseismology. Other inversion methods which do not need a reference model exist, such as the asymptotic inversion first applied to helioseismology by Gough (1984). A complete description of the inversion methods applied to helioseismology can be found in Christensen-Dalsgaard et al. (1990) and Gough and Thompson (1991).

3.3   Inference on Structure

Historically the first approach to the problem was the direct one. One of the main successes of this approach was the spectacular overall agreement of the theoretical $k_h - \omega$ diagram, produced by a standard solar model, with the observed one (Libbrecht 1988a). The differences between the observed and calculated eigenfrequencies ranges from a few μHz, for low-ℓ, low frequency modes (Paternò 1996), to several tens of μHz, for high-ℓ, high frequency modes (Bachmann et al. 1995).

The direct approach has been used for evaluating from which regions in the Sun's interior the discrepancies between theory and observation arise, denoting that the physics applied there is inadequate for describing the relevant phenomena. Dziembowski et al. (1988) have shown that the major inconsistencies derive from the treatment of the surface layers, in particular the superadiabatic ones, casting serious doubts on the validity of the mixing-length theory as a tool for describing the surface turbulent convection in stars.

Paternò et al. (1993) have applied a new treatment of stellar convection, proposed by Canuto and Mazzitelli (1992), to a standard solar model and compared its eigenfrequencies with those calculated for a model which made use of the same microphysics, but whose convective transport was modelled by means of the classical mixing-length theory. The results showed overall better agreement with observations of the frequencies calculated for the solar model which used the new approach for modelling stellar convection, compared to those which used the mixing-length theory.

The direct approach has also been used to test the first standard solar models constructed using helium diffusion in the interior (Bahcall and Pinsonneault 1992). Guenther et al. (1993) have shown that diffusion produces systematic shifts of the order $1 - 5\,\mu$Hz in the p-mode eigenfrequencies, and, in particular, it increases the eigenfrequencies at low ℓs while decreases them at high ℓs, going in the direction of a better agreement with observations.

The first significant results concerning the application of the inversion technique to the Sun were obtained by Christensen-Dalsgaard et al. (1985) who determined the sound speed from a set of about 2800 observed eigenfrequencies, and by Gough (1986) who determined the location of the base of the convection zone at $r_c \simeq 0.7\,R_\odot$. Since then, several efforts have been done for inverting data in order to test the correctness of the standard solar models in view of the improvements achieved in the description of the relevant physics (Gough and Kosovichev 1988; Dziembowski et al. 1990, 1992, 1994, 1995; Voronstov and Shibahashi 1991). The results indicate the substantial correctness of the standard solar models which are improved by the use of the recent equation of state (MHD, EOS/OPAL95) and opacity (OPAL92, OPAL95) tables.

A significant improvement in the standard solar models has been achieved with the inclusion of elemental diffusion. Basu et al. (1996) have compared two models only differing between them by the inclusion or not of the diffusion of helium and other elements heavier than helium. The result of their comparison is shown in Fig. 1, where the square of the sound speed relative difference between the Sun and models, both constructed with MHD equation of state and OPAL95 opacities, is given as a function of the Sun's fractional radius. It appears that the deviations of the real (seismic) Sun

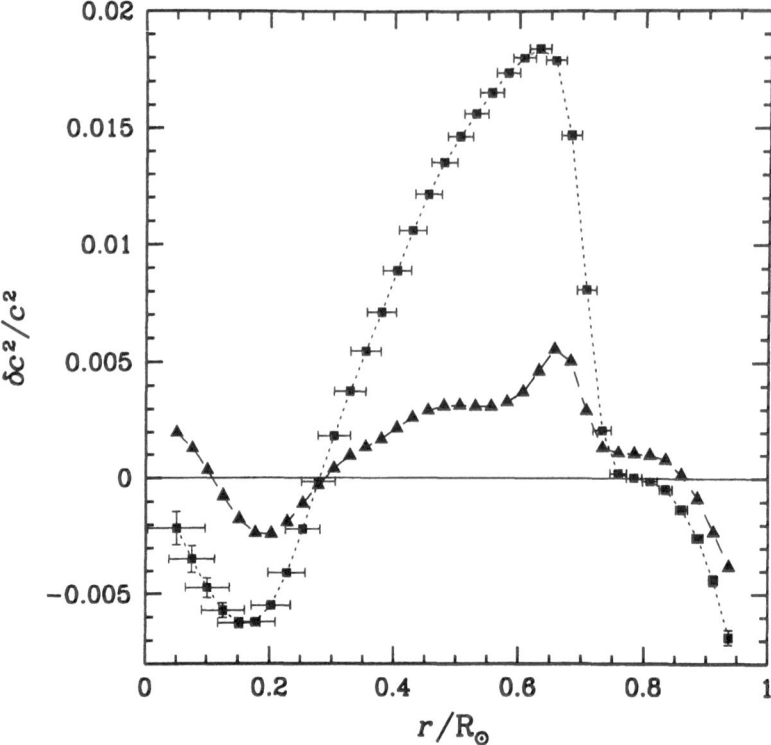

Fig. 1. The square of the sound speed relative difference between the Sun and two standard solar models without diffusion (squares) and with He and metal diffusion (triangles) as obtained from helioseismic data inversion by Basu et al. (1996).

from the model are much larger in the case of no diffusion (squares) than in the case of diffusion (triangles). The same authors have also compared the MHD model with two EOS/OPAL95 models with diffusion and found that the latter models give a better agreement with the real Sun.

There is a clear indication that the central sound speed is underestimated in the models, while the central mass concentration is overestimated (Basu et al. 1995). Since in the solar core the gas is fully ionized so that $\Gamma_1 \simeq 5/3$ and the perfect gas law is a good approximation, a larger sound speed means that either the central temperature or the hydrogen abundance of the real Sun are larger than those of the models, or both. The models with diffusion, which are closer to the real Sun than the classical ones without diffusion, have indeed larger central temperatures (see Table 2) and neutrino fluxes (see Table 3). In any case small discrepancies of the order of 1% still remain between the stratification of the real Sun and that predicted by the models.

Dziembowski et al. (1994) pointed out that, with the quoted observational errors, it is possible to achieve a precision of about 10^{-3} in the sound speed determination through most of the Sun's interior, except in the innermost layers ($r \leq 0.05 R_\odot$) where the precision is about 10^{-2}. At this level of accuracy, the seismic model of the Sun poses severe constraints to the accuracy with which the microphysics should describe the relevant phenomena.

3.4   Inference on Dynamics

As discussed in Sect. 3.1, rotation removes the mode degeneracy and the eigenfrequencies split into multiplets whose separation in frequency depends on the rotational speed. The frequency splittings can be expressed as expansions, generally truncated at the 5th term, in terms of Legendre polynomial in m/L, where m is the azimuthal order of the mode and $L = \sqrt{\ell(\ell+1)}$ (Libbrecht and Morrow 1991):

$$\Delta\omega_{n\ell m} = L \sum_i a_i \mathcal{P}_i(m/L) \qquad (18)$$

The odd coefficients of the expansion (18) are a measure of the rotational field, while the even coefficients are a measure of the Sun's asphericity. The 1st coefficient of the expansion reflects the radial dependence of the angular velocity, while the 3rd and 5th its latitudinal dependence, and the sum $a_1 + a_3 + a_5$ the equatorial rate of rotation. For each p-mode multiplet (ℓ and n fixed) the rotational splitting $\Delta\omega_{n\ell m}$ reflects the rotation rate of the regions in which the mode propagates; therefore the splitting of each multiplet gives a weighted average of the rotation rate over the region probed by the mode. It has been found (Libbrecht 1988b) that a_1, a_3 and $a_1 + a_3 + a_5$ are almost constant, with their surface values, for the modes which are trapped in the convection zone, and decrease with ℓ for the modes which penetrate below.

The forward (Morrow 1988) as well as the inverse (Christensen-Dalsgaard and Schou 1988, Libbrecht 1988b) analyses gave a surprising result which contradicted all the numerical models of internal dynamics already constructed, showing that isorotation surfaces in the convection zone were mainly radial, with the surface differential rotation persisting at all the depths. Below the convection zone to a depth of about $0.4 R_\odot$ the rotation appeared to be rigid at a rate corresponding to the surface rotation observed at a latitude of about $30°$. There is a shallow transition layer located at the base of the convection zone in which the radial and latitudinal rotational stresses are mostly concentrated. More recent work (Tomczyk et al. 1995), based on the inversion of 673 p-mode multiplets with $1 \leq \ell \leq 80$, explored the inner rotation deep down to $0.2 R_\odot$ with an accuracy from about 2% to 10% for the innermost layers. The authors, in agreement with previous measurements, found a weak radial dependence in the convection zone, no latitudinal dependence and no important radial variations of the angular velocity below $0.65 R_\odot$.

The rotation of the Sun's core can only be inferred by means of very accurate measurements of the lowest degree p-mode splittings, namely $\ell = 1, 2, 3$. However, since the eigenfunctions of these modes in the central regions have small amplitude and slow radial variation, so reflecting the average dynamical properties of a broad region, it is impossible to deduce the rotation of the layers below $0.15 - 0.2\,R_\odot$ by data inversion. Recently, the helioseismic networks BISON (Elsworth et al. 1995) and IRIS (Lazrek et al. 1996) have provided the p-mode splittings for $\ell = 1, 2, 3$ with an accuracy better than 5%. Surprisingly the BISON splittings are some 8% smaller than the IRIS splittings, and this discrepancy cannot be reconciled by the error bars of the measurements. The results of the inversion of the two sets of data (Di Mauro 1996), using also the BBSO data set for $5 \leq \ell \leq 60$ (Libbrecht 1989), are shown in Fig. 2 where the behaviour of the equatorial angular velocity in the solar interior is given. The significant discrepancy between the low ℓ splitting values is particularly amplified by inversion in the core. IRIS splittings indicate that the core rotation is only slightly faster than the surface equatorial rotation, while BISON splittings are consistent with a solar core rotating at a rate even slower than the surface polar rate. However, a scenario whereby the core rotates more slowly than the surface cannot easily be reconciled with our understanding of the rotational history of the Sun, namely that the angular momentum is transferred outwards from the more rapidly rotating central regions. This occurs both because the Sun's outer regions are torqued down by the magnetized solar wind and the solar core contracts relative to the rest of the Sun during the evolution. Paternò et al. (1996) and Di Mauro et al. (1996) also found that the recent determinations of the solar oblateness outside the Earth's atmosphere by the Solar Disk Sextant (Lydon and Sofia 1996) lead to a value of the quadrupole moment consistent with the rotational law deduced from the IRIS data.

4   The Neutrino Problem

The neutrino problem came out in the 70s when the first results indicated that the neutrino flux measured by the Homestake detector was about one third of the predicted one (Bahcall and Davis 1976). In the late 80s the Kamiokande results indicated that ^{8}B neutrinos were about the 50% of those predicted (Hirata et al. 1990) and, in 1992, the first GALLEX experiment results (GALLEX coll. 1992) confirmed the neutrino deficit. Many attempts were made to find an astrophysical solution to the problem by conjecturing non-standard solar models, none of them sufficiently convincing (Bahcall 1989, Castellani et al. 1994). On the other hand, the theoretical estimate of the neutrino flux expected to be measured in the different experiments has been remarkably stable over the past 25 years, ranging, for the chlorine experiment, from 7 SNU to the present value of 9.3 SNU, and, for the gallium experiment, from 125 SNU to the present value of 137 SNU, and it appears

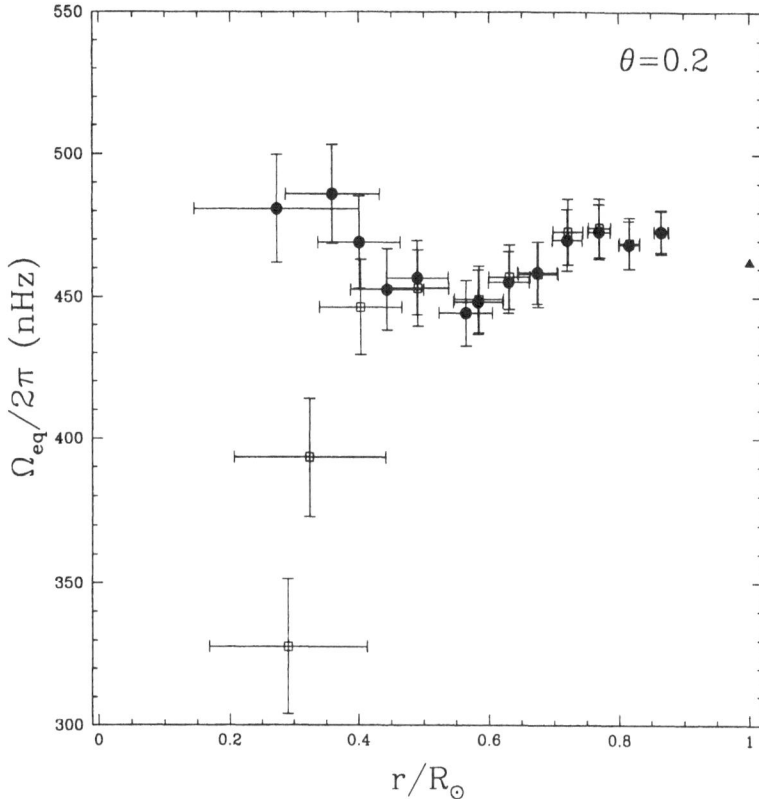

Fig. 2. Behaviour of the solar equatorial angular velocity as deduced from inversion of the IRIS plus BBSO (filled circles) and BISON plus BBSO (open squares) splitting data. The triangle indicates the surface equatorial rate of rotation and θ is the inversion regularization parameter (courtesy of M. P. Di Mauro).

to be particularly robust (Bahcall 1996).

The new standard solar models constructed using element diffusion appear to be more consistent with helioseismic data than the previous standard models without diffusion (see Sect. 3.3), even if refinements are necessary to reduce the still remaining small discrepancies. However there is a clear trend in the attempt to match the solar models to the seismic Sun, which tends to increase the divergence of the predicted from the measured neutrino flux. Therefore the neutrino problem, which existed at times when the solar models predicted a lower neutrino flux, is now more striking.

Table 4. Results of the four neutrino experiments in operation. The measuring reaction (col. 2) and the sensitivity to the solar reactions (col. 3), as numbered in Table 1, are given together with the measured (col. 4) and predicted (col. 5) neutrino fluxes by one of the best helioseismic models (Bahcall et al. 1995, 5th col. of Tables 2 and 3). The fluxes are given in SNU except for Kamiokande for which the flux is given in $10^6 \, cm^{-2} \, s^{-1}$ The errors are at 1σ, and statistical and systematic errors are added in quadrature. Theoretical uncertainties derive from the uncertainties in describing the relevant physical processes.

experiment	reaction	ν-s	flux (meas.)	flux (pred.)
Homestake	$\nu_e(^{37}Cl, e^-)^{37}Ar$	$2 \to 8$	2.55 ± 0.25	$9.3^{+1.2}_{-1.4}$
Kamiokande	$\nu_e(e^-, e^-)\nu_e$	$3, 5$	2.70 ± 0.36	$6.62 \times \left(1.00^{+0.14}_{-0.17}\right)$
GALLEX	$\nu_e(^{71}Ga, e^-)^{71}Ge$	$1 \to 8$	69.7 ± 8	137^{+8}_{-7}
SAGE	$\nu_e(^{71}Ga, e^-)^{71}Ge$	$1 \to 8$	69 ± 13	137^{+8}_{-7}

4.1 Results of the Neutrino Experiments

The updated results of the neutrino experiments [Homestake (Cleveland 1995), Kamiokande (Suzuki 1995), GALLEX (GALLEX coll. 1996), SAGE (SAGE coll. 1994)] and their comparison with one of the best helioseismic models (Bahcall et al. 1995) are given in Table 4. The GALLEX and SAGE experiments have an energy threshold of 0.233 MeV, Homestake of 0.814 MeV and Kamiokande of 7.5 MeV. The different thresholds reflect the sensitivity of the detectors to the solar reactions as listed in Table 1. Kamiokande practically measures only the ^{8}B neutrinos, Homestake essentially the ^{8}B plus a fraction of the ^{7}Be and all the CNO neutrinos and GALLEX and SAGE all emitted neutrinos, especially the most abundant low energy ones produced by the $p(p, e^+\nu_e)D$ reaction.

The average neutrino fluxes, as listed in Table 4, derive from more than 100 runs, in almost 25 years, of the Homestake detector operation, from about 50 runs, in almost 5 years, of the GALLEX and SAGE detector operation, and about 2500 days of the Kamiokande detector operation. Two of the experiments, GALLEX and SAGE, were calibrated with a very intense ($\simeq 7 \, L^\nu_\odot$) ^{51}Cr neutrino source, and the results of calibration indicated that the efficiency of the detectors approached 100% (GALLEX coll. 1995, Kirsten 1996), thus validating the response of the detectors to solar neutrinos.

4.2   Seismic vs. Neutrino Sun

The mutual consistency of the neutrino experiments and how they compare with the standard solar models can be discussed by plotting the neutrino fluxes of the ^{7}Be and CNO reactions $[\phi(^7\text{Be} + \text{CNO})]$ vs. the ^{8}B reaction neutrino flux $[\phi(^8\text{B})]$, which is the only signal present in the Kamiokande experiment (see also Castellani et al. 1996). The total flux $\phi(^7\text{Be} + \text{CNO})$ comes essentially from ^{7}Be reaction, since the CNO reactions give a contribution of 5% for chlorine and 7% for gallium experiments. In order to produce Fig. 3 for the three experiments Kamiokande, Homestake and GALLEX (SAGE gives the same results as GALLEX), we used the neutrino cross-sections as given by Bahcall (1989) and assumed that all the neutrinos produced by the basic $p(p, e^+\nu_e)$D reaction are present in the GALLEX (or SAGE) signal. This assumption is firmly based on the fact that the above reaction is basic for the maintenance of the hydrostatic structure of the Sun and provides almost all the neutrinos which are constrained by the solar luminosity [see Eq.(10)]. The effective cross-section for ^{7}Be plus CNO neutrino capture has been calculated as a weighted average over the fluxes of the cross-sections of the individual reactions. In Fig. 3 the allowed domains of the experiments are represented by strips $\pm 1\sigma$ wide, centered on the mean experimental values. The Kamiokande domain is a strip parallel to the $\phi(^7\text{Be} + \text{CNO})$ axis; the Homestake domain is an oblique strip with two limit cases, $\phi(^8\text{B}) = 0$ and $\phi(^7\text{Be} + \text{CNO}) = 0$; the GALLEX domain is a strip analogous to the Homestake strip, but with a different slope.

It appears from Fig 3 that the results of the experiments, taken at their mean values, are mutually exclusive and they are only marginally consistent when taken at 1σ. The GALLEX mean signal, as the result of the average over the three exposure periods for a total amount of 53 runs (69.7 SNU, GALLEX coll. 1996), leaves no room for the ^{7}Be, CNO and ^{8}B neutrinos, but the Kamiokande experiment indicates that about a half of the ^{8}B emitted neutrinos are measured. This means about 8 SNU in the GALLEX signal. However, if we consider the combined results of the previous two GALLEX exposure periods (39 runs, GALLEX coll. 1992, 1994) which gave 77.8 $\pm$ 10 SNU, or if we take the recent results at 1σ, GALLEX and Kamiokande are still compatible with the presence of all the low energy neutrinos emitted by the $p(p, e^+\nu_e)$D reaction and some 50% of the ^{8}B neutrinos. In any case the experiments indicate that the ^{7}Be neutrinos are practically absent. Only the future Borexino experiment, which measures the ^{7}Be principal line at 0.86 MeV, will be able to reveal the nature of this problem. In the neutrino flux diagram of Fig. 3 the standard solar model (SSM, Bahcall et al. 1995) appears to be very far from the small region, formed by the area above the zero line limited by the mutual intercepts of the three strips, which defines the domain of the mutual consistency of the experiments, namely the position of the neutrino Sun.

Since the main reactions which produce the emitted neutrinos depend

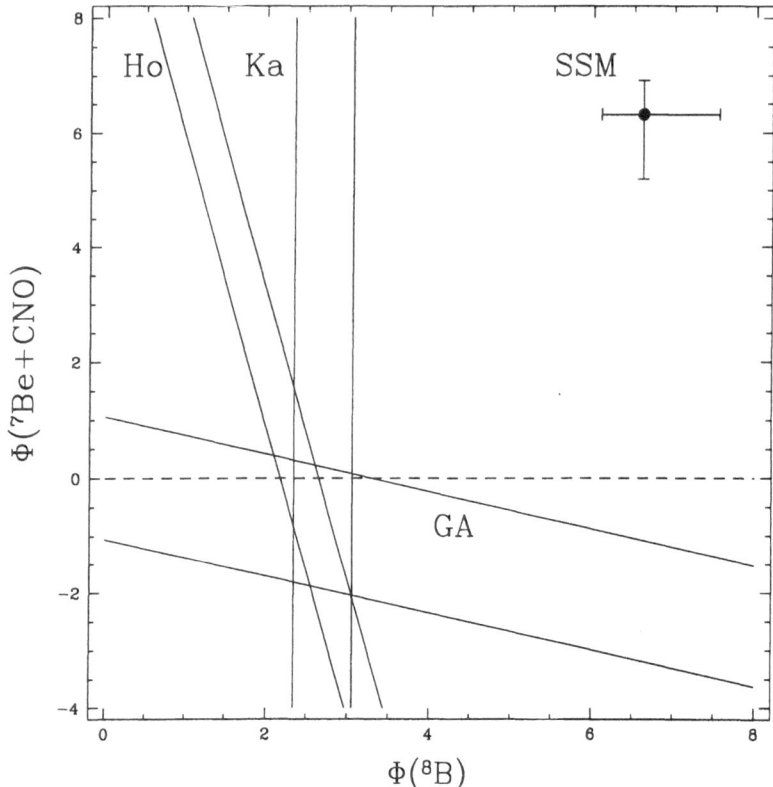

Fig. 3. The ^{8}B $[\phi(^8B)$ in $10^6\,\mathrm{cm}^{-2}\,\mathrm{s}^{-1}]$ and ^{7}Be plus CNO $[\phi(^7\mathrm{Be} + \mathrm{CNO})$ in $10^9\,\mathrm{cm}^{-2}\,\mathrm{s}^{-1}]$ neutrino fluxes from the results of the Homestake (Ho), Kamiokande (Ka) and GALLEX (GA) experiments. The allowed domains are indicated by strips $\pm 1\sigma$ wide centered on the mean experimental values, and it has been assumed that the GALLEX signal contains all the neutrinos emitted by the $\mathrm{p}(\mathrm{p}, e^+\nu_e)\mathrm{D}$ reaction (69.7 SNU). For comparison, the position of the standard solar model (SSM) of Bahcall et al. (1995) is given with its theoretical uncertainties.

on temperature with very different power laws $[\phi(\mathrm{pp}) \propto T^{-1}, \phi(^7\mathrm{Be}) \propto T^8, \phi(^8B) \propto T^{23}]$, attempts to reconcile the SSM with the neutrino Sun have recently been made by lowering artificially the opacity in the core, which means lowering the central temperature. A reduction of the opacity by about 40%, which reflects on a 6% reduction of the central temperature, has the consequence of reducing only by a factor of 2 the ^{7}Be neutrinos, but by factor of 6 the ^{8}B neutrinos (Castellani et al. 1996), in disagreement with the neutrino experimental results. From a nuclear point of view, an artificial increase by a factor of 14 of the astrophysical factor S_{33}, which determines the rate of the ^{3}He(^{3}He, 2p)^{4}He neutrinoless reaction, produces almost the same effect, while a change of S_{17}, which determines the production of ^{8}B, reflects only on the prodution of these neutrinos (Castellani et al. 1996). These authors found that, in order to approach closely the SSM to the neutrino Sun, it is necessary to combine together the astrophysical and nuclear changes in such a way that opacity should be reduced by some 50%, and S_{33} and S_{17} should be increased by a factor of 14 and 5 respectiveley. They concluded that, due to the large and unjustified changes in the physical processes, there is little chance of an astrophysical (or/and nuclear) solution to the neutrino problem.

On the other hand, since the helioseismic data inversion seems to indicate that the core of the real Sun should be somewhat hotter than that of the model Sun, any adjustment of the physical processes which produce a better agreement with helioseismic data goes in the direction of increasing the ^{7}Be and ^{8}B neutrino fluxes which are produced by reactions which strongly depend on temperature. In fact, as can be seen from Table 3, the models which are more consistent from a helioseismic point of view are those which use EOS/OPAL95 equation of state and OPAL95 opacities with He and metal diffusion (the SSM of Fig. 3, Bahcall et al. 1995), and produce more ^{7}Be and ^{8}B neutrinos than the previous solar models constructed with older equation of state and opacities or without diffusion or eventually with only He diffusion. As discussed in Sect. 3.3, the higher sound speed in the Sun's central regions required by helioseismology might also be explained in terms of a larger hydrogen abundance there, but considerations on the solar age seem to exclude this possibility.

Non-standard neutrino properties, which require massive neutrinos, have been proposed as a suitable solution for the observed neutrino deficit (Bahcall and Bethe 1990). The solution relies upon the change of the neutrino flavour in the path from the solar core to the Earth, which renders the emitted electron neutrinos undetectable to the experiments. Among the proposed mechanisms, such as the revisited vacuum oscillations (Berezhiani and Rossi 1996) and spin flip in the solar magnetic field (Cisneros 1971), the most attractive and elegant one is the MSW resonant conversion in matter (Mikheycv and Smirnov 1986). The four neutrino experiments indicate with a high degree of confidence that the two main parameters of the MSW effect, the squared mass difference between two neutrino flavours Δm^2 and the flavour mixing angle in the matter $\sin^2 2\theta_M$, have the values $7 \times 10^{-6}\,\mathrm{eV}^2$ and 5×10^{-3} respectively. On using the above values with a standard solar model, it is deduced that the MSW effect converts almost all the ^{7}Be and about 50% of the ^{8}B electron neutrinos into different flavour neutrinos, but it leaves unaltered the neutrino flux emitted by the p(p, $e^+\nu_e$)D reaction. This is exactly what is demanded by the results of the four neutrino experiments. An experimental proof of the MSW effect in laboratory is very difficult, but the forthcoming data of the Super-Kamiokande experiment, which is able to measure the shape of the ^{8}B neutrino spectrum, can show the signature of the MSW effect in the spectrum distortion and enlighten us on the nature of the neutrino.

5 Acknowledgements

I am most grateful to S. Basu for providing Fig. 1 and to M. P. Di Mauro for inverting the IRIS and BISON helioseismic data and providing Fig. 2.

References

Anders E., Grevesse N.: 1989, Geochim. Cosmochim. Acta **53**, 197

Bachmann K. T., Duvall T. L., Jr., Harvey J. W., Hill F.: 1995, in *GONG 1994: Helio- and Astero-Seismology from the Earth and Space*, R. K. Ulrich, E. J. Rhodes Jr. and W. Däppen eds., Astr. Soc. Pacific Conference Series, Vol. **76**, p. 156

Backus G. E., Gilbert F.: 1968, Geophys J. R. Astron. Soc. **16**, 169

Bahcall J. N.: 1989, *Neutrino Astrophysics*, Cambridge Univ. Press, Cambridge

Bahcall J. N.: 1996, ApJ **467**, 475

Bahcall J. N., Davis R., Jr.: 1976, Science **191**, 264

Bahcall J. N., Bethe H. A.: 1990, Phys. Rev. Lett. **65**, 2233

Bahcall J. N., Pinsonneault M. H.: 1992, Rev. Mod. Phys. **64**, 885

Bahcall J. N., Pinsonneault M. H., Wasserburg G. J.: 1995, Rev. Mod. Phys. **67**, 781

Basu S., Christensen-Dalsgaard J., Schou J., Thompson M. J., Tomczyk S.: 1995, ApJ **460**, 1064

Basu S., Christensen-Dalsgaard J., Schou J., Thompson M. J., Tomczyk S.: 1996, Bull Astron. Soc. India **24**, 147

Berezhiani Z. G., Rossi A.: 1996, Phys. Lett. B **367**, 219

Canuto V. M., Mazzitelli I.: 1992, ApJ **389**, 724

Castellani V., Degl'Innocenti S., Fiorentini G., Lissia M., Ricci B.: 1994, Phys. Rev. D **50**, 4749

Castellani V., Degl'Innocenti S., Fiorentini G., Lissia M., Ricci B.: 1996, Phys. Reports, in press

Chandrasekhar S.: 1964, ApJ **139**, 664

Christensen-Dalgaard J., Schou J.: 1988, in *Seismology of the Sun and Sun-like Stars*, E. J. Rolfe ed., ESA-SP 286, Paris, p. 149

Christensen-Dalsgaard J., Berthomieu G.: 1991, in *Solar Interior and Atmosphere*, A. N. Cox, W. C. Livingston and M. S. Matthews eds., The University of Arizona Press, Tucson, p. 401

Christensen-Dalsgaard J., Schou J., Thompson M. J.: 1990, MNRAS **242**, 353

Christensen-Dalsgaard J., Duvall T. L., Jr., Gough D. O., Harvey J. W., Rhodes E. J., Jr.: 1985, Nat **315**, 378

Cisneros A.: 1971, Ap&SS **10**, 87

Cleveland B.: 1995, Nucl. Phys. B (Proc. Suppl.) **38**, 47

Craig I. J. D., Brown J. C.: 1986, *Inverse Problems in Astronomy*, Hilger, Bristol

Deubner F.-L.: 1975, A&A **44**, 371

Di Mauro M. P.: 1996, private communication

Di Mauro M. P., Paternò L., Sofia S.: 1996, in *Sounding Solar and Stellar Interiors*, IAU Symp. 181, poster papers, in press

Dziembowski W. A., Paternò L., Ventura R.: 1985, A&A **151**, 47

Dziembowski W. A., Paternò L., Ventura R.: 1988, A&A **200**, 213

Dziembowski W. A., Pamyatnykh A. A., Sienkiewicz R.: 1990, MNRAS **244**, 542

Dziembowski W. A., Pamyatnykh A. A., Sienkiewicz R.: 1992, Acta Astron. **42**, 5

Dziembowski W. A., Goode P. R., Pamyatnykh A. A., Sienkiewicz R.: 1994, ApJ **432**, 417

Dziembowski W. A., Goode P. R., Pamyatnykh A. A., Sienkiewicz R.: 1995, ApJ **445**, 509

Elsworth Y., Howe R., Isaak G. R., McLeod C. P., Miller B. A., New R., Wheeler S. J., Gough D. O.: 1995, Nat **376**, 669

Evans J. W., Michard R.: 1962, ApJ **136**, 493

GALLEX coll.: 1992, Phys. Lett B **285**, 376

GALLEX coll.: 1994, Phys. Lett B **327**, 377

GALLEX coll.: 1995, Phys. Lett B **342**, 440

GALLEX coll.: 1996, Phys. Lett. B, in press

Goldreich P., Keeley D. A.: 1977, ApJ **212**, 243

Goldreich P., Murray N., Kumar P.: 1994, ApJ **424**, 466

Gough D. O.: 1984, Phil. Trans. Roy. Soc. London, Ser. A **313**, 27

Gough D. O.: 1986, in *Seismology of the Sun and Distant Stars*, D. O. Gough ed., NATO-ASI Ser., Vol. **169**, Reidel, Dordrecht, p. 125

Gough D. O., Kosovichev A. G.: 1988, in *Seismology of the Sun and Sun-like Stars*, E. J. Rolfe ed., ESA-SP 286, Paris, p. 195

Gough D. O., Thompson M. J.: 1991, in *Solar Interior and Atmosphere*, A. N. Cox, W. C. Livingston and M. S. Matthews eds., The University of Arizona Press, Tucson, p. 519

Grevesse N., Noels A.: 1993, in *Origin and the Evolution of the Elements*, N. Prantzos, E. Vangioni-Flam and M. Cassé eds., Cambridge Univ. Press, Cambridge, p. 15

Guenther D. B., Pinsonneault M. H., Bahcall J. N.: 1993, ApJ **418**, 469

Hansen C. J., Cox J. P., Van Horn H. M.: 1977, ApJ **217**, 151

Hirata K. S. et al.: 1990, Phys. Rev. Lett. **65**, 1297

Iglesias C., Rogers F. J.: 1996, ApJ **404**, 943

Kamionkowski M., Bahcall J. N.: 1994a, ApJ **420**, 884

Kamionkowski M., Bahcall J. N.: 1994b, Phys Rev. C **49**, 545

Kirsten T: 1996, in *New Trends in Solar Neutrino Physics*, V. Berezinsky and G. Fiorentini eds., LNGS - INFN, L'Aquila, p. 3

Lazrek M., Pantel A., Fossat E., Gelly B., Schmider F. X., Fierry-Fraillon D., Grec G., Loudagh S., Ehgamberdiev S., Khamitov S., Hoeksema T., Pallé P. L., Régulo C.: 1996, Solar Phys. **166**, 1

Leibacher J. W., Stein R. F.: 1971, Astrophys. Lett. **7**, 191

Leighton R. B.: 1961, Nuovo Cimento Suppl. Ser. **10 22**, 321

Libbrecht K. G.: 1988a, Space Science Reviews **47**, 275

Libbrecht K. G.: 1988b, in *Seismology of the Sun and Sun-like Stars*, E. J. Rolfe ed., ESA-SP 286, Paris, p. 131

Libbrecht K. G.: 1989, ApJ **336**, 1092

Libbrecht K., G., Morrow C. A.: 1991, in *Solar Interior and Atmosphere*, A. N. Cox, W. C. Livingston and M. S. Matthews eds., The University of Arizona Press, Tucson, p. 479

Lydon T. J., Sofia S.: 1996, Phys. Rev. Lett. **76**, 177

Mihalas D., Däppen W., Hummer D. G.: 1988, ApJ **331**, 815

Mihalas D., Hummer D. G., Mihalas B W., Däppen W.: 1990, ApJ **350**, 300

Mikheyev S. P., Smirnov A. Yu.: 1986, Nuovo Cimento C **9**, 17

Morrow C. A.: 1988, in *Seismology of the Sun and Sun-like Stars*, E. J. Rolfe ed., ESA-SP 286, Paris, p. 91

Paternò L.: 1996, in *Highlights of European Astronomy*, M. Rodonò and S. Catalano eds., Mem. Soc. Astron. Ital., in press

Paternò L., Sofia S., Di Mauro M. P.: 1996, A&A **314**, 940

Paternò L., Ventura R., Canuto V. M., Mazzitelli I.: 1993 ApJ **402**, 733

Rogers F. J., Iglesias C.: 1992, ApJS **79**, 507

Rogers F. J., Swenson F. J., Iglesias C.: 1996, ApJ **456**, 902

SAGE coll.: 1994, Phys. Lett. B **328**, 234

Suzuki Y.: 1995, Nucl. Phys. B (Proc. Suppl.) **38**, 54

Tomczyk S., Schou J., Thompson M. J.: 1995, ApJ **448**, L57

Ulrich R. K.: 1970, ApJ **162**, 993

Unno W., Osaki Y., Ando H., Shibahashi H.: 1989, *Nonradial Oscillations of Stars*, 2nd ed., University of Tokyo Press, Japan

Voronstov S. V., Shibahashi H.: 1991, PASJ **43**, 739

Modulation of Solar and Stellar Activity Cycles

N. O. Weiss[1] and S. M. Tobias[1,2]

[1] Department of Applied Mathematics and Theoretical Physics,
   University of Cambridge, Cambridge CB3 9EW, UK
[2] Present address: Joint Institute for Laboratory Astrophysics,
   University of Colorado, Boulder, CO 80309–0440, USA

Abstract. Cyclic magnetic activity in the Sun and other similar stars is inter-rupted by episodes of reduced activity such as the Maunder Minimum. This pattern is reproduced in mean field ($\alpha\omega$) dynamo models, where growth of the field is limited by the nonlinear action of the Lorentz force on differential rotation. The observed aperiodicity can be ascribed to deterministic rather than stochastic processes, and chaotic modulation is demonstrated for a low-order system, for two-dimensional dynamo waves and for a simple global model. Amplitude modulation that leads to grand minima is distinct from modulation associated with changes in the symmetry of the field.

1 Introduction

In this review we shall focus on a particular aspect of magnetic activity in the Sun, namely the long-timescale (~ 200 yr) modulation of the solar cycle that is associated with grand minima. Recent work on simplified models of the solar dynamo has led to a much clearer understanding of the origin of this behaviour. Fortunately, it is possible to extend the record of solar activity, which has only been observed systematically since telescopes were invented at the beginning of the seventeenth century, back for almost 10,000 years by using proxy data. However, our Sun is just one star and its properties only change on a slow evolutionary timescale. To explain its magnetic behaviour it is necessary to consider the Sun in the context of other similar stars, for the effects of varying a parameter such as rotation can then be inferred from observations of stars with different rotation rates. We shall therefore exploit the solar-stellar connection and discuss the past and future evolution of the solar dynamo, with the aim of providing convincing evidence for the existence of chaotically modulated stellar dynamos.

The basic features of solar and stellar dynamos have been discussed, for example, by Weiss (1994). Our treatment here is organised as follows. First we summarize the observed patterns of cyclic behaviour in the Sun, empha-sizing the Maunder Minimum and other grand minima that are deduced from changes in the abundances of cosmogenic isotopes. Then solar activity is re-lated to cyclic activity in other late-type stars, where the key parameter is the rotation rate, Ω. Section 3 provides a brief survey of dynamo theory as applied to stars. We argue that there is a transition from periodic to modulated to chaotic behaviour as Ω increases. This sequence of transitions is then demonstrated in Section 4 for a simple toy model of the dynamo process. More elaborate models, involving solutions of nonlinear partial differential equations, are discussed in Sections 5 and 6. These calculations show that two types of modulation can occur. The first is associated with deviations from dipole symmetry, while the other allows grand minima at fixed parity (as in the simple model). Finally, in Section 7, we summarize these results and comment briefly on the connection between solar activity and climatic change.

2   Observational Background

The systematic properties of the solar cycle are well known (see e.g. Stix 1989; Foukal 1990; Wilson 1994). The principal feature is the 11–year activity cycle. The sunspot fields have opposite senses in each hemisphere and reverse direction after each 11–year cycle, so the magnetic cycle has a 22–year period. These cycles vary irregularly in duration and in strength, as shown in Figure 1. Although the mean period is well-defined, there is a tendency for longer cycles to be correlated with weak activity.

This pattern has persisted through fifteen activity cycles. For a period of 30 years at the beginning of the nineteenth century, however, the cycles were extremely weak and during the latter part of the seventeenth century sunspots almost disappeared. Careful studies of the records have conclusively shown that the lack of sunspots was not caused by inadequate observations (Ribes & Nesme–Ribes 1993; Hoyt & Schatten 1996), and the reality of the Maunder Minimum (1645–1715) is beyond dispute. Moreover, the few spots that did appear between 1680 and 1700 were almost all confined to the Sun's southern hemisphere (Ribes & Nesme–Ribes 1993).

Although systematic observations of sunspots only extend back to the invention of the telescope, the pattern of activity can be followed back through proxy data. Unstable isotopes such as ^{10}Be and ^{14}C are produced in the earth's atmosphere by galactic cosmic rays, and the incidence of cosmic rays is affected by magnetic fields in the solar wind. Thus the production rates of ^{10}Be and ^{14}C are reduced when the Sun is magnetically active. These isotopes are preserved in polar icecaps and tree trunks, respectively, and anomalies in their abundances can be precisely dated. Careful measurements have confirmed that results for these two isotopes are consistent, and the ^{14}C record (which is used for radiocarbon dating) has now been carried back for almost 10,000 years. After removing slow trends caused by changes in the Earth's magnetic field and in oceanic circulation the remaining variations in ^{14}C abundance can be ascribed to changes in solar activity. Thus the wiggles in Figure 2 (Stuiver & Braziunas 1989) represent the envelope of the magnetic

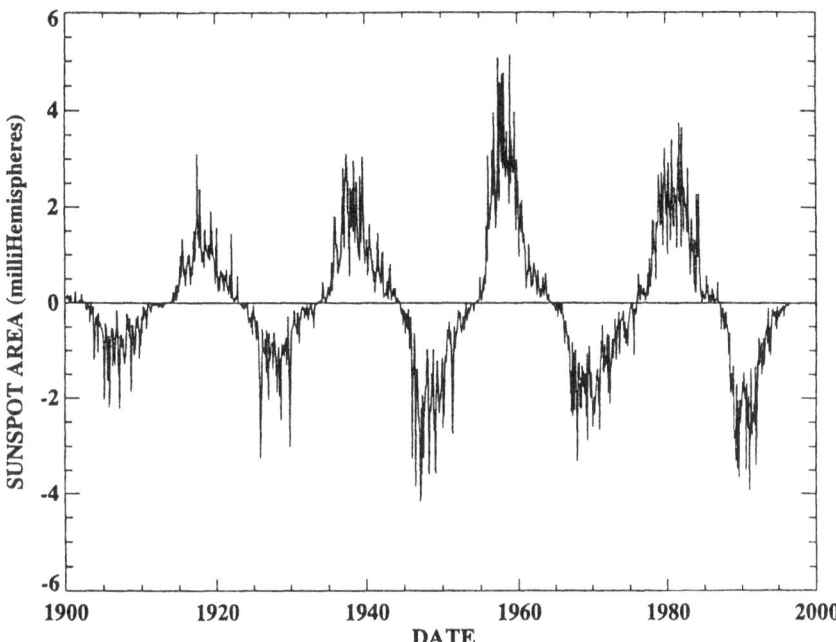

Fig. 1. The 22–year magnetic cycle (1900–1996), represented by sunspot areas with the sign flipped at sunspot minimum. (Courtesy of D.H. Hathaway.)

cycles: they demonstrate that irregular grand minima, with a characteristic timescale of around 200 years, are a persistent feature of magnetic activity in the Sun. This recurrent modulation is what we wish to explain.

To establish the origin of modulated cyclic activity we need to consider other stars on the main sequence (Soderblom & Baliunas 1988; Noyes, Baliunas & Guinan 1991). Stars with masses less than 1.2 $M_\odot$ are cool and have deep outer convection zones. They are also magnetically active. The fields can be measured directly (Saar 1996), and their presence can also be inferred from X ray, radio or photometric observations. From these measurements it is clear that there are many stars that are much more active than the Sun. The most significant results have been obtained by monitoring chromospheric lines. Ca^+ H and K emission is known to be correlated with magnetic fields on the Sun, and stellar Ca^+ emission has been studied systematically over the last 30 years (Baliunas *et al.* 1995). These observations show that, for stars of given mass, magnetic activity depends on rotation and that the most rapid rotators are the most active.

Figure 3 illustrates the evolution of the angular velocity, Ω, in a star like the Sun. In the pre-main sequence T Tauri phase there is a distinction between fast and slow rotators (depending perhaps on the presence of an accretion disc and planetary formation). As stars approach the main sequence they contract, while conserving their angular momentum, and consequently

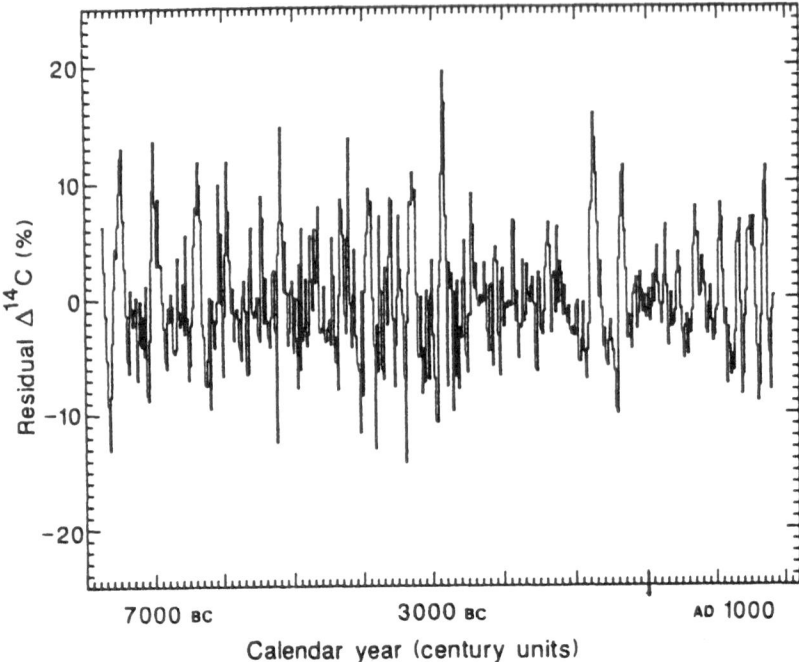

Fig. 2. Variations in ^{14}C production over the past 10 000 years; the maxima correspond to grand minima of solar activity. (After Stuiver & Braziunas 1989.)

spin up, so that the most rapid rotators in very young clusters rotate near break-up speed, with periods of around 12 hr. The resulting magnetic fields are responsible for heating stellar coronæ and driving stellar winds, which in turn remove angular momentum. So stars gradually spin down owing to magnetic braking, and the rate of deceleration decreases with decreasing Ω.

Magnetic cycles appear in slowly rotating, middle-aged stars like the Sun. Furthermore, there is evidence to suggest that activity in these stars is modulated (Baliunas & Jastrow 1990). Figure 4 shows the behaviour of two very similar stars: the first exhibits an irregular pattern of cyclic variation, with a period of about 9 years, while the second is quiescent, suggesting that it has been caught in a grand minimum. We conclude therefore that stars whose mass and age are similar to the Sun's display a similar pattern of magnetic activity, and that solar activity has dropped as the Sun evolved and will continue to decline during the rest of its stay on the main sequence.

3   Stellar Dynamos

Cyclic activity can only be maintained by a dynamo that relies on two distinct physical processes. A poloidal magnetic field, lying in meridional planes, is drawn out by differential rotation to produce a strong toroidal (or azimuthal)

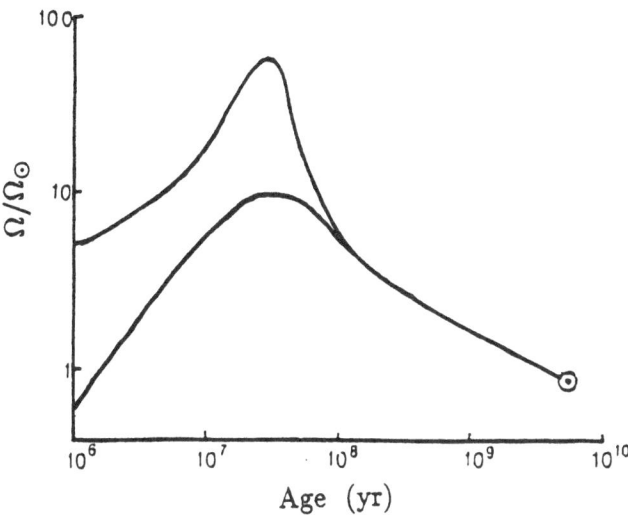

Fig. 3. Sketch showing the rotational history of stars like the Sun: ratio of surface angular velocity Ω to the rotation rate $\Omega_\odot$ of the present Sun as a function of age. The two curves correspond to initially fast and slow rotators.

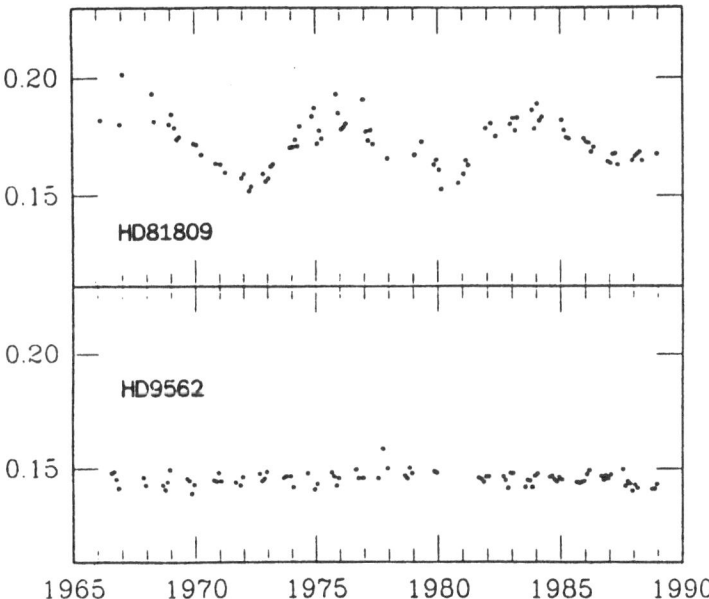

Fig. 4. Chromospheric Ca^+ emission for two stars like the Sun. The star in the upper panel shows cyclic activity with a period of 8.3 yr, while that in the lower panel is apparently inactive. (After Baliunas & Jastrow 1990.)

29

field that is antisymmetric about the equator. This field is then acted on by cyclonic eddies, whose systematic helical motion comes from the action of the Coriolis force on convective plumes, thus generating a reversed poloidal field (Parker 1979) and so the cycle proceeds. The effect of the cyclonic eddies – the key to the whole process – can be represented by the parameter α, which is proportional to the mean helicity $< \mathbf{u}\cdot\mathrm{curl}\ \mathbf{u} >$, where $\mathbf{u}$ is the turbulent velocity. Since α depends on the Coriolis force, it is antisymmetric about the equator.

It is widely accepted that the solar dynamo is located near the base of the convection zone and that the strong toroidal fields, which emerge through the photosphere to form active regions and sunspots, are anchored in a region of weak convective overshoot. This region, at the interface between the radiative and convective zones, is also the site of the strong radial shear in angular velocity that has been revealed by helioseismology (Tomczyk et al. 1995; Thompson et al. 1996). The cyclonic eddies, which are responsible for regenerating the poloidal field through the α–effect, occur, however, in the convection zone itself.

Although the dynamo process has been demonstrated in some fully consistent three-dimensional computations, most work has involved simplified models. These come in two forms: either it is assumed that the relevant magnetic flux is all confined to isolated flux ropes (e.g. Ferriz Mas, Schmitt & Schüssler 1994) or else it is supposed that there is a smoothly varying mean field. The truth probably lies in between but we shall follow the latter approach. Mean field $(\alpha\omega)$ dynamos, in which the axisymmetric poloidal field $\mathbf{B}_\mathrm{P}$ is generated by the α–effect, while the toroidal field $\mathbf{B}_\mathrm{T}$ is produced by differential rotation (the ω–effect), have been widely studied. The relevant parameter is the dynamo number

$$D = \frac{\alpha\Omega' L^4}{\eta^2} \propto \Omega^2 \ , \tag{1}$$

where Ω' is a measure of differential rotation, η is a turbulent diffusivity and L is an appropriate length scale. As $|D|$ (or Ω) is increased from a small value, cyclic dynamo action sets in as a magnetic instability, at an oscillatory (or Hopf) bifurcation when $D = D_c$. Thereafter the field oscillates with an amplitude that is determined by nonlinear effects, as either the α–effect or differential rotation (or both) are modified by the Lorentz force. In general, however, the toroidal field will increase with increasing Ω, in accord with stellar observations.

As we have seen, the solar cycle is actually aperiodic and is aperiodically modulated. This behaviour can be ascribed either to the influence of stochastic disturbances or to deterministic chaos. Unfortunately, the available records are not long enough to distinguish between these two possibilities (Weiss 1990; Morfill, Scheingraber & Sonett 1991). Stochastic modulation of cyclic activity can be produced if the disturbances have the same magnitude

as, for example, the α–effect itself (Schmitt, Schüssler & Ferriz Mas 1996)
Weak stochastic fluctuations will only have a significant effect if D is very
close to D_c (Hoyng 1988) – although the Sun is unlikely to be in such a
delicate position – or if the system is represented by trajectories in phase
space that approach some invariant manifold (Tobias 1996b). Otherwise,
mildly stochastic interference allows a system to evolve along trajectories
that "shadow" those on a chaotic attractor (Ott 1993), so that behaviour is
qualitatively unaffected. In what follows, we shall assume that the apparently
chaotic behaviour of the solar cycle is indeed deterministic.

A plausible explanation for modulation of the magnetic cycle is that there
are two states of convection and differential rotation in a star, of which the
first is hydrodynamically stable but magnetically unstable, while the second is magnetically unstable but hydrodynamically unstable (Parker 1979;
Zel'dovich, Ruzmaikin & Sokoloff 1983). The Lorentz force acts so as to drive
the system from the first state to the second, and grand minima are associated with the transition from the second to the first (Tobias, Weiss & Kirk
1995).

Mathematically, it is convenient to consider the sequence of bifurcations
as Ω (and hence $|D|$) is increased, so *reversing* the actual evolution of the
Sun's rotation in Figure 3. Then we expect to find the bifurcation sequence
shown schematically in Figure 5. Initially, the field-free state is stable to magnetic perturbations; the Hopf bifurcation H1 at $D = D_c$ gives rise to periodic
oscillations and the secondary Hopf bifurcation at H2 leads to doubly periodic (quasiperiodic) solutions, corresponding to regularly modulated cyclic
activity. Finally, there is a sequence of bifurcations that leads to chaotically
modulated behaviour. These transitions are demonstrated in the next section
for a simple model.

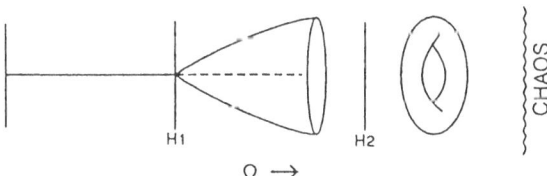

Fig. 5. Schematic bifurcation diagram, showing transitions from the field-free state
to periodic and then to doubly periodic activity, followed by a transition to chaos,
as the rotation rate Ω is increased. The bifurcations at H1 and H2 create a limit
cycle and a torus (strictly, a two-torus), respectively.

4 Modulation in a Low–order Model

Both cyclic activity and grand minima can be reproduced in solutions of coupled nonlinear ordinary differential equations. We shall construct a minimal system that displays these features. In order to obtain doubly periodic solutions (with trajectories that lie on a two-torus in phase space) it is necessary to consider a third-order system and this is also sufficient for chaos to occur. The motivation for this approach is the need to study generic properties of chaotically modulated cycles; such models also help to explain the origin of the modulation, although they lack predictive power. The simplest procedure for obtaining evolution equations is to truncate the partial differential equations that describe a mean field dynamo (e.g. Weiss, Cattaneo & Jones 1984; Schmaltz & Stix 1991; Covas *et al.* 1996). This is, however, open to the criticism that the interesting behaviour may be caused by arbitrary truncation and may vanish as more terms are included (as happens with the Lorenz equations, regarded as a description of convection). So it is preferable to use normal form equations and to rely on the associated mathematical theory, which guarantees that qualitative behaviour is robust (Tobias *et al.* 1995).

Let us therefore consider a third-order system in which all the hydrodynamic behaviour (convection and differential rotation) is collapsed onto the z-axis of cartesian co-ordinates. Since field-free solutions can persist this axis has to be invariant. The two hydrodynamic states mentioned in the previous section appear in a saddle-node bifurcation. To allow for a Hopf bifurcation the magnetic field is represented by two variables, x and y, corresponding to toroidal and poloidal fields, as indicated in Figure 6a. The normal form equations for a saddle-node/Hopf bifurcation take the form

$$\dot{r} = \lambda r + arz + cr^2 z \cos \phi \ , \tag{2}$$

$$\dot{\phi} = \omega - crz \sin \phi \ , \tag{3}$$

$$\dot{z} = \mu - z^2 - r^2 - bz^3 \ , \tag{4}$$

referred to cylindrical polar co-ordinates with $r^2 = x^2 + y^2$, $\tan \phi = y/x$. Here λ and μ are control parameters and a, b, c and ω are real constants (Guckenheimer & Holmes 1986; Tobias *et al.* 1995); the cubic terms in (2) and (4), together with the quadratic term in (3), have been added to remove degeneracies. If $c = 0$ the system is axisymmetric: then the primary Hopf bifurcation is followed by a secondary Hopf bifurcation that gives rise to quasiperiodic modulated solutions with trajectories that lie on a two-torus which encloses the unstable limit cycle. This torus expands until it eventually collides with the z–axis and is destroyed in a heteroclinic bifurcation. For $c \neq 0$ the modulation can become chaotic and the torus disappears in a heteroclinic tangle (Kirk 1991).

Tobias *et al.* (1995) illustrate a sequence of solutions for suitably chosen parameter values. As μ is increased, the bifurcation sequence is indeed as

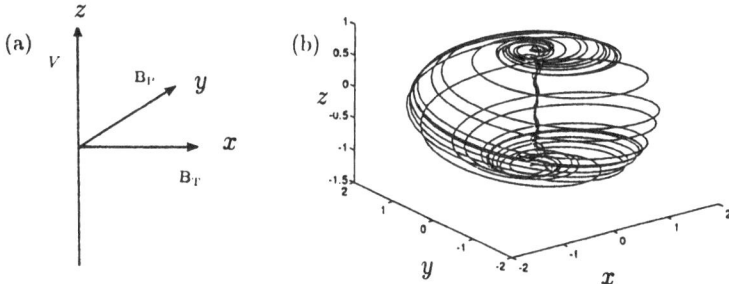

Fig. 6. (a) The co-ordinate system for the minimal third-order model. (b) The chaotic attractor in the three-dimensional phase space. (After Tobias *et al.* 1995.)

sketched in Figure 5. The z–axis is initially stable; then the magnetic instability appears at an oscillatory bifurcation (H1), giving rise to a stable limit cycle that corresponds to periodic cycles. The second oscillatory bifurcation (H2) leads to the appearance of a two-torus, corresponding to periodically modulated cycles. Finally, there is a transition to chaos. Figure 6b shows the chaotic attractor (which resembles a cored apple). Trajectories spiral out from the upper fixed point and around the outside of the attractor until they close in on the lower fixed point and slowly wind their way upwards in a tight spiral around the z–axis. The associated behaviour is represented in Figure 7, which shows the behaviour of the magnetic field, $x(t)$, to be compared with Figure 1, and also the envelope of activity, given by r^2, which can be compared with Figure 2.

Thus chaotic modulation is a feature of solutions of the toy system (4.1)–(4.3). Furthermore, since these are normal form equations, this behaviour is robust. Indeed, one can argue that the presence of cyclic magnetic activity implies that, at least in some stars, the activity should be modulated, and that chaotic modulation is associated with heteroclinicity.

5   Nonlinear Dynamo Waves

The next stage is to introduce spatial structure and to construct a system that is governed by partial differential equations. The simplest example of such a system is the infinite one-dimensional cartesian model originally introduced by Parker (1955, 1979), in which the magnetic field $\mathbf{B}(x,t) = (0, B, \partial A/\partial x)$, while the vector potential A and the toroidal field B satisfy the linear equations

$$\frac{\partial A}{\partial t} = \alpha B + \eta \frac{\partial^2 A}{\partial x^2} \ , \quad \frac{\partial B}{\partial t} = V' \frac{\partial A}{\partial x} + \eta \frac{\partial^2 B}{\partial x^2} \ , \tag{5}$$

where α and the velocity shear V' are both constant. These equations allow the growth of sinusoidal dynamo waves, propagating without change of form,

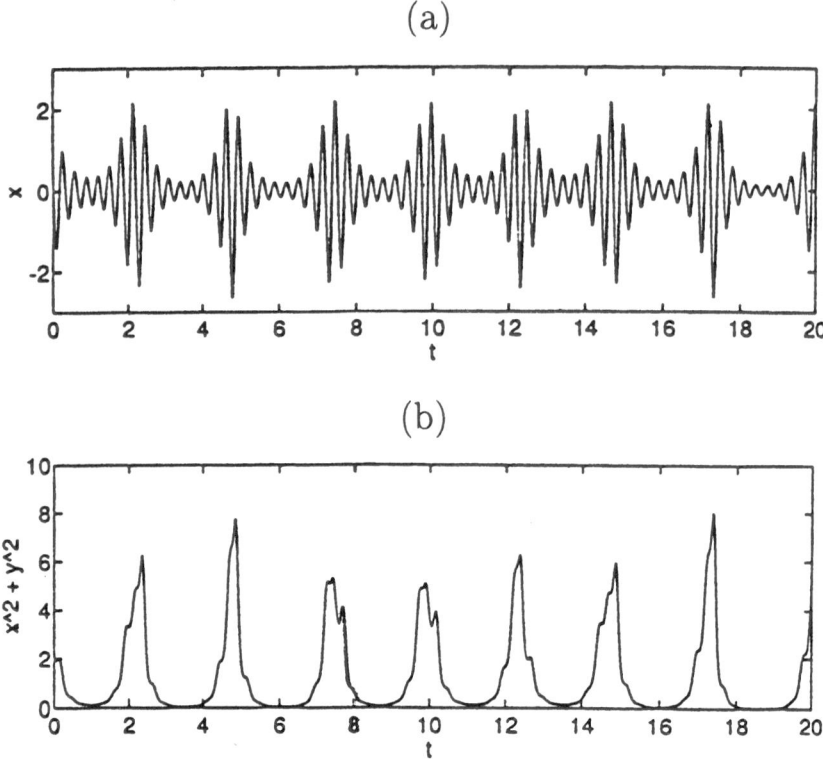

Fig. 7. Chaotically modulated behaviour in the low-order model, with parameters chosen to mimic the record of solar activity, showing (a) the toroidal field $x(t)$ and (b) the envelope of activity, $r^2(t)$. (After Tobias *et al.* 1995.)

when the dynamo number $D \equiv \alpha V'/2\eta^2 k^3 > 1$. This growth is halted by nonlinear effects. Of the various mechanisms that have been considered, the most plausible is the back-reaction of the Lorentz force, $\mathbf{j} \times \mathbf{B}$, exerted by the macroscopic field on differential rotation – often referred to as the Malkus–Proctor effect (Malkus & Proctor 1975). Since this force is quadratic in $\mathbf{B}$, there will be a contribution with twice the frequency of the dynamo wave – analogous to the "torsional oscillations" that are observed on the Sun (e.g Ulrich *et al.* 1988). A truncated nonlinear model (Weiss *et al.* 1984; Jones, Weiss & Cattaneo 1985) reveals a regime in which the initial Hopf bifurcation at $D = 1$ is followed by two further Hopf bifurcations, giving rise to quasiperiodically modulated waves with episodes of reduced activity that resemble grand minima, before solutions eventually become chaotic. (Since the waves are initially steady in a uniformly moving frame, an extra bifurcation is needed before dynamically complicated behaviour can appear.)

Parker (1993) has recently developed a more realistic two-dimensional model that mimics the structure of a dynamo at the base of the convection zone. The geometry remains cartesian, with the x, y, and z axes corresponding to the directions of increasing colatitude, longitude and radius, respectively, in a sphere. In this system the α–effect is confined to the upper half-space ($z > 0$), which represents the turbulent convection zone, while the sheared velocity $V(z)$, corresponding to differential rotation in a region of convective overshoot, is only present for $z < 0$. Dynamo action then leads to exponentially growing surface waves, whose amplitude is greatest at the interface.

If nonlinear effects are included this amplitude will saturate at some finite level. The most obvious process is suppression of the α–effect by the magnetic field. This happens while the mean field within the turbulent convection zone is still relatively weak (Vainshtein & Cattaneo 1992; Cattaneo & Hughes 1996), so raising a major difficulty for mean field dynamos. The problem can, however, be avoided if strong fields are confined to the region of weak convective overshoot, as in Parker's (1993) model. Tobias (1995) has extended this model to include the nonlinear back-reaction of the macroscopic magnetic field on the sheared velocity. The total velocity in the y–direction is then $v(x, z, t) = V(z) + u(x, z, t)$, where the perturbed component u evolves under the action of the Lorentz force and a turbulent viscosity $\nu(z) = \tau\eta(z)$, and the governing equations, in non-dimensional form, are

$$\frac{\partial A}{\partial t} = \alpha B + \eta\nabla^2 A \ , \tag{6}$$

$$\frac{\partial B}{\partial t} = D\left[v'\frac{\partial A}{\partial x} - \frac{\partial u}{\partial x}\frac{\partial A}{\partial z}\right] + \eta\nabla^2 B + \frac{\partial\eta}{\partial z}\frac{\partial B}{\partial z} \ , \tag{7}$$

$$\frac{\partial u}{\partial t} = \text{sgn}(D)\left[\frac{\partial A}{\partial x}\frac{\partial B}{\partial z} - \frac{\partial B}{\partial x}\frac{\partial A}{\partial z}\right] + \tau\left[\eta\nabla^2 u + \frac{\partial\eta}{\partial z}\frac{\partial u}{\partial z}\right] \ . \tag{8}$$

These equations are solved numerically in the finite region $\{0 < x \leq 2\pi; -1 < z < 1\}$, subject to the boundary conditions $A = B = u = 0$ at $z = 1$, $\partial A/\partial z = B = \partial u/\partial z = 0$ at $z = -1$, and to periodic lateral boundary conditions. Both α and V' are adjusted to vary smoothly with z, as shown in Figure 8, and η varies in the same way as α.

As $|D|$ is increased, with τ sufficiently small and other parameters suitably chosen, solutions of these equations show the same sequence of transitions as the low-order model. Travelling waves appear at $D = D_c$, with the form shown in Figure 9a; a second Hopf bifurcation then gives rise to periodically modulated waves, as illustrated in Figure 9b. These give way to waves that are quasiperiodically modulated and, finally, to chaotically modulated waves. Moreover, the modulation becomes much more significant when τ is very small. Since the waves initially travel without change of form, the value of $\langle B^2\rangle$, where the average is taken over the whole domain, is constant. At subsequent bifurcations $\langle B^2\rangle$ becomes modulated first periodically and then quasiperiodically. Figure 10a shows $\langle B^2\rangle$ for a chaotically modulated solu-

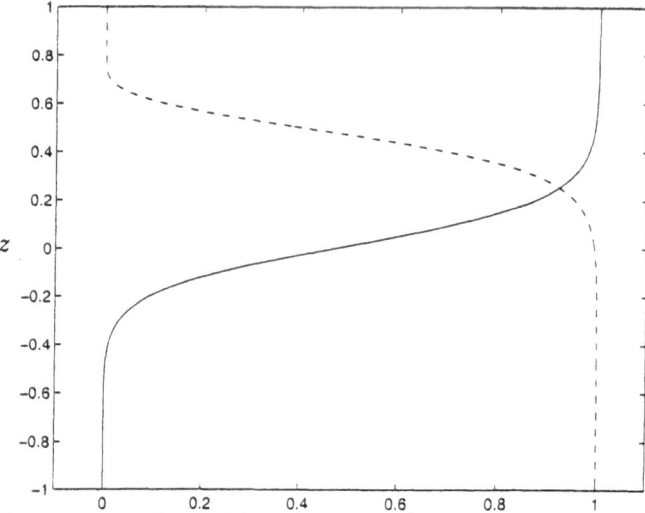

Fig. 8. The variation of α (solid line) and of V' (broken line) with depth z across the layer; notice the region of overlap at the interface.

tion: episodes of reduced activity are clearly visible in the time series, while the phase portrait in Figure 10b shows that trajectories approach the purely hydrodynamic state $(A = B = 0, v \neq 0)$, as they did in Figure 6. All this behaviour is consistent with that found for the low-order model of the previous section.

6   Modulation in a Global Model

Any recognizable model of a stellar dynamo must be able to reproduce the butterfly diagram, with oppositely directed toroidal fields on either side of the equator and boundaries at the poles. The dynamo waves of Section 5 represent local behaviour only: to obtain a global model we restrict attention to a bounded region with $0 < x < 2L$, and impose the boundary conditions $A = B = u = 0$ at $x = 0, 2L$. In a spherical system the sense of the cyclonic eddies, and hence the sign of the helicity, changes at the equator, while the azimuthal velocity vanishes at the poles. In a cartesian model the α–effect should therefore be antisymmetric about the equator $(x = L)$ and the ω– effect should drop to zero at the poles $(x = 0, 2L)$. Thus equations (6) and (7) are modified to become

$$\frac{\partial A}{\partial t} = \alpha \cos\left(\frac{\pi x}{2L}\right) B + \eta \nabla^2 A \tag{9}$$

and

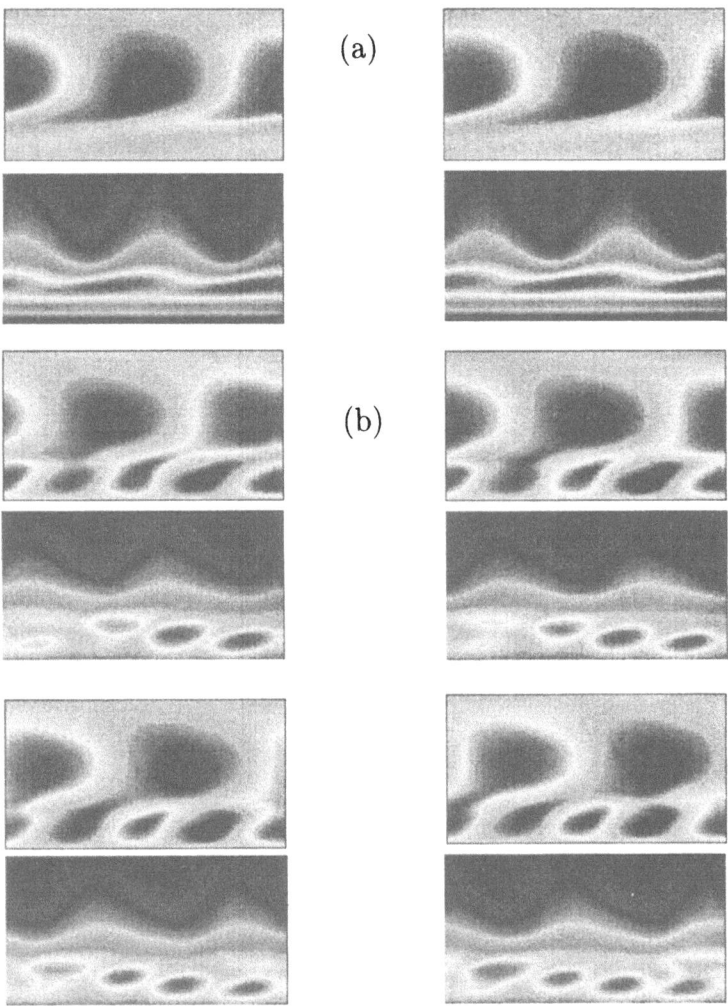

Fig. 9. Modulation of two-dimensional dynamo waves. (a) Rightward travelling waves at two successive times. (b) Modulated waves at four successive times. The upper and lower panels show the toroidal field and the velocity perturbation, respectively. (After Tobias 1995.)

$$\frac{\partial B}{\partial t} = D\left[\left\{V'\sin\left(\frac{\pi x}{2L}\right) + \frac{\partial u}{\partial z}\right\}\frac{\partial A}{\partial x} - \left\{\frac{\pi}{2L}V\cos\left(\frac{\pi x}{2L}\right) + \frac{\partial u}{\partial x}\right\}\frac{\partial A}{\partial z}\right] + \eta\nabla^2 B$$
(10)

where η is taken to be constant, and $\alpha(z)$ and $V'(z)$ are again chosen as in Figure 8, and $V(x,1) = 0$ (Tobias 1996a). Comparisons between two-dimensional cartesian and spherical models indicate that they have similar

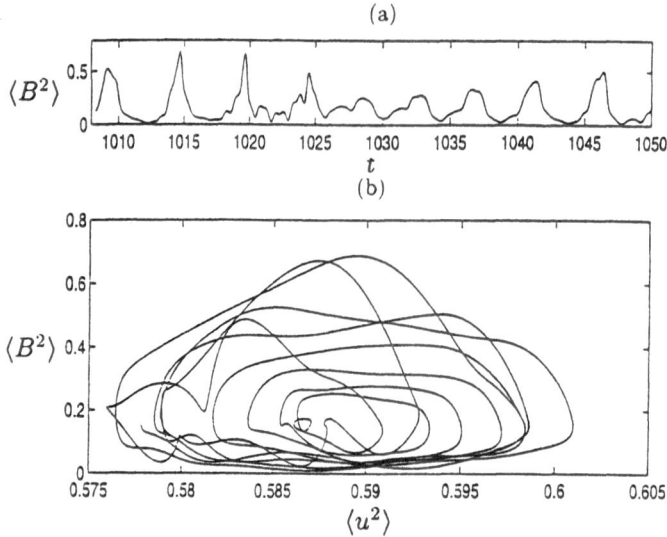

Fig. 10. Chaotically modulated dynamo waves. (a) Averaged magnetic energy $\langle B^2 \rangle$ as a function of time for $D = -2500, \tau = 0.01$. (b) Projection of the chaotic attractor onto the $\langle u^2 \rangle$–$\langle B^2 \rangle$ phase plane. (After Tobias 1995.)

bifurcation structures (Jennings *et al.* 1990) and this system has the computational advantage that variables can be expanded as sine series in the x-direction.

Bifurcations from the trivial field-free solution give rise to two distinct families of solutions with different symmetries about the equator. *Dipole* solutions, with B antisymmetric, have

$$\partial A/\partial x = B = \partial u/\partial x = 0 \text{ at } x = L \ , \tag{11}$$

while *quadrupole* solutions, with B symmetric, have

$$A = \partial B/\partial x = \partial u/\partial x = 0 \text{ at } x = L \ . \tag{12}$$

For $D < 0$, so that dynamo waves travel from the poles towards the equator, the first two bifurcations are at $D = -279$ and $D = -325$ and they give rise to oscillatory dipole and oscillatory quadrupole solutions respectively. Solutions on the branches that emerge from these bifurcations retain their symmetries, which can only be broken at a subsequent bifurcation (cf. Jennings & Weiss 1991).

6.1   Grand Minima with Dipole Symmetry

Although there are various spherical dynamo models in which symmetries are indeed broken in Hopf or pitchfork bifurcations (e.g. Brandenburg *et al.* 1989; Schmitt & Schüssler 1989; Kitchatinov, Rüdiger & Küker 1994), the

appropriately averaged fields on the real Sun do not normally deviate significantly from dipole symmetry, except near sunspot minimum (Yau 1988;
Watari 1996). Hence it is appropriate to investigate behaviour when the
dipole symmetry (11) is imposed, so that symmetry-breaking bifurcations
are suppressed.

Tobias (1996a) has obtained solutions for $L = 4$ and $\tau = 0.01$, which
follow the sequence of bifurcations sketched in Figure 5. The initial Hopf
bifurcation (H1) is followed by finite amplitude oscillations and the butterfly
diagram for $D = -500$, in Figure 11a, provides a plausible representation of
cyclic activity in the southern hemisphere of the Sun. As $|D|$ is increased there
is a secondary Hopf bifurcation (H2), leading to modulated cycles. Figure 12a
shows weakly modulated activity for $D = -860$ but the cycles for $D = -1500$,
in Figure 12b, are much more strongly modulated, with prolonged episodes
of reduced activity that correspond to grand minima. The corresponding
butterfly diagram is shown in Figure 11b. These quasiperiodic solutions have
trajectories that lie on a two-torus in phase space. Figure 13a shows the
projection onto a phase plane of the torus for $D = -1500$. As $|D|$ increases
the torus swells and approaches closer to the axis of zero magnetic energy,
so that the grand minima last longer. For yet larger values of $|D|$ the torus
breaks down and solutions become chaotic. A projection of the attractor for
$D = -2000$ is shown in Figure 13b.

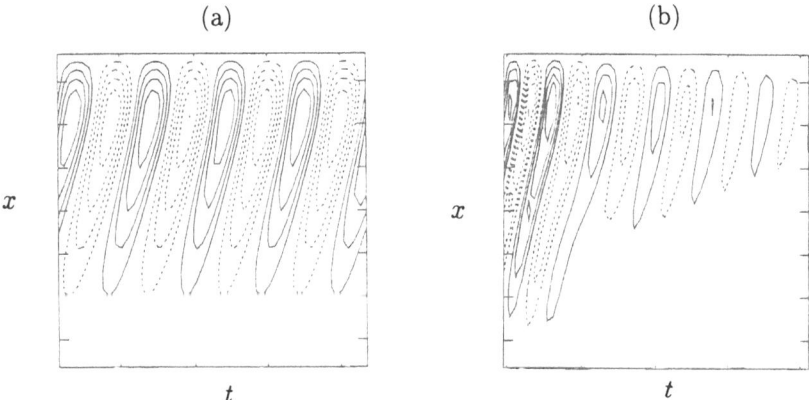

(a)   (b)

x x

t t

Fig. 11. Butterfly diagrams in the southern hemisphere for solutions with dipole
symmetry. (a) Periodic cycles for $D = -500, \tau = 0.01$. (b) Modulated cycles for
$D = -1500$. Here, and in subsequent figures, time is measured in arbitrary units.
(After Tobias 1995.)

All this behaviour follows the pattern found for the low-order model in
Section 4, thus implying that chaotic modulation is once again associated
with heteroclinicity. Furthermore, the ratio of the modulation period to the

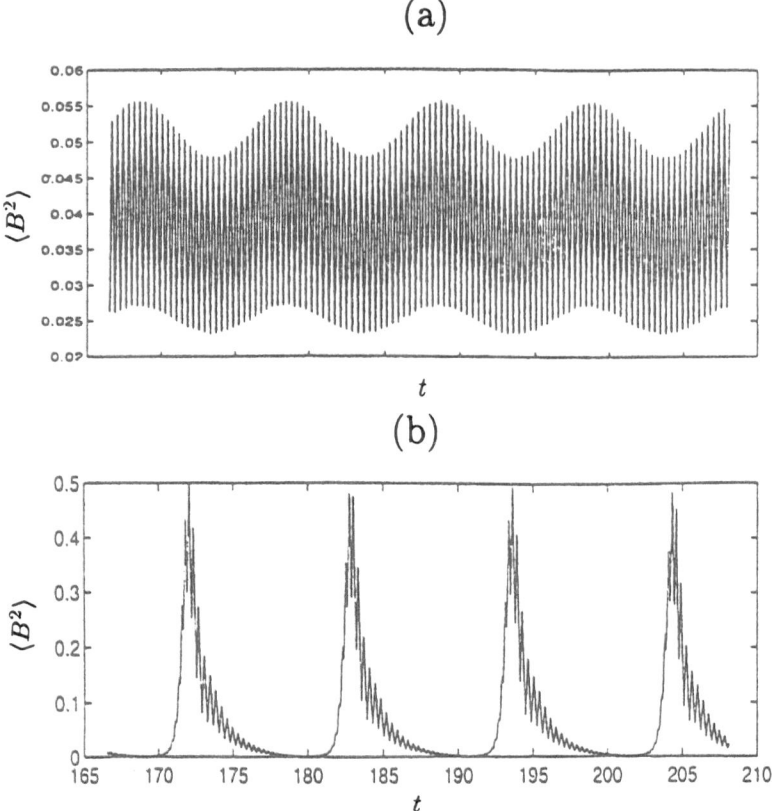

Fig. 12. Quasiperiodic cycles with dipole symmetry. $\langle B^2 \rangle$ as a function of time for (a) a weakly modulated solution ($D = -860$) and (b) a strongly modulated solution ($D = -1500$). (After Tobias 1995.)

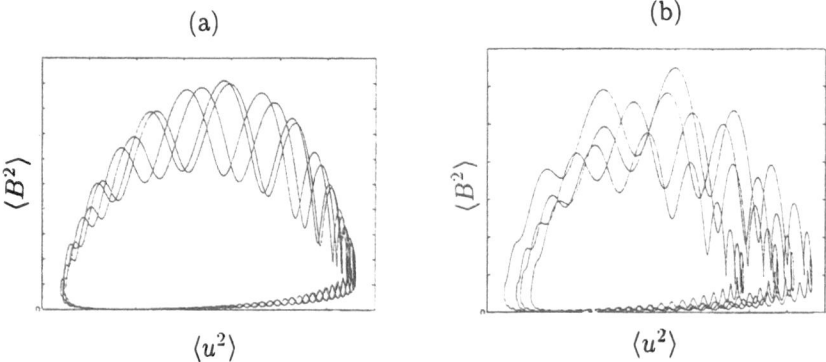

Fig. 13. Attractors for solutions with dipole symmetry. (a) Projection of a two-torus onto the $\langle u^2 \rangle$–$\langle B^2 \rangle$ phase plane for $D = -1500$. (b) Chaotic attractor for $D = -2000$. (After Tobias 1995.)

cycle period depends on the diffusivity ratio τ, apparently varying as $\tau^{-\frac{1}{2}}$; thus this ratio can be increased by reducing τ. This is not the only context in which a small value of τ is preferred (cf. Schlichenmaier & Stix 1995), though the physical justification for this preference remains unclear.

6.2 Symmetry-breaking and Modulation

It is important to establish whether the dipole solutions are stable to perturbations that break their symmetry. Tobias (1997) has therefore explored the behaviour of the system (8)–(10) when the symmetry condition (11) is *not* imposed. The numerical treatment only reveals branches with solutions that are stable, though the underlying bifurcation structure can only be fully understood if both stable and unstable branches are located (Jennings & Weiss 1990; Jennings 1991).

As before, there is a branch of stable periodic cycles with dipole symmetry emerging from the initial bifurcation at H1. With the diffusivity ratio fixed at $\tau = 0.1$ (for computational convenience) these dipole oscillations are stable at $D = -300$. Figure 14a shows the full butterfly diagram, extending from pole to pole, with toroidal fields that are antisymmetric about the equator. This symmetry is then broken at a secondary Hopf bifurcation, which gives rise to doubly periodic mixed-mode oscillations. The lack of symmetry is apparent in the butterfly diagram for $D = -400$, in Figure 14b. The symmetry of the system can be described by the parity $P = (\langle B_q^2 \rangle - \langle B_d^2 \rangle)/(\langle B_q^2 \rangle + \langle B_d^2 \rangle)$, where $B = B_q + B_d$ and B_q, B_d are symmetric and antisymmetric, respectively, about the equator. Thus $P = 1$ for a pure quadrupole field and $P = -1$ for a pure dipole field. Figure 15 shows how $\langle B^2 \rangle$ and P vary with time for this weakly modulated solution. This behaviour, which is associated with symmetry-breaking and is quite different from the strong modulation in Figure 12b, is referred to as Type 1 modulation (cf. Knobloch & Landsberg 1996). By $D = -500$ stability has been transferred to the branch of pure quadrupole oscillations, so B is symmetric and strictly periodic, as can be seen in Figure 14c. These quadrupole cycles persist until they in turn undergo a Hopf bifurcation when $D \approx 700$. After an interval with complicated quasiperiodic behaviour, stability is transferred back to the pure dipole solutions by $D - -900$.

Meanwhile, however, the oscillations have undergone a different Hopf bifurcation that gives rise to modulation without loss of symmetry, as in the runs with $\tau = 0.01$ that have already been described. This behaviour, with a timescale determined by τ, is modulation of Type 2 (Tobias 1997). As the modulation increases, giving rise to grand minima, the symmetry is again broken but in such a way that quadrupole fields only become significant near grand minima. Figure 16 shows the variation of $\langle B^2 \rangle$ and P with time for a chaotically modulated solution at $D = -1200$: when the magnetic energy is large the field is close to being a pure dipole but P increases rapidly during episodes of reduced activity. The butterfly diagram in Figure 14d illustrates

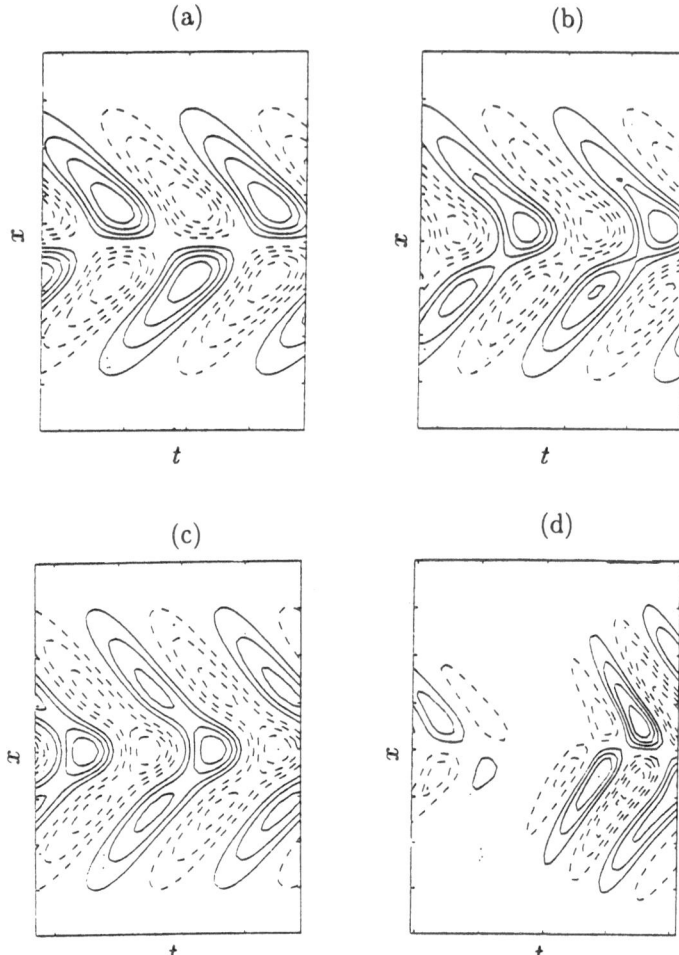

Fig. 14. Butterfly diagrams showing modulation and symmetry-breaking. (a) Periodic dipole for $D = -300, \tau = 0.1$. (b) Type 1 modulation: mixed-mode solution for $D = -400$. (c) Periodic quadrupole for $D = -500$. (b) Type 2 modulation: mixed-mode solution for $D = -1200$. Note the marked loss of symmetry at the grand minimum. (After Tobias 1997.)

the loss of symmetry near a grand minimum, when the activity is concentrated in a single hemisphere. This is precisely what happened during the Maunder Minimum, when there there was an interval of 40 years during which virtually all the sunspots that were observed were in the southern hemisphere (Ribes & Nesme–Ribes 1993).

Taken together, the results in this section show that there are two possible types of modulation in a nonlinear dynamo. Modulation of Type 1, which first appears at lower dynamo numbers, involves nonlinear interactions between

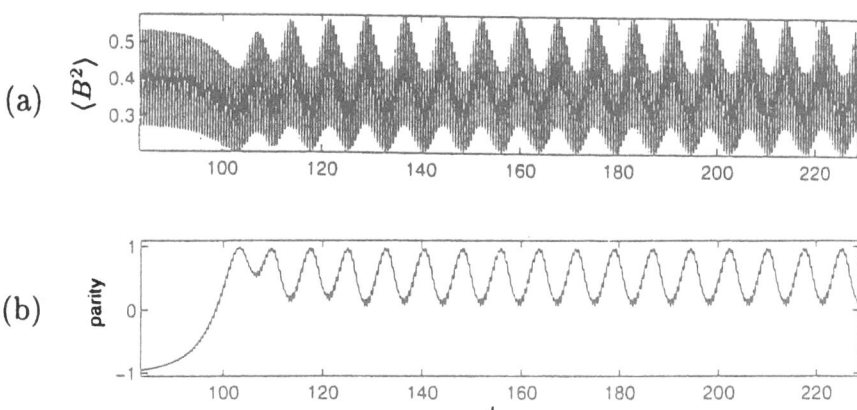

Fig. 15. Type 1 modulation: mixed-mode solution for $D = -400$. (a) $\langle B^2 \rangle$ and (b) the parity P as functions of time. (After Tobias 1997.)

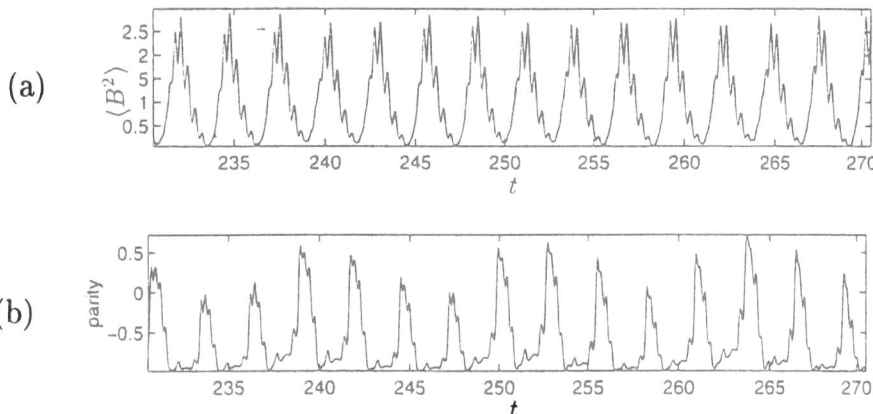

Fig. 16. Type 2 modulation: mixed-mode solution for $D = -1200$. (a) $\langle B^2 \rangle$ and (b) the parity P as functions of time. (After Tobias 1997.)

the dipole and quadrupole oscillations. Type 2 modulation, which leads to grand minima with only incidental loss of symmetry, occurs for higher values of $|D|$ but is a robust effect. The trajectories in phase space lie on a two-torus when the modulation first appears but as $|D|$ increases they approach the field-free subspace. Chaotic behaviour results from the breakdown of the torus and is associated with a heteroclinic region. Thus the low-order model of Section 4 illuminates the properties of the richer system studied here.

7 Modulation of the Solar Cycle

Grand minima recur throughout the record of magnetic activity on the Sun. To understand the origin of such behaviour in the solar dynamo, we have to look to other similar stars, with different rotation rates and different values of the dynamo number D. Our aim has been to demonstrate that a deterministic description suffices to explain the observed aperiodic modulation, and we have shown that episodes of reduced activity are associated with chaotic behaviour in a suite of simple models. What all these models have in common, mathematically, is the destruction of a two-torus as it approaches an invariant hydrodynamic subspace.

In Section 6.2 we emphasized the distinction between two types of modulation. Type 1 is associated with symmetry-breaking and parity interactions but does not normally produce grand minima; it too can be described by a low-order system (Knobloch & Landsberg 1996). The amplitude of activity cycles is substantially affected by modulation of Type 2, which has been our principal concern. Both types occur in mean-field dynamo models, though Type 2 requires a separation of timescales, such as is produced when the macroscopic field affects differential rotation. (Moreover, the viscosity has to be low to yield a modulation period that is an order of magnitude longer than the cycle period.)

The models suggest that both types of modulation should be present in stellar dynamos. The order in which they appear may well depend on details of the dynamo process: although symmetry-breaking happens first in simple models, Type 2 modulation may predominate for stars like the Sun. Nevertheless, the fact that the Sun's magnetic field deviates only slightly from dipole symmetry does not mean that other stars, with different rotation rates, should have toroidal fields that are antisymmetric about their equators.

Although we have focused our attention on deterministic mechanisms, there are various other explanations for intermittent cyclic activity. Most relate to the effects of large-scale turbulent convection, which can itself be regarded as a deterministic process, giving rise to on/off intermittency (Platt, Spiegel & Tresser 1994), or else as a source of stochastic disturbances that modulate magnetic cycles (Schmitt et al. 1996; Ossendrijver & Hoyng 1996). Stochastic effects can also be combined with deterministic modulation (Tobias 1996b).

Most treatments of the solar dynamo rely on mean field electrodynamics and assume that nonlinear saturation depends on the macroscopic field (for instance, through the Malkus–Proctor effect that we have invoked here). An alternative approach is through the back-reaction of the fluctuating microscopic fields on helicity and turbulent transport of angular momentum (Kitchatinov et al. 1994), which may well be important in some stars. The Malkus–Proctor effect should nevertheless predominate if strong fields are segregated from the region with vigorous convection, and confined to a layer with weak turbulence and strong shear, as postulated for the Sun. This is the situation represented in Parker's (1993) model and in the nonlinear results presented here. In the long run, such issues can only be settled by constructing fully consistent three-dimensional models of nonlinear stellar dynamos. At the moment, however, such computations (even with turbulent transport represented by laminar diffusivities) still pose a considerable challenge.

In conclusion, we draw attention to the influence of the Sun's magnetic activity on the terrestrial climate. There is increasing evidence that solar activity is correlated with temperature variations on the Earth. On a 200-year timescale, there are tantalising coincidences between climatic change (the Medieval Warm Period and the Little Ice Age) and modulation of the solar cycle (the Medieval Maximum followed by the Spörer and Maunder Minima). Although any changes in solar irradiance associated with grand minima are likely to be small there are other mechanisms – such as resonant coupling, or the effects of ultraviolet emission or cosmic rays on the atmosphere and cloud formation – that could well be important. So the issues discussed here may have immediate practical relevance.

References

Baliunas S.L. and 26 others (1995): Chromospheric variations in main-sequence stars II. ApJ **438**, 269–287

Baliunas S.L., Jastrow R. (1990): Evidence for long-term brightness changes of solar-type stars. Nature **348**, 520–523

Brandenburg A., Krause F., Meinel R., Moss D., Tuominen I. (1989): The stability of nonlinear dynamos and the limited role of kinematic growth rates. A&A **213**, 411–422

Cattaneo F., Hughes D.W. (1996) On the nonlinear saturation of the turbulent α–effect. Phys. Rev. E, in press

Covas E., Tworkowski A., Brandenburg A., Tavakol R. (1996): Robustness of truncated $\alpha\Omega$ dynamos with a dynamic α. Solar Phys., in press.

Ferriz Mas A., Schmitt D., Schüssler M. (1994): A dynamo effect due to instability of magnetic flux tubes. A&A **289**, 949–956

Foukal P.V. (1990): *Solar Astrophysics* (Wiley–Interscience: New York)

Guckenheimer J., Holmes P. (1986): *Nonlinear Oscillations, Dynamical Systems and Bifurcations of Vector Fields* (2nd Impression, Springer: New York)

Hoyng P. (1988): Turbulent transport of magnetic fields. III. Stochastic excitation of global magnetic modes. ApJ **332**, 857–871

Hoyt D.V., Schatten K.H. (1996): How well was the Sun observed during the Maunder minimum? Solar Phys. **165**, 181–192

Jennings R.L. (1991): Symmetry breaking in a nonlinear $\alpha\omega$-dynamo. Geophys. Astrophys. Fluid Dyn. **57**, 147–189

Jennings R.L., Brandenburg A., Moss D., Tuominen I. (1990): Can stellar dynamos be modelled in less than 3 dimensions? A&A **230**, 463–473

Jennings R.L., Weiss N.O. (1991): Symmetry breaking in stellar dynamos. MNRAS **252**, 249–260

Jones C.A., Weiss N.O., Cattaneo F. (1985): Nonlinear dynamos: a complex generalization of the Lorenz equations. Physica **14D**, 161–176

Kirk V. (1991): Breaking of symmetry in the saddle-node-Hopf bifurcation. Phys. Lett. A **154**, 243–248

Kitchatinov L.L., Rüdiger G., Küker M. (1994): Λ–quenching as the non-linearity in stellar turbulence dynamos. A&A **292**, 125–132

Knobloch E., Landsberg A.S. (1996): A new model of the solar cycle. MNRAS **278**, 294–302

Malkus W.V.R., Proctor M.R.E. (1975): The macrodynamics of the alpha-effect in rotating fluids. J. Fluid Mech. **67**, 417–444

Morfill G.E., Scheingraber H., Sonett C.P. (1991): Sunspot number variations: stochastic or chaotic? In *The Sun in Time*, ed. C.P. Sonett, M.S. Giampapa, M.S. Matthews (University of Arizona Press: Tucson), pp. 30–58

Noyes R.W., Baliunas S.L., Guinan E.F. (1991): What can other stars tell us about the Sun? In *Solar Interior and Atmosphere*, ed. A.N. Cox, W.C. Livingston, M.S. Matthews (University of Arizona Press: Tucson), pp. 1161–1185

Ossendrijver A.J.H., Hoyng P. (1996): Stochastic and nonlinear fluctuations in a mean field dynamo. A&A **313**, 959–970

Ott E. (1993): *Chaos in Dynamical Systems* (Cambridge University Press)

Parker E.N. (1955): Hydromagnetic dynamo models. ApJ **122**, 293–314

Parker E.N. (1979): *Cosmical Magnetic Fields: their origin and activity* (Clarendon Press: Oxford)

Parker E.N. (1993): A solar dynamo surface-wave at the interface between convection and nonuniform rotation. ApJ **408**, 707–719

Platt N., Spiegel E.A., Tresser C. (1994): The intermittent solar cycle. Geophys. Astrophys. Fluid Dyn. **73**, 147–161

Ribes J.C., Nesme–Ribes E. (1993): The solar cycle in the Maunder minimum AD1645 to AD1715. A&A **276**, 549–563

Saar S.H. (1996): Recent measurements of stellar magnetic fields. In *Magnetodynamic Phenomena in the Solar Atmosphere – Prototypes of Stellar Activity*, ed. Y. Uchida, T. Kosugi, H.S. Hudson (Kluwer: Dordrecht), in press.

Schlichenmaier R., Stix M. (1995): The phase of the radial mean field in the solar dynamo. A&A **302**, 264–270

Schmaltz S., Stix M. (1991): An $\alpha\Omega$ dynamo with order and chaos. A&A **245**, 654–661

Schmitt D., Schüssler M. (1989): Nonlinear dynamos I. One-dimensional model of a thin layer dynamo. A&A **223**, 343–351

Schmitt D., Schüssler M., Ferriz Mas A. (1996): Intermittent solar activity by an on–off dynamo. A&A **311**, L1–L4

Soderblom D.R., Baliunas S.L. (1988): The Sun among the stars: what the stars indicate about solar variability. In *Secular Solar and Geomagnetic Variations in the last 10,000 Years*, ed. F.R. Stephenson, A.W. Wolfendale (Kluwer: Dordrecht), pp. 25–49

Stix M. (1989): *The Sun: an Introduction* (Springer: Berlin)

Stuiver M., Braziunas T.F. (1989): Atmospheric ^{14}C and century-scale solar oscillations. Nature **338**, 405–408

Tobias S.M. (1995): *Nonlinear Solar and Stellar Dynamos* (Ph.D. Dissertation, University of Cambridge)

Tobias S.M. (1996a): Grand minima in nonlinear dynamos. A&A **307**, L21–L24

Tobias S.M. (1996b): Simple models of grand minima in the solar activity cycle. In *Magnetodynamic Phenomena in the Solar Atmosphere - Prototypes of Stellar Activity*, ed. Y. Uchida, T. Kosugi, H.S. Hudson, (Kluwer: Dordrecht), in press.

Tobias S.M. (1997): The solar cycle: parity interactions and amplitude modulation. A&A (to be submitted)

Tobias S.M., Weiss N.O., Kirk V. (1995): Chaotically modulated stellar dynamos. MNRAS **273**, 1150–1166

Tomczyk S., Schou J., Thompson M.J. (1995): Measurement of the rotation rate in the deep solar interior. ApJ **448**, L57–L60

Thompson M.J. and 25 others (1996): Differential rotation and dynamics of the solar interior. Science **272**, 1300–1305

Ulrich R.K., Boyden J.E., Webster L., Snodgrass H.B., Podilla S.P., Gilman P., Shieber T. (1988): Solar rotation measurements at Mount Wilson V. Reanalysis of 21 years of data. Solar Phys. **117**, 291–328

Vainshtein S.I., Cattaneo F. (1992) Nonlinear restrictions on dynamo action. ApJ **393**, 165–171

Watari S. (1996): Chaotic behavior of the north-south asymmetry of sunspots? Solar Phys. **163**, 259–266

Weiss N.O. (1990): Periodicity and aperiodicity in solar magnetic activity. Phil. Trans. R. Soc. London A **330**, 617–625

Weiss N.O. (1994): Solar and stellar dynamos. In *Lectures on Solar and Planetary Dynamos*, ed. M.R.E. Proctor, A.D. Gilbert, (Cambridge University Press), pp. 59-95

Weiss N.O., Cattaneo F., Jones C.A. (1984): Periodic and aperiodic dynamo waves. Geophys. Astrophys. Fluid Dyn. **30**, 305–341

Wilson P.R. (1994): *Solar and Stellar Activity Cycles* (Cambridge University Press)

Yau K.K.C. (1988): Analysis of pre-telescopic and telescopic sunspot observations. In *Secular Solar and Geomagnetic Variations in the last 10,000 Years*, ed. F.R. Stephenson, A.W. Wolfendale (Kluwer: Dordrecht), pp. 161–185

Zel'dovich Ya.B., Ruzmaikin A.A., Sokoloff D. (1983): *Magnetic Fields in Astrophysics* (Gordon and Breach: New York)

Dynamics of Flux Tubes
in the Solar Atmosphere: Observations

S.K. Solanki

Institute of Astronomy, ETH-Zentrum, CH-8092 Zürich, Switzerland

Abstract. Ground-based observations of the dynamics of solar magnetic flux tubes are reviewed. First a brief overview of the range of dynamic phenomena observed in both the largest and smallest flux tubes, i.e. in sunspots and magnetic elements, is given. Then three such phenomena — the Evershed effect, steady flows in small-scale magnetic features and oscillations in such features — are selected and discussed in detail.

1 Introduction

Flux tubes are, roughly speaking, bundles of magnetic field lines, which at the solar surface are observed as concentrations of more or less vertical magnetic field. Most of the observed flux and almost all the magnetic energy at the solar surface is concentrated into flux tubes. Since flux tubes emerge almost vertically through the solar surface we basically observe their cross-sections in the photospheric layers. These range in size from sunspots down to magnetic elements. Sunspots, the largest solar flux tubes are spatially easily resolved and their internal structure and dynamics can be studied in great detail. Magnetic elements, on the other hand, are in general spatially unresolved and indirect techniques must often be employed in order to observe their internal dynamics.

From an observational point of view the clearest evidence for flux tubes as magnetically distinct structures exists in the solar photosphere. For this reason the present review deals almost exclusively with observations of flux-tube dynamics in the photosphere. In this volume related reviews have been provided by Schmieder (1996) and Roberts and Ulmschneider (1996). The former, deals with dynamical phenomena in the upper atmosphere, while the theory behind flux-tube dynamics is the subject of the latter

Observations of dynamics related to flux tubes have recently been reviewed by, among others, Moore & Rabin (1985), Zwaan (1987, 1992), Stenflo (1989b, 1994), Schüssler (1990, 1992), Solanki (1990, 1993, 1995), Spruit et al. (1992), Lites (1992), Thomas & Weiss (1992) and Martin (1990).

A vast variety of phenomena fall under the heading of dynamics of flux tubes. In Sect. 2 some examples are listed. In Sects. 3 to 6 the following three topics are considered in greater detail: (1) observations of the Evershed effect, (2) observations of 'stationary' flows in and around magnetic elements, and

(3) observations of oscillations and waves in magnetic elements. Sects. 4–6 closely follow Solanki (1993).

2   Overview of dynamic phenomena

In this section I list the main dynamic phenomena found in sunspots and magnetic elements, with a brief description and selected references.

2.1   Observations of dynamics in sunspots

1. *Solar rotation*: Sunspots rotate at a different rate than the non-magnetic plasma (Ternulo et al. 1981, Gilman & Howard 1985, Balthasar et al. 1986). This difference can, with some theoretical input, provide information on the rotation law at the anchoring depth of the flux tube forming the sunspot.

2. *Meridional motions:* Sunspots can be used to trace large-scale equatorward or poleward flows (e.g. Kambry et al. 1991), which have been proposed as drivers of solar differential rotation.

3. *Motions of individual sunspots*: Such motions, i.e. local deviations from the rotation law, are observed as part of the evolution of active regions (e.g. Leka et al. 1994, Bumba et al. 1996). For example, preceding and following spots move away from each other (Gilman & Howard 1985) and at different speeds from the neutral line (Van Driel & Petrovay 1990), which gives information on the shape and the dynamics of the large flux tube underlying the spots and emerging through the solar surface in the form of an Ω loop (Petrovay et al. 1990, Moreno-Insertis et al. 1994).

4. *"Shearing" motions of sunspots relative to each other, or flux emergence within sunspots*: Such motions and processes can lead to energy build-up in the overlying magnetic arcade and its subsequent release in, e.g., flares (Hagyard et al. 1984, Kurokawa 1991, Wang 1994).

5. *Moat flow*: Small magnetic features and sometimes granules are seen to stream away from a sunspot (Harvey & Harvey 1973, Brickhouse & LaBonte 1988).

6. *Evershed effect*: An outward directed flow is observed in the photospheric layers of penumbrae and an inward directed flow in the chromospheric layers (see Sect. 3). The inward motion of bright and possibly dark features in the inner penumbra is also seen, sometimes accompanied by the outward motion of brighter structures in the outer penumbra (Muller 1973, 1976, Molowny-Horas 1994).

7. *Transition-region downflows*: Strong downflows are detected in transition-region spectral lines, mainly above sunspot umbrae. These downflows are probably related to the chromospheric inverse Evershed effect (Alissandrakis et al. 1988, Dere et al. 1990, Kjeldseth-Moe et al. 1993).

8. *Oscillations*: These are observed both in the umbra and the penumbra, with a 5 min period in photospheric layers and a 3 min period in chromospheric layers (Lites 1992).

9. *Running penumbral waves*: Time series in the Hα line of the chromospheric layers reveal waves travelling out from the umbra through the penumbra (Giovanelli 1972, Zirin & Stein 1972).

10. *Absorption of p-modes by sunspots*: The energy flux carried by p-modes towards a sunspot is considerably larger than the energy flux carried away from the sunspot. This absorption (and the scattering) can be used to obtain information on the subsurface structure of sunspots (Braun et al. 1987, 1992, Bogdan et al. 1993, cf. Bogdan 1992).

11. *Downflows below sunspots*: Recent helioseismological (tomographic) measurements by Duvall et al. (1996) have led to the discovery of a large-scale downflow below sunspots. Such a downflow had been proposed by Parker (1992) on theoretical grounds.

12. *Siphon flow*: A supersonic siphon flow has been observed across the neutral line of a δ-spot (Martínez Pillet et al. 1994).

2.2   Observations of dynamics of magnetic elements

1. *Solar rotation*: As in the case of sunspots the rotation law of magnetic patterns and features differs from that of the non-magnetic plasma (e.g., Snodgrass 1983, Stenflo 1989a, 1990, Howard 1990). The patterns are created mainly by large numbers of magnetic elements.

2. *Meridional motions:* Magnetic patterns also exhibit meridional motions (Howard 1991 and references therein).

3. *Evolution of active region plage*: The plages of active regions form, evolve, change shape and diffuse with time, implying motions of the underlying magnetic elements (e.g. Martin et al. 1985b, Strous et al. 1996).

4. *Horizontal displacement of quiet sun flux*: The supergranular flow determines the horizontal displacement of magnetic flux on the relevant scales in the quiet sun, with magnetic features being pushed to the supergranule boundaries (Livi et al. 1985, Zirin 1987, Martin 1990).

5. *Small-scale horizontal displacement*: On small scales magnetic elements are buffeted by the surrounding granules and undergo a random walk (Muller et al. 1994).

6. *Downflows*: Downflows within magnetic elements are either absent, or at the most small (Stenflo & Harvey 1985, Solanki 1986, Martínez Pillet et al. 1996). Magnetic elements are located in intergranular lanes and are surrounded by strong downflows (Title et al. 1987 Bünte et al. 1993a).

7. *Oscillations and waves*: There is evidence for longitudinal waves. The main frequency is five minutes, although shorter periods also seem to have been observed (Gionavelli et al. 1978, Fleck & Deubner 1991, Volkmer et al. 1995). There is mainly indirect evidence for transverse waves.

8. *Siphon flow*: A supersonic siphon flow has been detected across the neutral line of an active region (Rüedi et al. 1992b, Degenhardt et al. 1993).

3   The Evershed effect

3.1   The Evershed and the inverse Evershed effect

The dominant signature of photospheric dynamics in sunspots is the Evershed effect, named after its discoverer, J. Evershed. It is composed of a blueshift and (usually) a blueward asymmetry (i.e. an enhanced blue wing) of photospheric spectral lines in the discward part of the penumbra and a corresponding redshift and redward asymmetry in the limbward part of the penumbra (Evershed 1909, St John 1913, Maltby 1964, Schröter 1965, Lamb 1975, Wiehr et al. 1984, Ichimoto 1987, Dere et al. 1990, Shine et al. 1994, Rimmele 1994, Wiehr 1995, 1996).

The Evershed effect is height dependent. The line shifts decrease rapidly with height of line formation (St John 1913, Maltby 1964, Börner & Kneer 1992). Above the temperature minimum the line shifts change sign, with the discward part of the penumbra now showing redshifts. This so-called inverse Evershed effect has been observed by, e.g., St John (1913), Maltby (1975), Alissandrakis et al. (1988), Dere et al. (1990).

The simplest interpretation of the Evershed effect is an outflow from the umbra in the photosphere with a velocity of up to 5 km s^{-1} and an inflow with a larger velocity in the chromosphere. Nevertheless, the mass flux producing the inverse Evershed effect is only a few percent that in the photospheric flow (due to the orders of magnitude lower gas density in the chromosphere).

3.2   Inclination to the horizontal

In the photosphere the flow is nearly horizontal (e.g., Maltby 1964, Lamb 1975, Alissandrakis et al. 1988, Adam & Petford 1991, Börner & Kneer 1992). The magnetic field observed at a spatial resolution of a few arc sec, however, is inclined by up to 50° to the horizontal in the penumbra (Kawakami 1983, Lites & Skumanich 1990, Solanki et al. 1992, Lites et al. 1993, etc.). Since the field lines are frozen into the plasma this situation is at first sight unexpected. This problem was recognized by Lamb (1975) and Adam & Petford (1991), who reconstructed both the magnetic and velocity vectors of sunspots. Possible scenarios to explain both observations simultaneously are:

1. The gas flows in field-free channels embedded in the otherwise magnetic penumbra. In this manner the flow and the magnetic field would not interfere with each other. The observations show, however, that it is indeed the magnetized gas which is flowing (Solanki et al. 1992, 1994), ruling out this hypothesis.

2. The component of the flow perpendicular to the field-lines drags these along with it (proposed by Lamb 1975, Arena et al. 1990, Adam & Petford 1991). Since sunspots do not 'explode', i.e. expand rapidly with time, this scenario appears unlikely. Setting an upper limit of 3″ per day on the expansion of sunspots we obtain an upper limit of 30 m s^{-1} on the cross-field flow component, roughly an order of magnitude smaller than the values deduced by Lamb (1975) and Adam & Petford (1991).[1]
3. The Evershed effect is not caused by a material flow at all, but is rather produced by waves (Maltby & Erikson 1967, Erikson & Maltby 1968, Bünte et al. 1993b). A detailed analysis shows, however, that the wave hypothesis is inconsistent with the observations (Bünte & Solanki 1994).
4. The penumbral magnetic field is inhomogeneous on a small scale, with a horizontal component parallel to the flow and an inclined component. In this picure the Evershed flow is restricted to the horizontal component, as first proposed by Title et al. (1993). High resolution observations and measurements of broad-band circular polarization support a complex magnetic structure of this type (Degenhardt & Wiehr 1991, Schmidt et al. 1992, Title et al. 1993, Solanki & Montavon 1993, Hofmann et al. 1994).

Even slight deviations of the velocity vector from the horizontal are important for understanding the nature of the Evershed effect and the basic structure of the sunspot magnetic field. If the flow is inclined downward near the outer penumbral edge it is suggestive of material and (by proxy) magnetic field lines returning into the solar interior at the sunspot boundary. The measurement of such small deviations is extremely difficult, since small false vertical velocities can easily be introduced by, e.g., errors in the zero-level of the velocity, which is typically uncertain by 200–300 m s^{-1} or more (Shine et al. 1994). Thus Rimmele (1995b) finds evidence for a downflow of 400–600 m s^{-1} in the outer penumbra (and an upflow of 200–300 m s^{-1} in the inner penumbra), whereas Shine et al. (1994) deduce a small upflow (200–400 m s^{-1}) near the outer penumbral edge. The zero level is fixed by assuming that either the quiet sun or the umbra is at rest, i.e. shows unshifted spectral lines. Its uncertainty is due to the granular blue-shift of spectral lines outside sunspots and due to scattered light from the inner penumbra and the photosphere affecting the umbra. In the absence of such scattered light the umbra is expected to show no time-averaged shift (Beckers 1977).

The observations of Rimmele (1995b) are particularly intriguing, since they are suggestive of a flow along a shallow Ω-loop. This result is based on the asssumption that all line-of-sight velocities are vertical, which in turn derives from the location of the sunspot, which was very close to disc centre.

---

[1] Even if the field-line motions are assumed to be vertical an expansion of the sunspot cannot be avoided, since the downward field-line motions in the penumbra pull the extensive superpenumbral magnetic canopy (Giovanelli 1980, Solanki et al. 1992) down into the continuum-forming layer.

Significantly, as it turns out, the spot was not exactly at disc centre, however. Rimmele reports that the sunspot moved from $\mu = \cos\vartheta = 0.999$ to 0.985 during the two hours over which he observed it. The average μ of his observations, $\langle\mu\rangle \approx 0.992$, corresponds to $\langle\theta\rangle = 7.3°$. Horizontal velocities (v_H, e.g. due to the Evershed flow) can thus contribute to the line-of-sight velocity, v_{LOS}, although reduced by a factor $\sin\langle\theta\rangle \approx 0.13$. For a typical $v_H \approx 3$–4 km s^{-1} this gives $v_{LOS} = 400$–500 m s^{-1}, which is of the correct magnitude. This interpretation of Rimmele's observations is strengthened by considering his Fig. 1. The blue- and red-shifts do not form complete, symmetric rings around the otherwise symmetric sunspot, but rather appear to be concentrated on opposite sides of the spot, as expected for the normal, horizontal Evershed effect.

Rimmele's observations do, however, contain an additional piece of information that is independent of the exact location of the sunspot. The blue-shift is located closer to the umbra than the red-shift. Consequently, the flow appears to follow a convex path (as seen from Earth), with the exact angles to the solar surface being uncertain, however. At first sight the larger magnitude of the redshift (than the blueshift) suggests that the flow does indeed curve downwards at the outer penumbral edge. The conflicting observations of Shine et al. (1994), the uncertainty in the zero-level, and the earlier conclusion reached by Ichimoto (1987) and Rimmele (1995a) that the Evershed flow is concentrated into elevated filaments near the outer penumbral edge make this conclusion less certain.

Even the convexity of the Evershed flow-lines is not unchallenged. For example, Stellmacher & Wiehr (1971) claim that the velocity-neutral line in a sunspot lies on the discward side of the umbra. This appears to be supported by the majority of the observations of Pevtsov (1992). If this result is correct then it implies that at least in the inner penumbra the flow lines are concave. Note, however, that these measurements depend critically on the rather uncertain velocity zero-level (e.g., Arena et al. 1990), and need to be redone with particular care regarding this parameter.

One commonly used method to determine all three Evershed velocity components, radial, vertical and azimuthal, is to measure Doppler shifts over the whole sunspot and to assume that the velocity structure is axially symmetric (Haugen 1969). Unfortunately, any departure from axial symmetry in the true velocity will induce errors in the deduced velocity structure. Thus the (normally small) vertical and azimuthal velocities derived in this manner in the photospheric layers of sunspots (Lamb 1975, Adam & Petford 1991) need to be treated with caution, although the larger radial velocities are probably relatively reliable. Chromospheric vertical velocities obtained with this technique (Dere et al. 1990), being considerably larger, are also more reliable.

In conclusion, the balance of the evidence is in favour of convex flow- and field-lines, but it is unclear from the observations whether the flowing material returns into the sun at the outer sunspot boundary or continues moving upwards.

3.3   Horizontal structuring and vertical gradient
of the Evershed effect

If the Evershed effect is indeed restricted to filaments with horizontal field, then it must be highly structured. A number of authors (e.g. Shine et al. 1994, Wiehr 1995) do find a correlation of the Evershed effect with the dark penumbral filaments, or argue for such a correlation based on indirect observational evidence.

On the other hand, some authors have not found any significant correlation between flow velocity and brightness (e.g. Wiehr & Stellmacher 1989, Lites et al. 1990). Observations which may reconcile the contradictory results have been presented by Rimmele (1995a). He finds a good anticorrelation between velocity and brightness as long as he compares similar levels of the atmosphere. The correlation disappears if two very different levels are compared. For example, the velocity measured in the core of stronger lines does not correlate well with continuum intensity. This result is in agreement with measurements which show that line core intensity also does not correlate with continuum intensity (Wiehr & Stellmacher 1989). An alternative explanation for the different results obtained by different groups has been suggested by Wiehr & Degenhardt (1992). They maintain that a good correlation of the Evershed effect with continuum intensity minima is only achieved if the spatial resolution is better than $0\rlap{.}''5$.

Additional evidence for the inhomogeneity of the flow is provided by line profiles exhibiting a kink (Bumba 1960, Holmes 1961, Wiehr 1995), which is usually interpreted in terms of two spatially separated velocity components producing a complex line profile (e.g., Stellmacher & Wiehr 1971).

Another diagnostic having a bearing on the spatial inhomogeneity of the Evershed effect is the line asymmetry. In principle it is possible to explain such an asymmetry with either a vertical or a horizontal velocity gradient or distribution. There is evidence for both a vertical velocity gradient (see below) and a horizontal velocity distribution (see above). The directly observed vertical velocity gradient (from core shifts of lines formed at different heights), however, appears to be too small to account for all the observed asymmetry (Stellmacher & Wiehr 1980), and certainly cannot reproduce observed spectral lines showing two distinct velocity components (Wiehr 1995).

A novel approach, based on the automated least-squares fitting (also called the inversion) of line profiles, has been taken by Del Toro Iniesta et al. (1994). They derive horizontal fluctuations and vertical gradients of the Evershed velocity along different slices through a sunspot penumbra, and find that both the velocity and its large vertical gradient fluctuate along a slice. Their result favours the role of the vertical gradient in producing the line asymmetry. It is, however, based on the inversion of a single spectral line, so that they probably cannot properly distinguish between vertical gradients and sub-resolution horizontal fluctuations. The inversion approach is quite promising, though, and should be applied to more suitable data, possibly allowing for a more flexible model.

Undoubted is the importance of a vertical velocity gradient for the production of the broad-band circular polarization observed in penumbrae (Illing et al. 1975, Sánchez Almeida & Lites 1992, Solanki & Montavon 1993). The sense of the velocity gradient is not determined by these observations, however. Only the relative sign of the gradient of the LOS velocity and the magnetic vector enters into the production of net circular polarization over a spectral line.

The importance of the model geometry has been demonstrated by Degenhardt (1993). He showed that the observed line shifts *and* asymmetries can be reproduced, contrary to conventional wisdom, by a velocity that *increases* with height, if the filling factor of the flowing material decreases with height (since the line shifts reflect the product of velocity and filling factor). Thus the same observations can be reproduced by a velocity increasing or decreasing with height.

3.4   The upper atmosphere

The inverse Evershed effect observed in the chromosphere has a significant vertical component. According to the reconstructions of Dialetis et al. (1985), Alissandrakis et al. (1988) and Dere et al. (1990) the flow is almost horizontal in the superpenumbra, but becomes increasingly vertical towards the umbra. At disc centre downflows — called 'coronal rain' — can be seen in Hα above umbrae. EUV spectra have allowed such downflows to be followed into the overlying transition zone (e.g. Foukal 1976, 1978; Bruner et al. 1976, Brueckner et al. 1978, Nicolas et al 1982). Abdussamator (1971) and Lites (1980), however, see no such downflows.

According to Nicolas et al. (1982) sub- and supersonic downflows can be present within the same umbra. The observed velocity structure changes with temperature. At relatively low temperatures (6000 – 10 000 K, probed by EUV lines of neutral atoms) only subsonic velocities are seen. With increasing temperature the maximum of these subsonic downflow velocities increases, from less than 5 km s^{-1} observed in O I to approximately 40 km s^{-1} in O V. At even higher, truly coronal temperatures no downflow greater than approximately 10 km s^{-1} are observed. At high spatial resolution separate line components with shifts corresponding to supersonic velocities are often seen in addition to these subsonic downflows (Kjeldseth-Moe et al. 1988, 1993). This supersonic component is shifted by approximately 100 km s^{-1}, irrespective of the temperature (within the typical transition region temperature range over which this velocity component is present). These supersonic flows are extremely small-scale. The inverse Evershed effect at transition-region temperatures is hence highly inhomogeneous and possibly strongly filamented.

Also, the absence of supersonic flows at chromospheric temperatures implies the presence of shocks at higher temperatures, or greater heights.

The velocity of the inverse Evershed flow may possess a significant azimuthal component, which increases with distance from the sunspot (e.g., Maltby 1975, Dere et al. 1990). Such a twist is also seen in the superpenumbral Hα fibrils themselves (e.g., Hale 1908, 1930, Richardson 1941, Dere et al. 1990) and sometimes in extrapolations of photospheric magnetic fields into the chromosphere (e.g., Schmieder et al. 1989). It has recently been explained in terms of cyclonic motions in the presence of a magnetic field (Peters 1996).

Finally, Maltby (1975), from Hα Dopplergrams, finds that the chromospheric inflow is concentrated into channels with apparent average widths and lengths of $1''.6$ and $14''$, respectively, but with a very large scatter in these values. He argues for a loop-like structure of these channels, with the outer, upflowing part lying in the superpenumbra, well outside the penumbra.

3.5   The Evershed effect outside the penumbra

There has been considerable controversy about the presence or not of the Evershed effect beyond the outer boundary of the penumbra. Some investigators have presented convincing observations that the Evershed effect ends abruptly at the sunspot boundary (e.g., Wiehr 1996), while others have argued equally convincingly that it continues well beyond the visible boundary of the sunspot (e.g., Dere et al. 1990, Rimmele 1994). Recently it could be shown that although the Evershed effect does continue outside the sunspot boundary, it does so only above the base of the magnetic canopy (i.e. in the upper half of the photosphere; Solanki et al. 1994). The results of previous attempts to observe the photospheric Evershed effect outside the visible sunspot thus depend strongly on the type of spectral line employed to measure it and the exact technique used to determine the line shift and/or asymmetry.

Wiehr (1996) argues against a continuation of the Evershed effect, even above the canopy base, on the grounds of the disappearance of the Stokes I shift of lines of different strength just outside the penumbral boundary (Stokes I is the normal unpolarized spectrum). Now, Stokes I profiles may have quite different formation heights in the presence of magnetic canopies than in the quiet sun. This and other problems makes such observations less than ideal for the detection of flows in the canopy. Needed are new Stokes V, Q and U (i.e. polarized) observations, as well as radiative transfer calculations to study the influence of the change in the atmosphere at the transition from the penumbra to the quiet sun on the line profiles.

Even if the flow continues into the superpenumbral canopy the mass flux there is much smaller than in the penumbra (Solanki et al. 1994). If this is correct then either the material flowing through the penumbra returns to the solar interior at its outer boundary, or the Evershed effect is not a steady flow.

3.6   The Evershed effect as a non-stationary phenomenon

Recent observations have provided evidence that the Evershed flow is not entirely stationary. Time series of Dopplergrams recorded by Shine et al. (1994) show an average outwardly directed velocity of 3–4 km s^{-1} which, however, exhibits peak-to-peak modulations of 1 km s^{-1}. This modulation has an irregular repetitive behaviour, with a typical interval between peaks of roughly 10 minutes. Rimmele (1994) also finds the Evershed effect to be composed of velocity packets that repeat irregularly on a time scale of 15 minutes and that propagate with speeds of 2–5.5 km s^{-1}. Earlier evidence for a time-dependent outflow in penumbrae had been found by Schröter (1967), who studied the outward motion of bright structures, and Maltby (1975), who analyzed the inverse Evershed effect observed in Hα and found that the flow channels have a half-life of roughly 5 minutes.

4   Stationary flows in magnetic elements seen in the Stokes V zero-crossing wavelength

A direct indicator of flows within magnetic elements is the zero-crossing wavelength λ_V of Stokes V. Stokes V is the difference between right and left circular polarization and λ_V corresponds to its central wavelength. Older observations of λ_V showed sizeable downflows ranging up to 1.6 km s^{-1}. Much of this work was done with the line-centre-magnetogram technique introduced by Giovanelli & Ramsay (1971) and extensively applied by Giovanelli & Brown (1977) and Giovanelli & Slaughter (1978). It is based on the idea that for an *antisymmetric* Stokes V profile the signal disappears if the single magnetograph slit used is centred at λ_V. Any shift of λ_V with respect to the slit position produces a signal whose strength is a measure of the shift. The weakness of this technique is that asymmetric Stokes V profiles produce spurious shifts. For photospheric spectral lines this technique gives λ_V shifts of 0.5 km s^{-1}. For stronger lines, with cores formed in the chromosphere, the shift disappears. Other measurements have also shown λ_V shifts. For example the Stokesmeter data of Harvey et al. (1972) and the magnetograms of Stenflo (1973) suggested downflows of 0.5 km s^{-1}. Harvey (1977a) published a value of 1.6 km s^{-1}. Wiehr (1985) found shifts ranging from 0 to 2 km s^{-1} and Scholier & Wiehr (1985) also found downflows in the majority of the spectra analysed by them.

   Since the mid 1980s evidence against the presence of λ_V shifts greater than 250 m s^{-1} has accumulated. Investigators not finding any significant λ_V shift are: Stenflo & Harvey (1985), Solanki (1986), Stenflo et al. (1987a), Muglach & Solanki (1992), Fleck (1991) and Martínez Pillet et al. (1996). The last mentioned authors do find evidence of a small redshift in the magnetic features of approximately 100 m s^{-1} at large μ, with indications of a small blueshift at small μ. They also found evidence for a decrease in the average λ_V shift with increasing magnetic filling factor.

Spatially localized snapshots of λ_V often show significant blue or redshifts (up to 1 km s^{-1}) relative to the local or spatially averaged Stokes I (Fleck 1991, Amer & Kneer 1993, Martínez Pillet et al. 1996). A part of the shifts relative to the local Stokes I profile may have to do with the strong variability of the wavelength of the Stokes I profile (Fleck 1991), but not the comparisons with the spatially averaged Stokes I. When averaged over spatial position (Solanki & Pahlke 1988, Amer & Kneer 1993, Martínez Pillet et al. 1996), or time (Fleck 1991) the λ_V shifts in general become less than approximately 0.2–0.3 km s^{-1}, both when compared to laboratory wavelengths (Solanki 1986, Martínez Pillet et al. 1996) or to Stokes I profiles, if the relevant corrections for solar rotation, convective blueshift, etc. have been carried out. This is true for all lines formed between the low photosphere and the lower to middle chromosphere. As pointed out by Solanki (1986) these observations do not contradict the large flows (10–20 km s^{-1}) observed in the C IV doublet in the transition zone of active regions (e.g. Doschek et al. 1976, Lites et al. 1976, Feldman et al. 1982), even in the presence of extensive magnetic canopies, such as those proposed by Gabriel (1976), or Solanki & Steiner (1990) and observed by Giovanelli (1980) and Jones & Giovanelli (1983).

Can the contradiction between the older and the newer measurements of downflow velocity be reconciled? The main difference between the observations that show strong downflows and those that do not is that the former have a lower *spectral* resolution. Since Stokes V profiles of almost all spectral lines are asymmetric, with stronger blue wings than red wings near disc centre, spectral smearing tends to shift the zero-crossing wavelength towards the red, thus simulating a downflow (Solanki & Stenflo 1986). The magnitudes of the redshifts observed by, e.g., Giovanelli & Slaughter (1978) and Wiehr (1985) can be reproduced by smearing Stokes V profiles observed with high spectral resolution (FTS data) by an amount appropriate to the instrumental parameters relevant to the downflow observations (Solanki & Stenflo 1986).

This explanation works for all observations in the visible of spatially moderate resolution showing large redshifts (cf., Grossmann-Doerth et al. 1996), except those of Scholier & Wiehr (1985) and those involving the Fe I 15648.5 Å line. The predominance of redshifts observed by Scholier & Wiehr (1985) turns out to be a selection effect. A larger sample of their observations was analysed by Solanki & Pahlke (1988), who found that averaged over all the profiles no net shift greater than 200 m s^{-1} is visible.

Therefore, only the Fe I 15648.5 Å line remains, with its chequered history of λ_V shifts. Harvey (1977a) found a zero-crossing redshift of 2.2 km s^{-1} with respect to the local Stokes I wavelength of this line. The convective blueshift affecting the Stokes I profile (Nadeau 1988) reduces the actual λ_V redshift to 1.6 km s^{-1}. This value played a pivotal role in the compilation of redshifts by Giovanelli & Slaughter (1978). The next measurement of this line was published ten years later by Stenflo et al. (1987b). At $\mu \approx 1$ they found a redshift of 1.1 km s^{-1} with respect to the local Stokes I. Although much smaller than the value measured by Harvey (1977a), this observation still suggested a downflow of approximately 0.5 km s^{-1} within magnetic features. Muglach & Solanki (1992), however, have demonstrated that if all the un-blended lines in the H-band are considered, then no residual downflow larger than 0.3 km s^{-1} is found in magnetic features. They also showed that for the Fe I 15648.5 Å line, the error in the λ_V position is particularly large by virtue of its large Zeeman splitting and the resulting minute $dV/d\lambda$ at λ_V (where V refers to Stokes V). Thus, a line that is a good magnetic diagnostic is not always a good velocity diagnostic. For the high S/N observations of Stenflo et al. (1987b) an uncertainty of approximately 0.6 km s^{-1} is deduced for the λ_V of this line. For the observations of Harvey (1977a), with their lower S/N, the uncertainty is proportionally larger.

The largest flow velocities within small-scale magnetic features are associ-ated with magnetic neutral lines. Close to the neutral line of an active region Rüedi et al. (1992b) observed an upflow and a lower field strength in one po-larity, while a downflow coupled to a higher field strength was present in the other polarity. This is exactly the signature predicted by siphon flow models. Subsequently Degenhardt et al. (1993), from a comparison of the observed V profiles with the profiles generated by a grid of siphon-flow models developed by Montesinos & Thomas (1993), concluded that the observed flow was su-personic near the top of the loop connecting the 2 polarities. Due to its low formation height the 15648 Å line is formed below the shock in the down-flowing leg and therefore only reveal line shifts corresponding to subsonic velocities. Martínez Pillet et al. (1994) later observed shifts of photospheric V profiles corresponding directly to supersonic velocities (up to 14 km s^{-1}). Their observations, however, were made in a δ-spot, where the physical con-ditions are rather different from those in magnetic elements. Finally, Rüedi et al. (1992a) also noticed the presence of relative shifts between the Stokes V profiles of 2 magnetic components in the same spatial element. The origin and implications of these shifts, which are seen only in a minority of the spectra, is still unclear. Also unclear is the origin of the line shifts (of both signs) seen in high resolution spectra. One possible explanation is the presence of siphon flows along loops connecting magnetic elements of opposite polarity, with upflowing and downflowing legs being randomly distributed (Degenhardt & Kneer 1992). Another possibility, namely that these shifts represent snap-shots of non-stationary motions, is discussed in Sect. 6. Finally, at present we also do not know the cause of the small shifts (≈ 100 m s^{-1}) of the spa-tially averaged V profiles observed by Martínez Pillet et al. (1996). Even such small velocities would drain the corona on a time scale of hours to days in open field regions unless sufficient matter can diffuse across the field lines into the flux tube. An alternative explanation involves a net Doppler shift produced by non-stationary velocities within the flux tube.

In conclusion, there is currently no compelling observational evidence for the presence of stationary flows $\gtrsim 250$ m s^{-1} within small-scale magnetic features if we average over many of them. Individual features can exhibit substantial line shifts. It is also to be expected that large-scale stationary flows with substantial velocities are present under some circumstances. Siphon flows fall under this category.

5 Stationary flows diagnosed from the Stokes V asymmetry

Another diagnostic of stationary flows is the blue-red asymmetry of Stokes V. Stenflo et al. (1984a) discovered that Stokes V profiles observed in faculae and the network near $\mu = 1$ are strongly asymmetric. Both the absolute value of the area, A_b, and amplitude, a_b, of the blue Stokes V wing differ from the absolute value of the red wing area, A_r, and amplitude, a_r, respectively. Quantitatively, this asymmetry may be expressed by the relative amplitude, δa, and area, δA, asymmetry defined as

$$\delta a = \frac{a_b - a_r}{a_b + a_r} \quad \text{and} \quad \delta A = \frac{A_b - A_r}{A_b + A_r}. \tag{1}$$

Close to solar disc centre both δa and δA are positive for most Fe I (Solanki & Stenflo 1984) and Fe II lines (Solanki & Stenflo 1985). For the weakest lines noise and blends do not allow the intrinsic asymmetry to be determined, while the strongest lines do not appear to show a significant δA. For photospheric lines of intermediate and large strength $\delta a > \delta A$.

These results have been confirmed for smaller samples of lines, but for many more solar regions by, e.g., Stenflo & Harvey (1985), Wiehr (1985), Grossmann-Doerth et al. (1996) and Martínez Pillet et al. (1996). Lines in the infrared H-band are also asymmetric in the same sense as the lines in the visible, but the magnitude of the asymmetry is smaller (Muglach & Solanki 1992, Rüedi et al. 1995), while in the IR J-band (1–1.3 μm), which harbours stronger lines, surprisingly large asymmetries have been observed (Rüedi et al. 1995). Closer to the limb both δa and δA change sign, i.e. the red Stokes V wing becomes stronger than the blue wing (Stenflo et al. 1987a, Pantellini et al. 1988, Grossmann-Doerth et al. 1996, Martínez Pillet et al. 1996). The observed centre-to-limb variation of δA has been used by Mürset et al. (1988) to reproduce the broad-band circular polarization measured by Kemp et al. (1987), which therefore represent an independent confirmation of the direct measurements of δA. Hence there is good qualitative agreement between the various measurements. Quantitatively, however, there are some differences in the centre-to-limb variation. Whereas according to Stenflo et al. (1987a) the δA of the medium-strong Fe I 5250.2 Å line changes sign at $\mu \approx 0.4$, Grossmann-Doerth et al. (1996) find this to be the case already at $\mu \approx 0.8$. An intermediate μ value (≈ 0.6–0.7) for the transition from positive to negative δA is favoured by Martínez Pillet et al. (1996) for Fe I 6301.5 and 6302.5 Å.

Due to the much larger sets of observations underlying their results the centre-to-limb variation of Grossmann-Doerth et al. (1996) and Martínez Pillet et al. (1996) are probably the more reliable. Note that for these intermediate strength lines δa does not become significantly negative near the limb.

Well before the first observations of asymmetric Stokes V profiles outside of sunspots, Illing et al. (1975) had already proposed a combination of magnetic field and velocity gradients to explain the broad-band circular polarization observed in sunspots (Illing et al. 1974a,b, 1975). Gradients of a cospatial velocity and magnetic field have been relatively successful in reproducing selected line parameters $(\delta A, \lambda_V)$ at disc centre of a number of lines, if a velocity decreasing with height and a magnetic field increasing with height is assumed (Sánchez Almeida et al. 1988, 1989). Both theory and observations, however, constrain the field strength to decrease with height in the main body of the flux tube forming a magnetic element (cf. Sects. 5.1.3 and 5.1.5 of Solanki 1993). If this constraint is accepted then the asymmetry and the absence of a λ_V shift cannot be simultaneously reproduced by flows within the magnetic features (Solanki & Pahlke 1988).

Van Ballegooijen (1985) pointed out that downflows outside the magnetic elements can produce an asymmetric Stokes V if the expansion of the magnetic features with height is taken into account, so that some rays pass through both magnetic and non-magnetic parts of the atmosphere. Grossmann-Doerth et al. (1988, 1989) showed that such downflows do not produce any zero-crossing shift. Solanki (1989) was able to reproduce the δA values of four widely different lines without shifting λ_V using a thin-tube model of a flux tube surrounded by a downflow. The observations require downflow velocities of 1-2 km s^{-1} in the immediate surroundings of magnetic features. In addition, the surroundings must be approximately 200–300 K cooler than the average quiet sun, highly suggestive of intergranular lanes. The observed δa values have so far not been reproduced (simultaneously with other line parameters) by any models incorporating only stationary flows (see Sect. 6 for more on the modelling of δa).

The centre-to-limb variation of δA has been modelled by Bünte et al. (1993a) and by Knölker et al. (1991). The latter authors analyse the Stokes V profiles resulting from dynamical flux-tube models incorporating a convective cell that resembles a granule in the surroundings of the flux tube. They find that the asymmetry of the calculated Fe I 5250.2 Å line changes sign near $\mu = 0.4$, in accordance with the observations of Stenflo et al. (1987a). Bünte et al. (1993a) used hydrostatic models of flux tubes with an external velocity field that is allowed to assume different forms. They could reproduce the observations of Stenflo et al. (1987a) only for a relatively sophisticated model incorporating overturning convective cells having hot up- and cool downflows, with the flux tubes positioned in the downflows. Their model turns out to be too simple to reproduce quantitatively the centre-to-limb variation of δA of a strong and weak Fe I line simultaneously, although it gives the correct qualitative dependence. This picture of magnetic elements located in the downflowing intergranular lanes is confirmed by direct imaging (e.g., Title et al. 1987).

A different view of the production of Stokes V asymmetry has been proposed by Sánchez Almeida et al. (1996). They argue that the magnetic features are composed of many narrow strands of field, each thinner than the horizontal photon mean-free path. These strands are interleaved with flowing material. Since they imply a completely different structure of solar magnetic elements compared to the generally accepted view of flux tubes with diameters of 100–500 km a demonstration is required that this concept can also successfully reproduce other observational constraints besides δA. Note that the increase of δa towards the edges of magnetic elements observed by Fleck (1991) at high spatial resolution supports the idea of a downflow in the surroundings of a "standard" flux tube.

In summary, the δA observed in magnetic elements appears to be largely a product of the granulation surrounding the magnetic elements. Conversely, δA can constrain the velocity and the temperature of the convection in active regions. It has the potential of becoming a powerful diagnostic of the properties of 'abnormal granulation'.

6   Observations of non-stationary velocities

A number of diagnostics support the presence of non-stationary velocities in magnetic elements: time-series of λ_V, spatial scatter of λ_V values, line widths of Stokes V and Stokes I, and the fraction of the Stokes V amplitude asymmetry, δa, which exceeds δA.

The most direct of these, time-series of λ_V, have so far mainly uncovered oscillations or waves with a period close to five minutes and an amplitude of 0.2–0.3 km s^{-1} in photospheric spectral lines (Giovanelli et al. 1978, Wiehr 1985, Fleck 1991, Muglach et al. 1995). The amplitude increases with height and reaches 0.75 km s^{-1} in Hα. Since the velocity fluctuations in the (lower) chromospheric Mg I b 5183 Å line occur on average 18 s later than in the photospheric Fe I 5166 Å line Giovanelli et al. (1978) conclude that the waves are propagating upwards with a phase velocity of approximately 100 km s^{-1}. Roberts (1983) has identified these with longitudinal tube waves (also called sausage modes) in the presence of radiative damping. Then 5 min oscillations, which lie below the cutoff frequency and should not propagate, have a propagating component even in linear theory. The oscillations within the magnetic elements (as seen in Stokes V) are almost in phase with oscillations in the non-magnetic atmosphere (as seen in Stokes I, Giovanelli et al. 1978, Wiehr 1985), so that either the coupling between the internal and external atmospheres is extremely strong, or both oscillations have a common excitation mechanism. The results appear to be relatively independent of the spatial resolution of the observations which varies between 1″ and 8″.

Oscillations with similar periods have also been observed in the Stokes V amplitude by Tanenbaum et al. (1971) and Dara et al. (1987). They either reflect the temperature perturbations associated with the velocity oscillations (e.g., in longitudinal tube waves, Herbold et al. 1984, Solanki & Roberts 1992) or instrumental effects, and are unlikely to represent actual changes in the magnetic flux, as suggested, possibly somewhat prematurely, by Dara et al. (1987).

There is one significant exception to the 5 minute periods, however. In an isolated magnetic feature near disc centre Volkmer et al. (1995) find 100s periods in λ_V, with a 280 m s^{-1} amplitude. According to linear theory this period corresponds to propagating longitudinal tube waves, i.e. acoustic waves propagating along the flux tube, modified by the magnetic field. Volkmer et al. (1995) estimate an energy flux of 1.6–2.3×10^7 erg cm^{-2} s^{-1}, sufficient to heat the associated chromospheric network if the wave is not significantly radiatively or acoustically damped on the way. Unfortunately, only a single magnetic feature was found to exhibit such oscillations and confirmation is important.

Another measure of possible non-stationary velocities is the deviation of the λ_V of individual profiles recorded at high spatial resolution from the spatially averaged value. Each such profile represents a snapshot of a small part of the solar surface. A group of such snapshots may show oscillations at different phases in different magnetic elements, resulting in the observed distribution of λ_V values of high resolution profiles. Such a distribution may, however, also only represent differences in stationary flows from one magnetic element to the next (e.g., siphon flows, Degenhardt & Kneer 1992). These velocity fluctuations were first stressed by Amer & Kneer (1993), who saw fluctuations in λ_V larger than 1 km s^{-1} for profiles with a spatial resolution of better than 1″. Newer high-resolution observations by Kneer & Stolpe (1996) in plages show somewhat larger fluctuations in λ_V (400 m s^{-1}) than in Stokes I (300 m s^{-1}). A similar rms value of λ_V is determined by Martínez Pillet et al. (1996) from a much larger set of observations with a spatial resolution of roughly 1″.

Stokes I time series, power and phase spectra also suggest differences between active regions and the magnetic network on the one hand and the supergranule cell interiors on the other (e.g., Deubner & Fleck 1990, Title et al. 1992a, Von Uexküll & Kneer 1995), although some observers do not find any significant differences (Mein & Mein 1976, Mein 1977). Low frequency oscillations ($\nu \lesssim 2.5$ mHz, i.e. periods in excess of roughly 6.5 minutes) are more prominent in plage and network, particularly at chromospheric heights, while higher frequencies, in particular the three minute peak (which was not noticed in Stokes V by Fleck 1991), are considerably stronger in the cell interiors. Furthermore, the power in the p-mode oscillations is significantly lower in network and plages than in the quiet sun. This is true even in the chromosphere, where most of the quiet-sun power resides in the 3 minute oscillations (Bocchialini et al. 1994). Considerable differences also exist between the phase spectra, which suggest downward propagating waves in the network (see Deubner & Fleck 1990 for details). Bocchialini & Baudin (1994), applying a wavelet analysis to the data of Bocchialini et al. (1994) confirmed this conclusion. The direct relation of such observations to mass motions within magnetic structures like flux tubes and canopies is difficult to establish, however.

The widths of the V profiles of relatively Zeeman-insensitive lines suggest that non-stationary velocities with much larger amplitudes than those deduced from time-series of λ_V are present in magnetic elements (e.g., Solanki 1986, Pantellini et al. 1988, Zayer et al. 1990). Typical rms values of such velocities lie between 1 and 3 km s^{-1}, depending on the spectral line and to some extent on the observed region. A NLTE analysis suggests that the Stokes V widths of photospheric lines can be reproduced by rms velocities of 1.5–2.5 km s^{-1} (Bruls & Solanki 1993), while the low chromospheric Mg I b-lines require a broadening of 1.6–1.8 km s^{-1} (Briand & Solanki 1995). These values are larger than those obtained from fits to quiet-sun Stokes I profiles. The Stokes V profile widths tell us that additional broadening velocities, somewhat larger than the 'turbulent velocities' in the quiet sun, are present in magnetic elements. In the quiet sun, however, most of the line broadening comes from the granulation (Nordlund 1984a), which is, of course, not present within the magnetic features. Therefore, the broadening must be due to an oscillatory or wave-like velocity field, or due to the presence of stationary up- and downflows within different spatially unresolved magnetic elements (e.g., siphon flows on small scales, as suggested by Degenhardt & Kneer 1992). A fraction of the broadening velocity is detected as the rms of the λ_V spatial fluctuations seen in high resolution observations (e.g., Amer & Kneer 1993, Martínez Pillet et al. 1996). It falls short of the line broadening velocity by a factor of 3–5, however (e.g., Kneer & Stolpe 1996).

Stokes I line profiles in active regions also support the large observed Stokes V widths. Cavallini et al. (1988), Immerschitt & Schröter (1989) and Brandt & Solanki (1990) find an increase in line width with increasing activity, respectively magnetic flux. This may be due to larger rms velocities in active regions, although Brandt & Solanki (1990) point out that a generally lower temperature in the non-magnetic part of active regions may also produce such an effect. The reduction in rms granular velocities in active regions, deduced by Nesis et al. (1989), Günther & Mattig (1991) and Title et al. (1992b) from high spatial resolution observations of line shifts and by Hanslmeier et al. (1991) from line bisectors, implies that any enhanced velocities in active regions must be distributed on small horizontal scales (possibly within the magnetic elements).

There is some evidence that the velocity amplitudes derived from the Stokes V widths decrease slightly with filling factor (Zayer et al. 1990). This is suggestive of a decrease in the excitation of tube waves and oscillations in regions with large filling factors, a view consistent with the increased lifetimes of granular cells in active regions (Title et al. 1989) and the decreased rms scatter of λ_V around its average with increasing filling factor (Martínez Pillet et al. 1996).

The velocity amplitudes deduced from line broadening near the limb are almost the same as near disc centre (Pantellini et al. 1988). Similarly, the rms of the λ_V of high resolution observations does not decrease towards the limb (Martínez Pillet et al. 1996). These two observations suggest that transverse wave modes (e.g., kink mode, torsional Alfvén mode) are just as strongly excited as longitudinal modes (e.g., sausage mode, overstable oscillations).

Another possible source of horizontal velocity visible in Stokes V is the motion of complete magnetic elements that are continually being shuffled around by the evolving granulation. It is still unclear, however, what magnitude this effect is expected to have. Whereas Title et al. (1990) mention *maximum* velocities of 3 km s^{-1} (but give no average value), Zirin (1985) reports an *average* horizontal velocity of only 0.06 km s^{-1}. His spatial resolution, however, is considerably lower, and he may well be missing much of the motion taking place on a small scale. De Boer & Kneer (1992) and Title et al. (1992a) mention the presence of strong horizontal motions of facular bright points, but give no numbers. Of interest in this respect are the short duration rapid horizontal movements of bright points observed by Muller et al. (1994). Identifying bright points with flux tubes Choudhuri et al. (1992, 1993) determine the amount of energy generated in the form of an upward propagating kink-wave pulse along a slender flux tube by the observed rapid horizontal movements. In the latter paper they include the effect of the temperature jump between the chromosphere and corona which reflects a part of the wave energy flux. They conclude that the energy flux is sufficient to heat the quiet solar corona, even in the presence of reflections at the transition zone. Finally, Volkmer et al. (1995) have presented the first evidence of quasi-periodic transverse motions of small-scale magnetic features (although only of a single feature).

The final parameter indicating non-stationary velocities in magnetic features is δa. Although Solanki (1989) could reproduce the observed δA values with an external downflow (Sect. 5), he was unable to reproduce the $\delta a/\delta A$ ratio. The δa values of the synthetic profiles were too small. He was able to increase $\delta a/\delta A$ by introducing, in an very simple model, an oscillatory motion with different weights for the up- and downflowing components. Solanki & Roberts (1992), however, found that the $\delta a/\delta A$ ratio of Fe I and II lines cannot be simultaneously reproduced if they consider a more sophisticated model, namely a linear tube wave calculated in the thin-tube approximation. Their result casts doubt on mechanisms of enhancing δa relative to δA that are based on a temperature-velocity correlation. Grossmann-Doerth et al. (1991) have, therefore, considered the influence of a simple representation of a non-linear wave with different up- and downflow velocity amplitudes.

They can roughly reproduce the δA, δa and λ_V values of five spectral lines observed near disc centre, if the upflow velocity is a factor of 3–4 smaller than the downflow velocity, whose amplitude may range from 2 to 4 km s^{-1}. Of course, their model also includes a downflow outside the flux tube.

Degenhardt & Kneer (1992) favour a superposition of differently directed siphon flows in different flux tubes as a method of enhancing δa. Basically, their proposal is a correlation between field strength and velocity. Another mechanism that may enhance δa, but which has not yet been analysed in detail, is the superposition of waves or oscillations with a distribution of amplitudes and/or phases across the cross-section of a magnetic element. Such oscillations have been seen in the self-consistent model calculations of Knölker et al. (1991) and Steiner et al. (1995, 1996). Martínez Pillet et al. (1996) recently found that δa decreases somewhat as the filling factor increases while δA remains essentially unchanged, suggesting that the influence of internal dynamics is also reduced. Grossmann-Doerth et al. (1996), on the other hand, report that both δa and δA decrease with increasing filling factor.

The difference $\delta a - \delta A$ remains positive even at very small μ, although it does decrease towards the limb (Martínez Pillet et al. 1996). Ploner & Solanki (1996), from an investigation of the line profiles produced by linear kink-mode waves, discovered that this wave mode produces δa and δA of similar magnitude, so that another source of the $\delta a - \delta A$ excess must be sought near the solar limb.

In summary, some diagnostics indicate the presence of broadening velocities up to 3 km s^{-1} (both Stokes V line broadening and δa give similar velocity amplitudes), but λ_V time series only show rather low amplitude perturbations, while snapshots at high spatial resolution yield λ_V with an intermediate rms. These results suggest that besides low amplitude tube waves excited by p-modes, other oscillations or waves are also present in magnetic features, but have rarely been visible in λ_V time series due to one or more of the following reasons: a) Many flux tubes with tube waves having different phases are present in the resolution element of a given observation. However, unless the flux tubes are very small (diameters $\lesssim 0.1$–$0.2''$) some residual signal of such waves should be visible in the high resolution observations of Fleck (1991). b) The tube waves have a very short wavelength (and consequently also a short period). If it becomes smaller than the half-width of the Stokes V contribution function of a particular line, the waves become invisible in the λ_V time series of that line (Solanki & Roberts 1992). Even in this case, however, the observational constraint that the time-averaged λ_V is close to the rest wavelength restricts the properties of any short period waves considerably. c) It is possible that the dynamical thin-tube approximation breaks down and most of the power is present in higher order modes having multiple nodes or at least different phases across the cross-section of the tube. Such waves are extremely difficult to detect, due to the cancellation of phases when averaging over even a single flux tube.

7 Conclusions

Observations of solar magnetic flux tubes have uncovered many and diverse dynamical phenomena. Also, a number of open problems associated with them have found a solution in recent years, but many remain, partly due to the difficulties faced by the observer, often associated with the fine-scale structure of the velocity field and the underlying magnetic field. With improving observational and analysis techniques we can look forward to the clarification of a number of the current uncertainties and questions, but probably also to the discovery of new and unexpected phenomena.

References

Abdussamatov H.I., 1971, *Solar Phys.* **16**, 384
Adam M.G., Petford A.D., 1991, *Solar Phys.* **135**, 319
Alissandrakis C.E., Dialetis D., Mein P., Schmieder B., Simon G., 1988, *Astron. Astrophys.* **201**, 339
Alissandrakis C.E., Dara H.C., Koutchmy S., 1991, *Astron. Astrophys.* **249**, 533
Amer M.A., Kneer F., 1993, *Astron. Astrophys.* **273**, 304
Arena F., Landi Degl'Innocenti E., Noci G., 1990, *Solar Phys.* **129**, 259
Balthasar H., Vázquez M., Wöhl H., 1986, *Astron. Astrophys.* **155**, 87
Beckers J.M., 1977, *Astrophys. J.* **213**, 900
Bocchialini K., Baudin F., 1995, *Astron. Astrophys.* **299**, 893
Bocchialini K., Vial J.-C., Koutchmy S., 1994, *Astrophys. J.* **423**, L67
Bogdan T.J., 1992, in *Sunspots: Theory and Observations*, J.H. Thomas, N.O. Weiss (Eds.), Kluwer, Dordrecht, p. 345
Bogdan T.J., Brown T.M., Lites B.W., Thomas J.H., 1993, *Astrophys. J.* **406**, 723
Börner P., Kneer F., 1992, *Astron. Astrophys.* **259**, 307
Brandt, P.N., Solanki, S.K.: 1990, *Astron. Astrophys.* **231**, 221
Braun D.C., Duvall T.J., Jr., LaBonte B.J., 1987, *Astrophys. J.* **319**, L27
Braun D.C., Duvall T.L., Jr., La Bonte B.J., Jefferies S.M., Harvey J.W., Pomerantz M.A., 1992, *Astrophys. J.* **391**, L113
Briand C., Solanki S.K., 1995, *Astron. Astrophys.* **299**, 596
Brickhouse N.S., LaBonte B.J., 1988, *Solar Phys.* **115**, 43
Brueckner G.E., Bartoe J.-D.F., Van Hoosier M.E., 1978, in *Proceedings OSO-8 Workshop*, E. Hansen, S. Schaffner (Eds.), University of Colorado Press, Boulder, CO, p. 380
Bruls J.H.M.J., Solanki S.K., 1993, *Astron. Astrophys.* **273**, 293
Bruner E.C., Jr., Chipman E.G., Lites B.W., Rottmann G.J., Shine R.A., Athay R.G., White O.R., 1976, *Astrophys. J.* **210**, L97
Bumba V., 1960, *Isv. Krim, Astr. Obs.* **23**, 212
Bumba V., Klvaña M., Kálman B., 1996, *Astron. Astrophys. Suppl. Ser.* **118**, 35
Bünte M., Solanki S.K., 1995, *Astron. Astrophys.* **297**, 861
Bünte M., Solanki S.K., Steiner O., 1993a, *Astron. Astrophys.* **268**, 736
Bünte M., Darconza G., Solanki S.K., 1993b, *Astron. Astrophys.* **274**, 478
Cavallini, F., Ceppatelli, G., Righini, A.: 1988, *Astron. Astrophys.* **205**, 278
Choudhuri A.R., Auffret H., Priest E.R., 1992, *Solar Phys.* **143**, 49

Choudhuri A.R., Dikpati M., Banerjee D., 1993, *Astrophys. J.* **413**, 811

Dara, H.C., Alissandrakis, C.E., Koutchmy, S.: 1987, *Solar Phys.* **109**, 19

De Boer, C.R., Kneer, F.: 1992, *Astron. Astrophys.* **264**, L24

Degenhardt D., 1993, *Astron. Astrophys.* **277**, 235

Degenhardt, D., Kneer, F.: 1992, *Astron. Astrophys.* **260**, 411

Degenhardt D., Wiehr E., 1991, *Astron. Astrophys.* **252**, 821

Degenhardt D., Solanki S.K., Montesinos B., Thomas J.H., 1993, *Astron. Astrophys.* **279**, L29

Del Toro Iniesta J.C., Tarbell T.D., Ruiz Cobo B., 1994, *Astrophys. J.* **436**, 400 penumbra.

Dere K.P., Schmieder B., Alissandrakis C.E., 1990, *Astron. Astrophys.* **233**, 207

Deubner, F.-L., Fleck, B.: 1990, *Astron. Astrophys.* **228**, 506

Dialetis D., Mein P., Alissandrakis C.E., 1985, *Astron. Astrophys.* **147**, 93

Doschek G.A., Feldman U., Bohlin J.D., 1976, *Astrophys. J.* **205**, L177

Duvall T.L., Jr., D'Silva S., Jefferies S.M., Harvey J.W., Schou J., 1996, *Nature* **379**, 235

Eriksen G., Maltby P., 1967, *Astrophys. J.* **148**, 833

Evershed J., 1909, *Monthly Notices Royal Astron. Soc.* **69**, 454

Feldman, U., Cohen, L., Doschek, G.A.: 1982, *Astrophys. J.* **255**, 325

Fleck, B.: 1991, *Rev. Mod. Astron.* **4**, 90

Fleck, B., Deubner, F.-L.: 1991, in *Mechanisms of Chromospheric and Coronal Heating*, P. Ulmschneider, E.R. Priest, R. Rosner (Eds.), Springer, Berlin, p. 19

Foukal P.V., 1976, *Astrophys. J.* **210**, 575

Foukal P.V., 1978, *Astrophys. J.* **223**, 1046

Gabriel A.H.: 1976, *Phil. Trans. Roy. Soc. London* **A281**, 339

Gilman P.A., Howard R., 1985, *Astrophys. J.* **295**, 233

Giovanelli R.G., 1972, *Solar Phys.* **27**, 71

Giovanelli, R.G.: 1980, *Solar Phys.* **68**, 49

Giovanelli, R.G., Brown, N.: 1977, *Solar Phys.* **52**, 27

Giovanelli, R.G., Ramsay, J.V.: 1971, in *Solar Magnetic Fields*, R. Howard (Ed.), Reidel, Dordrecht, *IAU Symp.* **43**, 293

Giovanelli, R.G., Slaughter, C.: 1978, *Solar Phys.* **57**, 255

Giovanelli, R.G., Livingston, W.C., Harvey, J.W.: 1978, *Solar Phys.* **59**, 49

Grossmann-Doerth, U., Schüssler, M., Solanki, S.K.: 1988, *Astron. Astrophys.* **206**, L37

Grossmann-Doerth, U., Schüssler, M., Solanki, S.K.: 1989, *Astron. Astrophys.* **221**, 338

Grossmann-Doerth U., Schüssler M., Solanki S.K., 1991, *Astron. Astrophys.* **249**, 239

Grossmann-Doerth U., Keller C.U., Schüssler M., 1996, *Astron. Astrophys.* in press

Günther, E., Mattig, W.: 1991, *Astron. Astrophys.* **243**, 244

Hagyard M.J., Smith Jr. J.B., Teuber D., West E.A., 1984, *Solar Phys.* **91**, 115

Hale G.E., 1908, *Astrophys. J.* **28**, 100

Hale G.E., 1930, *Astrophys. J.* **71**, 73

Hanslmeier, A., Mattig, W., Nesis, A.: 1991, *Astron. Astrophys.* **244**, 521

Harvey, J.W: 1977, in *Highlights of Astronomy*, E.A. Müller (Ed.), Vol. 4, Part II, p. 223

Harvey K.L., Harvey J.W., 1973, *Solar Phys.* **28**, 61

Harvey, J.W., Livingston, W., Slaughter, C: 1972, in *Line Formation in the Presence of Magnetic Fields*, High Altitude Obs., NCAR, Boulder, CO, p. 227

Haugen E., 1969, *Solar Phys.* **9**, 88

Herbold, G., Ulmschneider, P. Spruit, H.C. and Rosner, R.: 1985, *Astron. Astrophys.* **145**, 157

Hofmann J., Deubner F.-L., Fleck B., Schmidt W., 1994, *Astron. Astrophys.* **284**, 269

Holmes J., 1961, *Monthly Notices Royal Astron. Soc.* **122**, 301

Howard R., 1967, *Solar Phys.* **2**, 3

Howard R.F., 1990, *Solar Phys.* **126**, 299

Howard R.F., 1991, *Solar Phys.* **131**, 259

Ichimoto K., 1987, *Publ. Astron. Soc. Japan* **39**, 329

Illing, R.M.E., Landman, D.A., Mickey, D.L.: 1974a, *Astron. Astrophys.* **35**, 327

Illing, R.M.E., Landman, D.A., Mickey, D.L.: 1974b, *Astron. Astrophys.* **37**, 97

Illing, R.M.E., Landman, D.A., Mickey, D.L.: 1975, *Astron. Astrophys.* **41**, 183

Immerschitt, S., Schröter, E.H.: 1989, *Astron. Astrophys.* **208**, 307

Jones H.P., Giovanelli R.G., 1983, *Solar Phys.* **87**, 37

Kambry M.A., Nishikawa J., Sakurai T., Ichimoto K., Hiei E., 1991, *Solar Phys.* **132**, 41

Kawakami H., 1983, *Publ. Astron. Soc. Japan* **35**, 459

Kemp, J.C., Henson, G.D., Steiner, C.T., Powell, E.R.: 1987, *Nature* **326**, 270

Kjeldseth-Moe O., Brynildsen N., Brekke P., Engvold O., Maltby P., Bartoe J.-D.F., Brueckner G.E., Cook J.W., Dere K.P., Soker D.G., 1988, *Astrophys. J.* **334**, 1066

Kjeldseth-Moe O., Brynildsen N., Brekke P., Maltby P., Brueckner G.E., 1993, *Solar Phys.* **145**, 257

Kneer F., Stolpe F., 1996, *Solar Phys.* **164**, 303

Knölker, M., Grossmann-Doerth, U., Schüssler, M., Weisshaar, E.: 1991, *Adv. Space Res.* **11**, 285

Kurokawa H., 1991, in *Flare Physics in Solar Activity Maximum 22*, Y. Uchida et al. (Eds.), Lecture Notes in Physics, Vol. 387, p. 39

Lamb S.A., 1975, *Monthly Notices Royal Astron. Soc.* **172**, 205

Leka K.D., Van Driel-Gesztelyi L., Nitta N., Canfield R.C., Mickey D.L., Sakurai T., Ichimoto K., 1994, *Solar Phys.* **155**, 301

Lites B.W., 1980, *Solar Phys.* **68**, 327

Lites B.W., 1992, in *Sunspots: Theory and Observations*, J.H. Thomas, N.O. Weiss (Eds.), Kluwer, Dordrecht, p. 261

Lites B.W., Skumanich A., 1990, *Astrophys. J.* **348**, 747

Lites B.W., Bruner E.C., Jr., Chipman E.G., Shine R.A., Rottman G.J., White O.R., Athay R.G., 1976, *Astrophys. J.* **210**, L111

Lites B.W., Scharmer G.B, Skumanich A., 1990, *Astrophys. J.* **355**, 329

Lites B.W., Elmore D.F., Seagraves P., Skumanich A., 1993, *Astrophys. J.* **418**, 928

Livi S.H.B., Wang J., Martin S.F., 1985, *Australian J. Phys.* **38**, 855

Maltby P., 1964, *Astrophys. Norvegica* **8**, 205

Maltby P., 1975, *Solar Phys.* **43**, 91

Maltby P., Eriksen G., 1967, *Solar Phys.* **2**, 249

Martin S.F., 1990, in *Solar Photosphere: Structure, Convection and Magnetic Fields*, J.O. Stenflo (Ed.), Kluwer, Dordrecht, *IAU Symp.* **138**, 129

Martin S.F., Livi S.H.B., Wang J., 1985, *Australian J. Phys.* **38**, 929
Martínez Pillet V., Lites B.W., Skumanich A., Degenhardt D., 1994, *Astrophys. J.* **425**, L113
Martínez Pillet V., Lites B.W., Skumanich A., 1996, *Astrophys. J.* in press
Mein, N.: 1977, *Solar Phys.* **52**, 283
Mein, N., Mein, P.: 1976, *Solar Phys.* **49**, 231
Molowny-Horas R., 1994, *Solar Phys.* **154**, 29
Montesinos B., Thomas J.H., 1993, *Astrophys. J.* **402**, 314
Moore R.L., Rabin D., 1985, *Ann. Rev. Astron. Astrophys.* **23**, 239
Moreno Insertis F., Caligari P., Schüssler M., 1994, *Solar Phys.* **153**, 449
Muglach, K., Solanki, S.K.: 1992, *Astron. Astrophys.* **263**, 301
Muglach K., Solanki S.K., Livingston W.C., 1995, in *Infrared Tools for Solar Astrophysics: What's Next?*, J.R. Kuhn, M.J. Penn (Eds.), World Scientific, Singapore, p. 387
Muller R., 1973, *Solar Phys.* **29**, 55
Muller R., 1976, *Solar Phys.* **48**, 101
Muller R., Roudier Th., Vigneau J., Auffret H., 1994, *Astron. Astrophys.* **283**, 232
Mürset, U., Solanki, S.K., Stenflo, J.O.: 1988, *Astron. Astrophys.* **204**, 279
Nadeau, D.: 1988, *Astrophys. J.* **325**, 480
Nesis, A., Fleig, K.-H., Mattig, W.: 1989, in *Solar and Stellar Granulation*, R.J. Rutten, G. Severino (Eds.), Kluwer, Dordrecht, p. 289
Nicolas K.R., Kjeldseth-Moe O., Bartoe J.-D.F., Brueckner G.E., 1982, *Solar Phys.* **81**, 253
Nordlund, Å.: 1984a, in *Small–Scale Dynamical Processes in Quiet Stellar Atmospheres*, S.L. Keil (Ed.), National Solar Obs., Sunspot, NM, p. 181
Pantellini, F.G.E., Solanki, S.K., Stenflo, J.O.: 1988, *Astron. Astrophys.* **189**, 263
Parker E.N., 1992, *Astrophys. J.* **390**, 290
Peters H., 1996, *Monthly Notices Royal Astron. Soc.* **278**, 821
Petrovay K., Brown J.C., Van Driel-Gesztelyi L., Fletcher L., Marik M., Stewart G., 1990, *Solar Phys.* **127**, 51
Pevtsov A.A., 1992, *Solar Phys.* **141**, 65
Ploner S.R.O., Solanki S.K., 1996, *Astron. Astrophys.* submitted
Richardson R.S., 1941, *Astrophys. J.* **93**, 24
Rimmele T.R., 1994, *Astron. Astrophys.* **290**, 972
Rimmele T.R., 1995a, *Astron. Astrophys.* **298**, 260
Rimmele T.R., 1995b, *Astrophys. J.* **445**, 511.
Roberts, B.: 1983, *Solar Phys.* **87**, 77
Roberts B., Ulmschneider P., 1996, these proceedings
Rüedi, I., Solanki, S.K., Livingston, W., Stenflo, J.O.: 1992a, *Astron. Astrophys.* **263**, 323
Rüedi, I., Solanki, S.K., Livingston, W.: 1992b, *Astron. Astrophys.* **261**, L21
Rüedi I., Solanki S.K., Livingston W., 1995, *Astron. Astrophys.* **293**, 252
Sánchez Almeida J., Landi Degl'Innocenti E., 1996, *Solar Phys.* **164**, 203
Sánchez Almeida, J., Lites, B.W.: 1992, *Astrophys. J.* **398**, 359
Sánchez Almeida, J., Collados, M., Del Toro Iniesta, J.C.: 1988, *Astron. Astrophys.* **201**, L37
Sánchez Almeida, J., Collados, M., Del Toro Iniesta, J.C.: 1989, *Astron. Astrophys.* **222**, 311

Schmieder B., 1996, these proceedings

Schmieder B., Raadu M., Démoulin P., Dere K.P., 1989, *Astron. Astrophys.* **213**, 402

Schmidt W., Hofmann A., Balthasar H., Tarbell T.D., Frank Z.A., 1992, *Astron. Astrophys.* **264**, L27

Scholiers, W., Wiehr, E.: 1985, *Solar Phys.* **99**, 349

Schröter E.H., 1965, *Z. Astrophys.* **62**, 228

Schröter E.H., 1967, in *Solar Physics*, J.N. Xanthakis (Ed.), Interscience, London, p. 325

Schüssler M., 1990, in *Solar Photosphere: Structure, Convection and Magnetic Fields*, J.O. Stenflo (Ed.), Kluwer, Dordrecht, *IAU Symp.* **138**, 161

Schüssler M., 1992, in *The Sun — a Laboratory for Astrophysics*, J.T. Schmelz, J.C. Brown (Eds.), Kluwer, Dordrecht, p. 191

Shine R.A., Title A.M., Tarbell T.D., Smith K., Frank Z.A., 1994, *Astrophys. J.* **430**, 413

Snodgrass H., 1983, *Astrophys. J.* **270**, 288

Solanki, S.K.: 1986, *Astron. Astrophys.* **168**, 311

Solanki, S.K.: 1989, *Astron. Astrophys.* **224**, 225

Solanki S.K., 1990, in *Solar Photosphere: Structure, Convection and Magnetic Fields*, J.O Stenflo (Ed.), , *IAU Symp.* **138**, 103

Solanki S.K., 1993, *Space Sci. Rev.* **61**, 1

Solanki S.K., 1995, in *Infrared Tools for Solar Astrophysics: What's Next?*, J.R. Kuhn, M.J. Penn (Eds.), World Scientific, Singapore, p. 341

Solanki S.K., Montavon C.A.P., 1993, *Astron. Astrophys.* **275**, 283

Solanki, S.K., Pahlke, K.D.: 1988, *Astron. Astrophys.* **201**, 143

Solanki, S.K., Roberts, B.: 1992, *Monthly Notices Royal Astron. Soc.* **256**, 13

Solanki, S.K., Steiner, O.: 1990, *Astron. Astrophys.* **234**, 519

Solanki, S.K., Stenflo, J.O.: 1984, *Astron. Astrophys.* **140**, 185

Solanki, S.K., Stenflo, J.O.: 1985, *Astron. Astrophys.* **148**, 123

Solanki, S.K., Stenflo, J.O.: 1986, *Astron. Astrophys.* **170**, 120

Solanki S.K., Rüedi I., Livingston W., 1992, *Astron. Astrophys.* **263**, 339

Solanki S.K., Montavon C.A.P., Livingston W., 1994, *Astron. Astrophys.* **283**, 221

Spruit H.C., Schüssler M., Solanki S.K., 1992, in *Solar Interior and Atmosphere*, A.N. Cox, W. Livingston, M.S. Matthews (Eds.), University of Arizona press, Tucson, AZ, p. 890

St. John C.E., 1913, *Astrophys. J.* **37**, 322

Steiner O., Grossmann-Doerth U., Knölker M., Schüssler M., 1995, *Rev. Mod. Astron.* **8**, 81

Steiner O., Grossmann-Doerth U., Knölker M., Schüssler M., 1996, *Solar Phys.* **164**, 223

Stellmacher G., Wiehr E., 1971, *Solar Phys.* **17**, 21

Stellmacher G., Wiehr E., 1980, *Astron. Astrophys.* **82**, 157

Stenflo, J.O.: 1973, *Solar Phys.* **32**, 41

Stenflo, J.O.: 1976, in *Basic Mechanisms of Solar Activity*, V. Bumba, J. Kleczek (Eds.), Reidel, Dordrecht, *IAU Symp.* **71**, 69

Stenflo J.O., 1989a, *Astron. Astrophys.* **210**, 403

Stenflo J.O., 1989b, *Astron. Astrophys. Rev.* **1**, 3

Stenflo J.O., 1990, *Astron. Astrophys.* **233**, 220

Stenflo J.O., 1994, *Solar Magnetic Fields: Polarized Radiation Diagnostics*, Kluwer, Dordrecht

Stenflo, J.O., Harvey, J.W.: 1985, *Solar Phys.* **95**, 99

Stenflo, J.O., Harvey, J.W., Brault, J.W., Solanki, S.K.: 1984a, *Astron. Astrophys.* **131**, 333

Stenflo, J.O., Solanki, S.K., Harvey, J.W.: 1987a, *Astron. Astrophys.* **171**, 305

Stenflo, J.O., Solanki, S.K., Harvey, J.W.: 1987b, *Astron. Astrophys.* **173**, 167

Strous L.H., Scharmer G., Tarbell T.D., Title A.M., Zwaan C., 1996, *Astron. Astrophys.* **306**, 947

Tanenbaum, A.S., Wilcox, J.M., Howard, R.: 1971, in *Solar Magnetic Fields*, R. Howard (Ed.), Reidel, Dordrecht, *IAU Symp.* **43**, 348

Ternullo M., Zappala R.A., Zucarello F., 1981, *Solar Phys.* **74**, 111

Thomas J.H., Weiss N., 1992, in *Sunspots: Theory and Observations*, J.H. Thomas, N. Weiss (Eds.), Kluwer, Dordrecht, p. 3

Title A.M., Tarbell T.D., Topka K.P., 1987, *Astrophys. J.* **317**, 892

Title, A.M., Tarbell, T.D., Topka, K.P, Ferguson, S.H., Shine, R.A. and the SOUP Team: 1989, *Astrophys. J.* **336**, 475.

Title, A.M., Tarbell, T.D., Topka, K.P., Cauffman, D., Balke, C., Scharmer, G.: 1990, in *Phys. of Magn. Flux Ropes*, C.T. Russell, E.R. Priest, L.C. Lee (Eds.), Geophysical Monograph 58, American Geophys. Union, Washington, DC, p. 171

Title A.M., Topka K.P., Tarbell T.D., Schmidt W., Balke C., Scharmer G.: 1992a, *Astrophys. J.* **393**, 782

Title A.M., Frank Z.A., Shine R.A., Tarbell T.D., Topka K.P., Scharmer G., Schmidt W., 1992b, in *Sunspots: Theory and Observations*, J.H. Thomas, N.O. Weiss (Eds.), Kluwer, Dordrecht, p. 195

Title A.M., Frank Z.A., Shine R.A., Tarbell T.D., Topka K.P., Scharmer G., Schmidt W., 1993, *Astrophys. J.* **403**, 780

Van Ballegooijen, A.A., 1985, in *Theoretical Problems in High Resolution Solar Physics*, H.U. Schmidt (Ed.), Max Planck Inst. f. Astrophys., Munich, p. 177.

Van Driel-Gesztelyi L., Petrovay K., 1990, *Solar Phys.* **126**, 285

Volkmer R., Kneer F., Bendlin C., 1995, *Astron. Astrophys.* **304**, L1

von Uexküll M., Kneer F., 1995, *Astron. Astrophys.* **294**, 252

Wang J., 1994, *Solar Phys.* **155**, 285

Wiehr, E.: 1985, *Astron. Astrophys.* **149**, 217

Wiehr E., 1995, *Astron. Astrophys.* **298**, L17

Wiehr E., 1996, *Astron. Astrophys.* **309**, L4

Wiehr E., Degenhardt D., 1992, *Astron. Astrophys.* **259**, 313

Wiehr E., Stellmacher, G., 1989, *Astron. Astrophys.* **225**, 528

Wiehr E., Koch A., Knölker M., Küveler G., Stellmacher G., 1984 , *Astron. Astrophys.* **140**, 352

Zayer, I., Solanki, S.K., Stenflo, J.O., Keller, C.U.: 1990, *Astron. Astrophys.* **239**, 356

Zirin, H.: 1985, *Australian J. Phys.* **38**, 961

Zirin H., 1987, *Solar Phys.* **110**, 101

Zirin H., Stein A.,1972, *Astrophys. J.* **178**, L85

Zwaan C., 1987, *Ann. Rev. Astron. Astrophys.* **25**, 83

Zwaan C., 1992, in *Sunspots: Theory and Observations*, J.H. Thomas, N.O. Weiss (Eds.), Kluwer, Dordrecht, p. 75

Dynamics of Flux Tubes
in the Solar Atmosphere: Theory

B. Roberts[1] and P. Ulmschneider[2]

[1] School of Mathematical and Computational Sciences, University of St Andrews, St Andrews, Fife KY16 9SS, Scotland
[2] Institut für Theoretische Astrophysik, Universität Heidelberg, Tiergartenstr. 15, D-69121 Heidelberg, Germany

Abstract. The modes of oscillation of a photospheric magnetic flux tube are reviewed, taking into account both linear and nonlinear aspects. Analytical and computational developments are discussed, beginning with the basic characteristics of linear wave propagation and progressing to a consideration of nonlinearity and the question of the generation of tube waves and the energy flux they transport.

1 Introduction

The concept of a magnetic flux tube, which goes back to Michael Faraday, has proved to be particularly fruitful for solar studies. In the Sun's photosphere observations show that isolated small-scale magnetic flux tubes occur in the downdraught lanes between supergranules; these tubes are magnetically strong, with fields of about 2 kG confined to radii of about 100 km (see the reviews by Stenflo 1989, 1994; Solanki 1990, 1993, 1997; Schüssler 1991). Larger tube-like concentrations occur in the form of pores and sunspots, though the detailed internal structure of a sunspot is presently uncertain: a sunspot may be described as a uniform plug of magnetic field (a monolith) or alternatively as a conglomeration of individual small-scale magnetic tubes, assembled to form a sunspot, that retain their separate identities in the layers below the visible surface (see the reviews in Thomas & Weiss 1992). Observations of the absorption of *p*-modes in sunspots are likely to shed light on this question in due course (see the review by Bogdan 1992).

There are no *isolated* flux tubes in the corona, which is completely filled with magnetic field. But flux tubes still exist, corresponding to magnetised plasma regions that are delineated from their surroundings, appearing in the form of coronal loops. A coronal flux tube is primarily a region of high plasma density, although temperature and magnetic field differences may also arise.

Magnetic flux tubes are communication channels, linking one part of the Sun's atmosphere or interior with another. Their essential one-dimensionality means that the linkage is likely to be efficient. As such they are important conduits for momentum and energy transport, carrying flows or waves from one site to another. Although there are a number of similarities between photospheric and coronal flux tubes, there is an important distinction: whereas a photospheric tube is a region of enhancement in Alfvén speed, a coronal flux tube is a region of low Alfvén speed. Finally, it is interesting to note that there are a number of similarities in the basic structure of waves in isolated photospheric magnetic tubes and the waves in extragalactic jets (Roberts 1987; Bodo et al. 1989; Hardee 1995).

In this review we concentrate on the photospheric flux tube. Waves in magnetic flux tubes have also been reviewed in Roberts (1980, 1981, 1985a, 1986, 1990a, b, 1991a, b, 1992a, b), Spruit (1981a, b), Spruit & Roberts (1983), Thomas (1985, 1990), Hollweg (1986, 1990a), Edwin & Roberts (1987), Ryutova (1990a, b), and Edwin (1991, 1992).

2   Basic Modes of Oscillations

The basic modes of oscillation of an isolated magnetic flux tube are now well understood. Geometrically, there are *sausage* modes, *kink* modes and *fluting* modes. These geometrical forms are defined in terms of the patterns that the boundary of the flux tube makes when it is disturbed (see equation (13) below). If the displaced tube remains a circle centred on the axis of symmetry of the undisturbed tube, then this is the sausage mode; if the displaced tube remains circular, but is no longer about the axis of symmetry, then this is the kink mode. The various higher order distortions of the tube boundary, ranging from elliptical to highly castellated, are the fluting modes. The modes may be classified further according to their nature in the radial coordinate r: disturbances that inside the tube are *oscillatory* in r are referred to as *body* waves, and those that are *exponential* (or evanescent) in form are the *surface* waves. Basically, in a wide tube the surface waves are confined to near the boundary of the tube and do not penetrate far into the centre of the tube, whereas body modes disturb the centre of a wide tube; however, in a thin tube, the centre is disturbed for both modes. In addition to the modes that are trapped within a tube, disturbing the tube's environment only slightly, there arise *leaky* waves. Leaky waves are generated by motions within the tube that result in an outflow of wave energy: the tube is a generator for waves in its environment. These classifications may be applied to both fast and slow magnetoacoustic waves, so the description becomes complicated. Additionally, a tube may support torsional Alfvén waves.

The characteristic speeds that govern wave propagation in a magnetic flux tube are readily established. Consider a plasma of density ρ_0 and pressure p_0 within which is embedded a magnetic field $\mathbf{B_0}$ of strength B_0. The sound speed c_S and Alfvén speed c_A are defined by

$$c_S = \left(\frac{\gamma p_0}{\rho_0} \right)^{1/2}, \quad c_A = \left(\frac{B_0{}^2}{\mu_0 \rho_0} \right)^{1/2}, \tag{1}$$

where μ_0 is the magnetic permeability and γ the adiabatic index of the gas (taken to be 5/3). From these two speeds we may construct the *fast* magnetoacoustic speed, c_f, and the *slow* (or tube) magnetoacoustic speed, c_T:

$$c_f^2 = c_S^2 + c_A^2, \quad c_T^{-2} = c_S^{-2} + c_A^{-2}. \tag{2}$$

The fast speed c_f is super-sonic and super-Alfvénic, and the slow speed c_T is sub-sonic and sub-Alfvénic; the slow speed is particularly significant for waves in a thin magnetic flux tube (Defouw 1976; Roberts & Webb 1978). We may illustrate these speeds as follows. In a photospheric flux tube with magnetic field strength $B_0 = 2$ kG and plasma density $\rho_0 = 2.2 \times 10^{-4}$ kg m^{-3}, the Alfvén speed is $c_A = 12$ km s^{-1}; for a sound speed of $c_S = 8$ km s^{-1} we obtain $c_T = 6.7$ km s^{-1} and $c_f = 14.4$ km s^{-1}.

A speed equivalent to c_T is in fact common to a variety of elastic tubes, with the role of the Alfvén speed being played by the appropriate elastic speed of the physical situation. For example, in the case of a blood vessel the speed equivalent to c_A is the elastic speed in the membrane of the blood vessel. In this case, the speed of sound c_S in blood is much larger than the elastic speed, so effectively $c_S \gg c_A$ giving $c_T \approx c_A$; wave propagation in a blood vessel proceeds with a speed that is close to the elastic speed of the membrane walls. For water in a pipe, the relative magnitudes of the two basic speeds depends upon the material of the pipe. In a metal pipe, the elastic speed of the metal membrane is much larger than the speed of sound in water, and so the effective propagation speed is close to the sound speed in water (about 1.4 km s^{-1}). In a plastic pipe, the orderings in the two speeds are reversed and the effective propagation speed is close to the elastic speed in plastic (about 10 m s^{-1}), lying far below the speed of sound in water.

The kink mode disturbs both the tube and its environment, and so the characteristic speed for this wave involves both the density ρ_0 of the gas inside the tube and the density ρ_e of the gas in the environment. The kink mode is principally a result of the magnetic tension force, B_0^2/μ_0, in the magnetic field and so its speed of propagation, c_k, is given by (Ryutov & Ryutova 1976; Parker 1979; Spruit 1982)

$$c_k^2 = \frac{B_0^2/\mu_0}{\rho_0 + \rho_e} = \left(\frac{\rho_0}{\rho_0 + \rho_e}\right) c_A^2. \tag{3}$$

This speed too is sub-Alfvénic; for $\rho_e = 2\rho_0$ (consistent with a sound speed in the tube's environment of 9.6 km s^{-1}) we obtain $c_k = 6.9$ km s^{-1}, roughly 60% of the Alfvén speed.

To progress further it is necessary to consider the linear equations of ideal magnetohydrodynamics in some detail. Consider an equilibrium magnetic field $\mathbf{B_0} = B_0(r)\hat{\mathbf{z}}$ aligned with the z-axis of a cylindrical polar coordinate system r, θ, z. To begin with we will ignore the effects of gravity (see Section 4). Then the equilibrium gas pressure $p_0(r)$ and density $\rho_0(r)$ are structured by the magnetic field so as to maintain total pressure balance: the sum of the gas pressure and the magnetic pressure is a constant,

$$\frac{d}{dr}\left(p_0(r) + \frac{B_0^2(r)}{2\mu_0}\right) = 0. \tag{4}$$

Small amplitude motions $\mathbf{v}$ about the equilibrium (4) satisfy the wave equation

$$\frac{\partial^2 \mathbf{v}}{\partial t^2} - c_A^2 \frac{\partial^2 \mathbf{v}}{\partial z^2} = -c_A^2 \hat{\mathbf{z}} \frac{\partial}{\partial z} \text{div } \mathbf{v} - \frac{1}{\rho_0} \text{grad}\left(\frac{\partial p_T}{\partial t}\right). \tag{5}$$

Equation (5) follows from the time derivative of the momentum equation combined with the magnetohydrodynamic induction equation for an ideal medium. The total pressure, $p_T(r, \theta, z)$, the sum of the gas pressure perturbation p and the perturbation in the magnetic pressure, $B_0 B_z/\mu_0$, satisfies the evolution equation

$$\frac{\partial p_T}{\partial t} = \rho_0 c_A^2 \frac{\partial v_z}{\partial z} - \rho_0 (c_S^2 + c_A^2) \text{div } \mathbf{v}. \tag{6}$$

Equation (6) is a combination of the isentropic and induction equations.

For a flow $\mathbf{v}(r, \theta, z) = (v_r, v_\theta, v_z)$, the components of equation (5) give (Roberts 1986, 1992a)

$$\rho_0(r)\left(\frac{\partial^2}{\partial t^2} - c_A^2(r)\frac{\partial^2}{\partial z^2}\right)v_r + \frac{\partial^2 p_T}{\partial r \partial t} = 0, \tag{7}$$

$$\rho_0(r)\left(\frac{\partial^2}{\partial t^2} - c_A^2(r)\frac{\partial^2}{\partial z^2}\right)v_\theta + \frac{1}{r}\frac{\partial^2 p_T}{\partial \theta \partial t} = 0, \tag{8}$$

and

$$\left(\frac{\partial^2}{\partial t^2} - c_T^2(r)\frac{\partial^2}{\partial z^2}\right)v_z = -\left(\frac{c_S^2(r)}{c_f^2(r)}\right)\frac{1}{\rho_0(r)}\frac{\partial^2 p_T}{\partial z \partial t}, \tag{9}$$

with the evolution in $p_T(r, \theta, z)$ described by equation (6).

It is evident from the form of (7)–(9) that an inhomogeneous medium supports the phenomena of *phase-mixing* (Barston 1964; Heyvaerts & Priest 1983) and *resonant absorption* (Chen & Hasegawa 1974; Ionson 1978). Phase-mixing describes the process by which wave fronts become increasingly more corrugated as they propagate, in response to the fact that the Alfvén speed is different on different field lines. Resonant absorption occurs whenever the phase speed of an Alfvén wave matches the local Alfvén speed c_A or the phase speed of a slow wave matches the local slow speed c_T. Both processes produce small spatial scales *across* the magnetic field, the direction in which neither the Alfvén wave nor the slow wave is able to propagate in a uniform medium. Since both processes depend upon nonuniformity in the equilibrium state, they are are likely to occur preferentially where inhomogeneities are most pronounced, such as on the boundaries of isolated flux tubes.

Phase-mixing is evident when $\partial/\partial\theta = 0$, so that motions are symmetric about the axis of the tube. Then equation (8) decouples from (7)–(9) to give

$$\frac{\partial^2 v_\theta}{\partial t^2} = c_A^2(r)\frac{\partial^2 v_\theta}{\partial z^2}. \tag{10}$$

These are *torsional Alfvén waves* (e.g., Spruit 1982; Hollweg 1988, 1990b, 1991). Symmetric motions $v_\theta(r, z, t)$ take place independently of motions in the r and z-directions. The r-dependence in the torsional Alfvén wave satisfying (10) is arbitrary, being fixed by the means by which the wave is generated.

To illustrate the implications of the wave equation (10) consider the torsional oscillation

$$v_\theta = v_0 \sin[k_z z - k_z c_A(r)t]. \qquad (11)$$

Initially (at $t = 0$), this gives a disturbance of amplitude v_0 and wavelength $2\pi/k_z$. According to (11), the initial shape propagates forward with speed c_A, but its radial gradient, given by

$$\frac{\partial v_\theta}{\partial r} = -k_z v_0 \left(\frac{dc_A}{dr} \right) t \cos[k_z z - k_z c_A(r)t], \qquad (12)$$

grows secularly on a timescale of $(dc_A/dr)^{-1}$. This is phase-mixing; cross-field gradients are rapidly built-up, making any dissipative processes more efficient (Heyvaerts & Priest 1983; Ireland 1997). The phase-mixing timescale can be surprisingly short: for an isolated tube with c_A changing from about 10 km s^{-1} in the tube to zero outside it, over a distance of 10 km (one-tenth of a tube radius), it is 1 second. This suggests that any such modes, if generated, would rapidly be dissipated by non-adiabatic processes enhanced by phase-mixing (or resonant absorption), and so the edges of isolated flux tubes ought to be excessively hot, producing locally a bright ring around the tube. Much the same effect should operate on the edge of sunspots (Hollweg 1988; Roberts 1992a) or throughout the body of a spot, if spots are strongly inhomogeneous (Ryutova & Persson 1984). However, no such bright ring has been observed, either for thin tubes or for sunspots. Also, on theoretical grounds, we must note that unless there exists an ignorable coordinate (so that the assumption $\partial/\partial\theta = 0$ is valid), then the waves are coupled, through equations (7)–(9), and no simple phase-mixing occurs (Davila 1987, 1991; Parker 1991), though resonant absorption then takes place (eg., Goedbloed 1975, 1983; Rae & Roberts 1981; Lee & Roberts 1986; Poedts et al 1990). Resonant absorption would seem to be particularly important for coronal loops. The topic is reviewed by Hollweg (1990a, b) and Goossens (1991); see also Goossens, Ruderman & Hollweg (1995).

It is convenient to introduce the Fourier form of equations (7)-(8). Write

$$v_r(r, \theta, z, t) = v_r(r) \exp i(\omega t - n\theta - k_z z), \qquad (13)$$

with a similar form for p_T. The integer $n (= 0, \pm1, \pm2, \cdots)$ describes the geometrical form of the perturbations. The case $n = 0$ gives the *sausage* mode, and corresponds to *symmetric* motions of the tube. These are compressional oscillations ($p_T \neq 0$), in addition to the torsional Alfvén waves. The *kink* mode of the tube is given by setting $n = 1$ (or $n = -1$); in such modes the instantaneous motion of the tube resembles a snake, with the boundary of the tube a displaced circle. Finally, waves with $|n| \geq 2$ are the *fluting* modes of the tube.

With the Fourier form (13) for v_r and other variables, equations (6) - (9) yield ordinary differential equations. It proves convenient to work in terms of the total pressure perturbation, p_T, which satisfies (Edwin & Roberts 1983)

$$\rho_0(r)(k_z^2 c_A^2(r) - \omega^2) \frac{1}{r} \frac{d}{dr} \left\{ \frac{1}{\rho_0(r)(k_z^2 c_A^2(r) - \omega^2)} r \frac{dp_T}{dr} \right\} = \left(m^2(r) + \frac{n^2}{r^2} \right) p_T,$$

(14)

where

$$m^2(r) = \frac{(k_z^2 c_S^2(r) - \omega^2)(k_z^2 c_A^2(r) - \omega^2)}{(c_S^2(r) + c_A^2(r))(k_z^2 c_T^2(r) - \omega^2)}.$$

(15)

The radial velocity component follows from (7):

$$\rho_0(r)(k_z^2 c_A(r) - \omega^2) v_r = -i\omega \frac{dp_T}{dr}.$$

(16)

Equation (14) is singular at $\omega^2 = k_z^2 c_A(r)$ and at $\omega^2 = k_z^2 c_T(r)$ (Appert, Gruber & Vaclavik 1974; Goedbloed 1975, 1983); the first singularity is associated with the Alfvénic continuous spectrum and the second with the slow mode continuous spectrum. Both spectra have been investigated in detail for a number of equilibria (see the review by Goossens 1991), especially in connection with coronal heating (see the reviews by Narain & Ulmschneider (1990, 1996) and Browning (1991)).

Now we are interested in the solution of equation (14) appropriate for a magnetic flux tube. Consider, then, a uniform magnetic field $B_0 \hat{z}$ confined to a tube of radius $r = a$, embedded in a field-free environment:

$$B_0(r) = \begin{cases} B_0, & r < a, \\ 0, & r > a. \end{cases}$$

(17)

The interface $r = a$ is a current sheet across which conditions change discontinuously, while preserving total pressure balance with the external gas pressure p_e of the environment:

$$p_0 + \frac{B_0^2}{2\mu_0} = p_e.$$

(18)

Combined with the ideal gas law, pressure balance implies a connection between the density ρ_0, sound speed c_S and Alfvén speed c_A of the tube and the density ρ_e and sound speed c_{Se} in the environment:

$$\frac{\rho_e}{\rho_0} = \frac{c_S^2 + \frac{1}{2}\gamma c_A^2}{c_{Se}^2}.$$

(19)

The temperature structure of an isolated tube in the photosphere is complicated (Schüssler 1990; Solanki 1997), and temperature differences between the interior of a flux tube and its surroundings may arise (and in fact are required to reproduce observed chromospheric canopy heights (Solanki & Steiner 1990), as suggested by carbon monoxide observations (Ayres et al. 1986; Solanki et al. 1994)). But it is evident from (19) that we may expect $\rho_0 < \rho_e$: the tube is partially evacuated by the magnetic field.

Now the media inside and outside the flux tube are *uniform*, and this affords us some simplification. In a uniform medium, equation (14) for p_T inside the tube reduces to

$$r^2 \frac{d^2 p_T}{dr^2} + r \frac{dp_T}{dr} - (m_0^2 r^2 + n^2) p_T = 0, \tag{20}$$

where m_0^2 is the value of m^2 inside the tube. (We have cancelled a factor $(k_z^2 c_A^2 - \omega^2)$ corresponding to Alfvén waves.) Equation (20) is a form of Bessel's differential equation, with has solutions in form of the modified Bessel functions $I_n(m_0 r)$ and $K_n(m_0 r)$. The solution $K_n(m_0 r)$ is singular at $r = 0$ and so is rejected. Accordingly, we take

$$p_T = A_0 I_n(m_0 r), \qquad r < a, \tag{21}$$

where A_0 is an arbitrary constant.

In the field-free environment of the flux tube, where the sound speed is c_{Se} and the Alfvén speed is zero, solutions $p_T \propto I_n(m_e r)$ and $K_n(m_e r)$ arise; here m_e^2 is the value of $m^2(r)$ outside the tube,

$$m_e^2 = k_z^2 - \frac{\omega^2}{c_{Se}^2}. \tag{22}$$

If now we assume that $m_e^2 > 0$, corresponding to selecting waves that decay (in r) outside the tube, then the solution that is bounded for $r \to \infty$ is

$$p_T(r) = A_e K_n(m_e r), \quad r > a, \tag{23}$$

where A_e is an arbitrary constant.

It remains to match p_T and v_r across $r = a$. The result is the dispersion relation (Roberts & Webb 1978, 1979; Spruit 1982; Edwin & Roberts 1983; Cally 1985, 1986; Evans & Roberts 1990)

$$\frac{1}{\rho_0(k_z^2 c_A^2 - \omega^2)} m_0 \frac{I_n'(m_0 a)}{I_n(m_0 a)} + \frac{1}{\rho_e \omega^2} m_e \frac{K_n'(m_e a)}{K_n(m_e a)} = 0, \tag{24}$$

where a prime (') denotes the derivative of a modified Bessel function (e.g., $I_n'(m_0 a) \equiv dI_n(x)/dx$ evaluated at $x = m_0 a$, etc.). This is the dispersion relation governing magnetoacoustic waves in an isolated magnetic flux tube.

Equation (24) is valid whatever the nature of m_0, but it is written in a form that is particularly suitable for surface waves ($m_0^2 > 0$). For body waves ($m_0^2 < 0$) an alternative form is more convenient, obtained by writing $n_0^2 = -m_0^2$:

$$\frac{1}{\rho_0(k_z^2 c_A^2 - \omega^2)} n_0 \frac{J_n'(n_0 a)}{J_n(n_0 a)} + \frac{1}{\rho_e \omega^2} m_e \frac{K_n'(m_e a)}{K_n(m_e a)} = 0. \tag{25}$$

Here J_n denotes the Bessel function of order n. Both (24) and (25) are subject to the constraint that $m_e > 0$.

Solutions of the transcendental dispersion relations (24) and (25) are best obtained numerically (Edwin & Roberts 1983; Evans & Roberts 1990) or graphically, or analytically in certain limits. Their structure depends upon the ordering of the various speeds, c_S, c_{Se} and c_A. For an isolated photospheric tube we take $c_S < c_A$ and $c_{Se} < c_A$. Then there are slow body modes (sausage and kink) with phase-speeds ω/k_z that lie between c_T and c_S; these modes may be viewed as waves that are constrained within the tube, bouncing from side to side as they propagate along its interior. There are also *slow surface waves* with phase-speeds that are less than c_T. Finally, there are fast surface waves which have phase-speeds between c_k and c_{Se}.

Having established the basic structure of the linear modes of a flux tube in as simple a circumstance as may be envisaged, namely a uniform tube in the absence of gravity or flow, we turn now to a consideration of some of the complications that add to our picture.

3   Nonlinear Analytical Aspects

The case of most interest for photospheric tubes is the *long wavelength limit*, namely $k_z a \ll 1$. This restriction corresponds to longitudinal wavelengths $2\pi/k_z \gg 2\pi a$, which for a tube of radius a = 100 km means wavelengths much greater than 600 km. Granules have dimensions ranging from a few hundred km to 2000 km, with 1000 km being typical, and so tube waves generated by a typical granule just about satisfies this extreme. Supergranules, with scales of 3×10^4 km, certainly give long wavelength modes.

Consider, then, the long wavelength limit of the dispersion relation. For $c_{Se} = c_S$, the slow sausage mode ($n = 0$) gives (Roberts & Webb 1978, 1979; Edwin & Roberts 1983; Roberts 1985b)

$$\omega \sim k_z c_T \left[1 - \frac{1}{4} \left(\frac{\rho_e}{\rho_0} \right) \left(\frac{c_T}{c_A} \right)^4 k_z^2 a^2 K_0 \left(\lambda |k_z| a \right) \right] \qquad (26)$$

where $\lambda = c_T/c_A$, and the kink mode ($n = 1$) gives (Edwin & Roberts 1983)

$$\omega \sim k_z c_k \left[1 + \frac{1}{2} \left(\frac{\rho_e}{\rho_e + \rho_0} \right) \hat{\mu}^2 k_z^2 a^2 K_0(\hat{\mu}|k_z|a) \right] \qquad (27)$$

where $\hat{\mu} = (c_{Se}^2 - c_k^2)^{1/2}/c_{Se}$.

These approximate formulae for the wave speeds are important in nonlinear theories. Indeed, it has been shown that weakly nonlinear, weakly dispersive slow sausage surface waves have motions $v(z, t)$ along a thin tube which satisfy the nonlinear integrodifferential equation (Roberts 1985b)

$$\frac{\partial v}{\partial t} + c_T \frac{\partial v}{\partial z} + \beta_0 v \frac{\partial v}{\partial z} + \alpha_0 \frac{\partial^3}{\partial z^3} \int_{-\infty}^{\infty} \frac{v(s,t)ds}{[\lambda^2 a^2 + (z - s)^2]^{1/2}} = 0. \qquad (28)$$

The constant α_0 is directly connected with the dispersive correction in (26), and the coefficient β_0 of the nonlinear term is

$$\beta_0 = \frac{[(\gamma + 1)c_A^2 + 3c_S^2]c_A^2}{2(c_S^2 + c_A^2)^2}.$$

Equation (28) is sometimes referred to as the Leibovich–Roberts equation; it has been solved numerically (Weisshaar 1989) and exhibits solutions that are soliton-like in character. However, no analytical solution of equation (28) is known, though a number of its properties have been found (Bogdan & Lerche 1988). The derivation of (28) given by Roberts (1985b) rests upon the thin tube approximation (see below), but an alternative approach, directly from the full set of magnetohydrodynamic equations, yields the same result (Molotovshchikov & Ruderman 1987).

An equation of similar form to (28) arises in a magnetic *slab* (Roberts & Mangeney 1982; Roberts 1985b; Merzljakov & Ruderman 1985):

$$\frac{\partial v}{\partial t} + c_T \frac{\partial v}{\partial z} + \beta_0 v \frac{\partial v}{\partial z} + \alpha_1 \frac{\partial^2}{\partial z^2} \int_{-\infty}^{\infty} \frac{v(s,t)}{s - z} ds = 0. \tag{29}$$

This is the Benjamin–Ono equation; it has been studied extensively and has a soliton solution. The constant α_1 is a measure of dispersion.

Many other aspects of tube waves have recently been explored, in attempt to describe analytically the nonlinear behaviour of the rich spectrum of nonlinear waves that an isolated flux tube or slab may support; see Ferriz-Mas (1988), Ruderman (1993), Zhelyazkov et al. (1994) and Nakariakov, Zhugzhda & Ulmschneider (1996).

4   Gravitational Aspects

Stratification, so far ignored in our account, is in fact important in the photosphere. The pressure scale height at the temperature minimum falls to its lowest value, of about 100 km, and this is comparable with the radius of a tube. So stratification effects are important. (In the corona, where the scale height is large, such effects are of less consequence.) The addition of gravity to the description of modes given in Section 2 has not so far proved possible, mainly because the flux tube expands with height in a stratified atmosphere and this seriously complicates the description. Progress has been made by considering either a thin tube or a uniform unbounded field.

The case of a thin tube has been investigated by use of the so called *thin tube equations*. The sausage and kink modes are treated separately. Gravity renders an isolated flux tube non-uniform in height. The fall-off of the confining gas pressure in the environment of the tube forces the tube to expand outwards. Consider the *thin tube equations* for the *sausage* mode:

$$\frac{\partial}{\partial t} \rho A + \frac{\partial}{\partial z} \rho v A = 0, \tag{30}$$

$$\frac{\partial v}{\partial t} + v\frac{\partial v}{\partial z} = -\frac{1}{\rho}\frac{\partial p}{\partial z} + g, \tag{31}$$

$$\frac{\partial p}{\partial t} + v\frac{\partial p}{\partial z} = \frac{\gamma p}{\rho}\left(\frac{\partial \rho}{\partial t} + v\frac{\partial \rho}{\partial z}\right), \tag{32}$$

$$BA = \text{constant}, \tag{33}$$

$$p + \frac{B^2}{2\mu_0} = p_e. \tag{34}$$

In the above thin tube equations, $B(z,t)$ is the field-strength of a thin tube with cross-sectional area $A(z,t)$, $v(z,t)$ is the longitudinal flow speed within the tube, where the gas pressure and density are given by $p(z,t)$ and $\rho(z,t)$. The external gas pressure $p_e(z,t)$ is calculated on the boundary of the tube, with equation (34) embodying a boundary condition namely that the total pressure inside the tube is balanced at all times by the external pressure field p_e, which itself may vary in z and t in response to waves in the tube or to externally imposed motions. The z-axis is aligned with gravity $g\hat{z}$, pointing downwards. A derivation of equations (30)–(34) has been given by Roberts & Webb (1978), based upon expanding all dependent variables in Taylor series about the central axis ($r = 0$) of the tube, assuming symmetry and no motions in the θ-direction. Special cases of these equations were written down on physical grounds by Parker (1974), for an incompressible fluid, and by Defouw (1976) for an isothermal gas.

In equilibrium ($v = 0, \partial/\partial t = 0$) the thin tube equations yield

$$p_0(z) = p_0(0)e^N, \qquad \rho_0(z) = \rho_0(0)\frac{\Lambda_0(0)}{\Lambda_0(z)}e^N,$$

$$A_0(z) = A_0(0)e^{N/2}, \qquad B_0(z) = B_0(0)e^{N/2}, \tag{35}$$

where

$$N(z) = \int_0^z \frac{dz}{\Lambda_0(z)}$$

is the integrated pressure scale-height $\Lambda_0(z)\,(= p_0(z)/g\rho_0(z))$ inside the tube. For simplicity, we have taken the temperature inside the tube to be equal to that in the environment, assumed to be in hydrostatic equilibrium.

The linear form of the thin tube equations is readily found for the equilibrium (35). Consider first the $g = 0$ case, for which the equilibrium state is a uniform tube ($N = 0$). The linearised form of (30)–(34) with $g = 0$ yields (Roberts 1981):

$$\frac{\partial^2 v}{\partial t^2} - c_T^2\frac{\partial^2 v}{\partial z^2} = -\frac{1}{\rho_0}\frac{c_T^2}{c_A^2}\frac{\partial^2 p_e}{\partial z\partial t}. \tag{36}$$

To progress further requires consideration of $p_e(z,t)$, the external pressure field on the boundary of the oscillating tube. The simplest assumption to make is that the external pressure remains at its equilibrium value, a constant $p_e(z)$, in which case the above equation immediately yields $\omega^2 = k_z^2 c_T^2$, giving the slow sausage mode as expected. A more refined analysis, though, is to allow for disturbances in the environment of the tube, and then the external gas pressure satisfies the wave equation in r, z and t. By solving this equation for disturbances that decline exponentially as $r \to \infty$ we may recover the approximate dispersion relation (26), showing that *the dispersive correction is due to the tube wave disturbing the environment of the tube* (Roberts & Webb 1979; Roberts 1985b).

Returning to the stratified flux tube ($g \neq 0$), we progress by again assuming that p_e is simply the unperturbed external gas pressure, a function of z but not of time. Then, after some algebra, we obtain the Klein–Gordon equation (Roberts 1981; Rae & Roberts 1982; Roberts 1992b)

$$\frac{\partial^2 Q}{\partial t^2} - c_T^2(z)\frac{\partial^2 Q}{\partial z^2} + \Omega^2(z)Q = 0, \tag{37}$$

where $Q(z,t)$ is related to the flow $v(z,t)$ through

$$Q(z,t) = \left[\frac{\rho_0(z)A_0(z)c_T^2(z)}{\rho_0(0)A_0(0)c_T^2(0)}\right]^{1/2} v(z,t) \tag{38}$$

and Ω^2 is given by

$$\Omega^2(z) = \frac{c_T^2}{4\Lambda_0^2}\left[3\Lambda_0' + \frac{9}{4} - \frac{2}{\gamma} + \frac{4c_S^2}{\gamma c_A^2}\left(\frac{\gamma-1}{\gamma} + \Lambda_0'\right)\right]. \tag{39}$$

Here a prime denotes differentiation with respect to depth z.

In an *isothermal* atmosphere we obtain some simplification in the above: Λ_0, c_T and Ω^2 are constants, with Ω^2 reducing to (Defouw 1976; Roberts & Webb 1978)

$$\Omega^2 = \Omega_T^2 \equiv \frac{c_T^2}{4\Lambda_0^2}\left[\left(\frac{9}{4} - \frac{2}{\gamma}\right) + \frac{4c_S^2}{\gamma c_A^2}\left(1 - \frac{1}{\gamma}\right)\right]. \tag{40}$$

The presence of constant coefficients in the Klein–Gordon equation (37) applied to an isothermal atmosphere leads to a familiar dispersion relation,

$$\omega^2 = k_z^2 c^2 + \Omega_T^2, \tag{41}$$

where here the propagation speed c is the slow speed c_T. Equation (41) shows that Ω, given by (40), is the *cutoff frequency* for sausage modes in a thin tube. Much the same equation arises in the vertical propagation of sound waves in the absence of a magnetic field, where (41) applies with the c being the sound speed c_S and the cutoff frequency being $c_S/2\Lambda_0$. For the sausage mode, Ω^2 may be viewed as made up of two contributions, the first (corresponding to the first term on the right-handside of (40)) arising from the *geometrical shape* of the tube and the second (corresponding to the second term on the right of (40)) being determined by the tube's *elasticity*. A *rigid tube* ($c_A \gg c_S$), with exponential cross-sectional area determined according to the equilibrium (35) with Λ_0 constant, has cutoff frequency $(9/4 - 2/\gamma)^{1/2} c_S/2\Lambda_0$. A straight and vertical rigid tube has cutoff $c_S/2\Lambda_0$, the same as a vertically propagating sound wave. See also the discussion in Campos (1986).

In general the cutoff frequency for the sausage mode in a tube is less than the cutoff frequency of a rigid tube. In accordance with (41), an impulsively generated sausage wave results in a wave-front propagating with the tube speed c_T, trailing an oscillating wake which rises and falls with the frequency Ω (Rae & Roberts 1982).

The Klein–Gordon equation also describes the *kink* mode (Spruit & Roberts 1983; Roberts 1986). The linearised form of Spruit's (1981) thin tube equations for the kink mode lead to the wave equation

$$\frac{\partial^2 \xi}{\partial t^2} = c_k^2(z)\frac{\partial^2 \xi}{\partial z^2} + g\left(\frac{\rho_0 - \rho_e}{\rho_0 + \rho_e}\right)\frac{\partial \xi}{\partial z} \tag{42}$$

for transverse displacement $\xi(z,t)$. It is easy to cast equation (42) into an equation of the form (37), for suitable Q related to ξ; such a procedure allows us to compare acoustic waves, sausage tube waves and kink tube waves all in the one frame-work (Rae & Roberts 1982; Roberts 1986). With (42) cast in the form of (37), the speed c becomes the kink speed c_k and the square of the cutoff frequency becomes

$$\Omega^2 = \frac{c_k^2}{4\Lambda_0^2}(\frac{1}{4} + \Lambda_0'). \tag{43}$$

We can now compare the sausage and kink modes, and also a vertically propagating sound wave, by reference to the Klein–Gordon equation in an isothermal atmosphere, simply by noting their differing propagation speeds, cutoff frequencies and e-folding distances. Suppose that $c_A = 12$ km s^{-1}, $c_S = 8$ km s^{-1}, $\Lambda_0 = 140$ km. The sound wave has propagation speed 8 km s^{-1} and cyclic cutoff frequency, $(c_S/2\Lambda_0)/2\pi$, of 4.5 mHz (period 220 s), and the wave e-folds once in 280 km (two scale heights). By contrast, both the sausage and kink modes take 560 km to e-fold once (four scale heights), the sausage wave propagating with a speed $c_T = 6.7$ km s^{-1} and the kink wave with a speed $c_k = 6.9$ km s^{-1}. As noted by Spruit (1981b), the two waves have entirely different cutoff frequencies: the cyclic cutoff frequency $(\Omega/2\pi)$ for the sausage wave is 4.2 mHz (period 240 s), close to the sound wave, whereas the kink mode has a cyclic cutoff frequency of 2 mHz (period 500 s), about a factor of two different.

5   Nonlinear Numerical Computations

Using Eqs. (30) to (33), nonlinear time-dependent solutions for longitudinal wave propagation in thin magnetic flux tubes may be obtained numerically, through the method of characteristics. Radiation effects may also be included. For this, Eq. (32) is written in an entropy conservation form,

$$\frac{dS}{dt} = \frac{\partial S}{\partial t} + v\frac{\partial S}{\partial z} = \frac{dS}{dt}\bigg|_{Rad} , \tag{44}$$

where $(dS/dt)|_{Rad}$ is the radiative damping function. Eq. (33) follows from (44) on assuming $dS/dt|_{Rad} = 0$ and using the thermodynamic relations

$$\frac{d\rho}{\rho} = \frac{2}{\gamma - 1}\frac{dc_S}{c_S} - \frac{\mu}{\Re}dS , \qquad \frac{dp}{p} = \frac{2\gamma}{\gamma - 1}\frac{dc_S}{c_S} - \frac{\mu}{\Re}dS . \tag{45}$$

Here $\Re$ is the gas constant and μ the mean molecular weight. For one-dimensional computations it is convenient to use a Lagrangian frame (which follows the motion of a parcel of fluid) as this, when replacing differential equations by difference equations, permits us to more easily ensure mass and energy conservation in the numerical scheme. In place of the Eulerian scheme, which considers a fixed location, namely the height z, one now uses the Lagrangian height a as independent variable; a is the Eulerian height at time $t = 0$, the start of the calculation. Consider a mass element contained initially in the height interval da at height a which at a later time moves to the Eulerian height $z(a, t)$ and expands to the size dz. If $A_0(a)$ and $\rho_0(a)$ are, respectively, the cross-section and density of the element at time $t = 0$, while $A(a, t)$ and $\rho(a, t)$ are the same quantities at a later time, then mass conservation in a tube gives

$$\rho_0(a)A_0(a)da = \rho(a, t)A(a, t)dz , \tag{46}$$

from which one obtains the scale factor

$$l_a = \left(\frac{\partial z}{\partial a}\right)_t = \frac{\rho_0 A_0}{\rho A} . \tag{47}$$

Using the transformation equations between the Eulerian and Lagrangian frames,

$$\left(\frac{\partial f}{\partial t}\right)_a = \left(\frac{\partial f}{\partial t}\right)_z + v\left(\frac{\partial f}{\partial z}\right)_t , \qquad \left(\frac{\partial f}{\partial a}\right)_t = l_a\left(\frac{\partial f}{\partial z}\right)_t , \tag{48}$$

Eqs. (30) and (31), with the help of (34), are written in the forms

$$\frac{1}{l_a}\left(\frac{\partial v}{\partial a}\right)_t + \frac{2c_S}{\gamma - 1}\frac{1}{c_T^2}\left(\frac{\partial c_S}{\partial t}\right)_a - \frac{\mu}{\gamma\Re}\left(\frac{c_S^2}{c_A^2} + \gamma\right)\left(\frac{\partial S}{\partial t}\right)_a - \frac{v}{\rho c_A^2}\frac{dp_e}{dz} = 0 \tag{49}$$

$$\left(\frac{\partial v}{\partial t}\right)_a + \frac{1}{l_a}\left[\frac{2c_S}{\gamma - 1}\left(\frac{\partial c_S}{\partial a}\right)_t - \frac{c_S^2\mu}{\gamma\Re}\left(\frac{\partial S}{\partial a}\right)_t\right] + g = 0. \tag{50}$$

We have used Eq. (45) to eliminate the thermodynamic variables p and ρ in favour of the sound speed c_S and entropy S. Suitably combining Eqs. (49) and (50) with help of (44) brings these equations into characteristic form

$$dv \pm \frac{2}{\gamma - 1} \frac{c_S}{c_T} dc_S \mp \frac{\mu c_S^2}{\gamma \Re c_T} dS \mp \left[\frac{\mu c_T}{\gamma \Re} (\gamma - 1) \left. \frac{dS}{dt} \right|_{Rad} + \frac{v c_T}{\rho c_A^2} \frac{dp_e}{dz} \right] dt + g dt = 0,$$
(51)

along the two characteristics C^+, C^- given by

$$\left(\frac{da}{dt} \right)_\pm = \pm \frac{c_T}{l_a},$$
(52)

where the upper sign in these two equations is for the C^+ characteristic and the lower sign for the C^- characteristic. Instead of the two partial differential Eqs. (49) and (50) we now have four ordinary differential equations (51) and (52).

The numerical solution proceeds as follows. From a height point P at the new time level $t + \Delta t$, where one assumes preliminary values of the three unknowns $c_S(P)$, $S(P)$ and $v(P)$, one constructs the characteristics (52) in the t-a plane and their intersection points with the old time level t at which the solution is assumed to be given everywhere. From these intersection points using the finite difference forms of Eqs. (51) along the two C^+ and C^- characteristics, two of the three unknowns can be computed. The third unknown is obtained by integrating Eq. (44) along the fluid path $a = const.$ using $dS = (dS/dt)|_{Rad} dt$. In this way new estimates of $c_S(P)$, $S(P)$ and $v(P)$ are obtained and the process can be repeated until convergence.

This procedure is very accurate and stable and allows also one to treat the formation and propagation of shocks. Extensive longitudinal wave propagation calculations using this procedure have been performed by Herbold et al. (1985) and Rammacher & Ulmschneider (1989). In these calculations H^- continuum radiation was included in the optically thin approximation, while Mg II k and Ca II K line radiation were treated as optically thick. For these non-grey radiation treatments the NLTE statistical rate equations where solved together with the radiative transfer equations.

For the problem of coupled transverse and longitudinal tube waves the set of equations (30) to (34) have to be augmented by the transverse components. The tube is now considered to be inclined to the vertical. Let $\hat{e}_1$ be the unit vector along the tube, with direction cosines l_x, l_y and l_z with respect to the coordinate axes; the direction cosines are related by

$$l_x^2 + l_y^2 + l_z^2 = 1.$$
(53)

Then the longitudinal Eqs. (30) and (31) together with (34) can be written

$$\frac{\hat{e}_1}{l_a} \cdot \left(\frac{\partial \mathbf{v}}{\partial a} \right)_t + \frac{2 c_S}{\gamma - 1} \frac{1}{c_T^2} \left(\frac{\partial c_S}{\partial t} \right)_a - \frac{\mu}{\gamma \Re} \left(\frac{c_S^2}{c_A^2} + \gamma \right) \left(\frac{\partial S}{\partial t} \right)_a - \frac{v_z}{\rho c_A^2} \frac{dp_e}{dz} = 0,$$
(54)

$$\hat{e}_1 \cdot \left(\frac{\partial v}{\partial t}\right)_a + \frac{1}{l_a}\left[\frac{2c_S}{\gamma - 1}\left(\frac{\partial c_S}{\partial a}\right)_t - \frac{c_S^2\mu}{\gamma\Re}\left(\frac{\partial S}{\partial a}\right)_t\right] + gl_z = 0 \ . \tag{55}$$

These equations should be compared with (49) and (50). Combining the two equations together with (44) into characteristic form one obtains

$$\hat{e}_1 \cdot dv \pm \frac{2}{\gamma - 1}\frac{c_S}{c_T}dc_S \mp \frac{\mu c_S^2}{\gamma\Re c_T}dS$$

$$\mp \left[\frac{\mu c_T}{\gamma\Re}(\gamma - 1)\left.\frac{dS}{dt}\right|_{Rad} + \frac{v_z c_T}{\rho c_A^2}\frac{dp_e}{dz}\right]dt + gl_z\,dt = 0 \tag{56}$$

along the two characteristics C_1^+, C_1^- given by

$$\left(\frac{da}{dt}\right)_\pm = \pm\frac{c_T}{l_a} \ , \tag{57}$$

where the top sign in the last two equations is for the C_1^+ and the bottom sign for the C_1^- characteristic. Note that for purely vertical propagation, when $l_z = 1$, Eq. (56) reduces to (51).

The system (56) and (57) describes the longitudinal wave, where the signal propagation speed, as observed from the moving mass element, is $dz/dt = l_a\,da/dt = \pm c_T$.

For the kink mode the transverse components of the equation of motion and of the combined induction and continuity equations must be considered. From the equation of motion

$$\rho\frac{dv}{dt} = -\nabla p - \frac{1}{\mu_0}B \times (\nabla \times B) + \rho g \ , \tag{58}$$

we take the transverse component

$$\rho\left(\frac{dv}{dt}\right)^\perp = \frac{B^2}{\mu_0}\kappa\hat{e}_2 + \hat{e}_1 \times \left[\rho g - \nabla\left(p + \frac{B^2}{2\mu_0}\right)\right] \times \hat{e}_1 \ . \tag{59}$$

Here $\hat{e}_2$ is the unit vector in direction to the local center of curvature and $\kappa\hat{e}_2 = (\partial\hat{e}_1/l_a\partial a)_t$. Assuming horizontal pressure balance, and outside the tube hydrostatic equilibrium ($\nabla p_e = \rho_e g$), Eq. (59) is written

$$\rho\left(\frac{dv}{dt}\right)^\perp = \rho c_A^2\kappa\hat{e}_2 + (\rho - \rho_e)g^\perp \ . \tag{60}$$

This equation does not include forces due to the backreaction of the external medium. The complete Eq. (60) thus reads

$$\rho\left(\frac{dv}{dt}\right)^\perp = \rho c_A^2\kappa\hat{e}_2 + (\rho - \rho_e)g^\perp + f_{ext} \ . \tag{61}$$

In recent years there has been an extensive discussion about the correct form of the back-reaction force term f_{ext} (Spruit 1981; Choudhuri 1990;

Cheng 1992; Fan, Fisher & McClymont 1994; Moreno-Insertis, Ferriz-Mas & Schüssler 1994; Moreno-Insertis, Schüssler & Ferriz-Mas 1996). Osin, Volin & Ulmschneider (1996) take

$$\mathbf{f_{ext}} = -\rho_e \left(\frac{d_\perp \mathbf{v}^\perp}{dt} \right)^\perp . \tag{62}$$

Then, with Eqs. (61) and (62), the transverse component of the equation of motion can be written (Osin et al. 1996)

$$\left(\frac{\partial \mathbf{v}}{\partial t} \right)_a - \hat{\mathbf{e}}_1 \cdot \left(\frac{\partial \mathbf{v}}{\partial t} \right)_a \hat{\mathbf{e}}_1 - \frac{(\rho c_A^2 - \rho_e v^{\|2})}{(\rho + \rho_e) l_a} \left(\frac{\partial \hat{\mathbf{e}}_1}{\partial a} \right)_t -$$
$$\frac{2\rho_e v^{\|}}{(\rho + \rho_e)} \left(\frac{\partial \hat{\mathbf{e}}_1}{\partial t} \right)_a - \frac{(\rho - \rho_e)}{(\rho + \rho_e)} \mathbf{g}^\perp = 0 , \tag{63}$$

where $v^{\|} = \hat{\mathbf{e}}_1 \cdot \mathbf{v}$.

The transverse component of the combined induction and continuity equation (30) is given by

$$\left(\frac{\partial \hat{\mathbf{e}}_1}{\partial t} \right)_a = \frac{1}{l_a} \left[\left(\frac{\partial \mathbf{v}}{\partial a} \right)_t - \hat{\mathbf{e}}_1 \cdot \left(\frac{\partial \mathbf{v}}{\partial a} \right)_t \hat{\mathbf{e}}_1 \right] . \tag{64}$$

Combining Eqs. (63) and (64) into characteristic form one finds (after some algebra) the two equations

$$\left(1 - l_x^2 \right) dv_x - l_x l_y dv_y - l_x l_z dv_z - c_k^\pm dl_x$$
$$- \frac{\rho - \rho_e}{\rho + \rho_e} g l_x l_z dt = 0 , \tag{65}$$
$$\left(1 - l_y^2 \right) dv_y - l_x l_y dv_x - l_y l_z dv_z - c_k^\pm dl_y$$
$$- \frac{\rho - \rho_e}{\rho + \rho_e} g l_y l_z dt = 0 ,$$

both along the characteristics C_2^+, C_2^- given by

$$\left(\frac{da}{dt} \right)_\pm = \frac{c_k^\pm}{l_a} . \tag{66}$$

Here the kink speed $c_k^\pm$, the propagation speed of the pure transverse wave mode, is given by

$$c_k^\pm = -\frac{\rho_e}{\rho + \rho_e} v^{\|} \pm \sqrt{ \left(\frac{\rho_e}{\rho + \rho_e} \right)^2 v^{\|2} + \frac{\rho c_A^2 - \rho_e v^{\|2}}{\rho + \rho_e} } . \tag{67}$$

Note that the pure longitudinal wave propagates with the tube speed c_T and in the above approximation is not affected by the back-reaction of the external medium, whereas the propagation of the transverse mode is strongly affected by the back-reaction of the external fluid. Propagation of the transverse mode is in general asymmetric, $c_k^- \neq -c_k^+$. Symmetry is restored when the longitudinal flow speed $v^{\parallel}$ is negligible, and then

$$c_k^{\pm} = \pm c_k \, , \tag{68}$$

the kink speed introduced in Eq. (3). The square root in Eq. (67) points to the possibility of hyperbolicity violation and hence to the development of a wave instability. This will happen when the longitudinal fluid velocity $v^{\parallel}$ is large enough:

$$v^{\parallel} > \sqrt{\frac{\rho + \rho_e}{\rho_e}} c_A \, , \tag{69}$$

defining the threshold of the *fire-hose instability*, so-called because the tube experiences a kink-like disturbance driven by the flow (much as for water in a fire-hose).

In contrast with the case of purely longitudinal wave propagation, one now has 8 unknowns c_S, S, $\mathbf{v}$, $\hat{\mathbf{e}}_1$ in the coupled longitudinal-transverse tube wave problem. To solve for these unknowns we have the two relations (56) along the C_1^+, C_1^- characteristics given by Eq. (57) and two relations (65) each along the C_2^+, C_2^- characteristics given by Eq. (66), the entropy conservation relation (44) along the fluid path and Eq. (53). Using this procedure, longitudinal-transverse wave computations have been performed by Ulmschneider & Zähringer (1989), Ulmschneider, Zähringer & Musielak (1991) and Osin et al. (1996). Moreover, as shown by Zhugzhda, Bromm & Ulmschneider (1995), the above procedure also allows to compute shock formation and propagation.

Although presently a procedure which incorporates all three coupled modes, the longitudinal, transverse and torsional wave propagation using the thin flux tube approximation, has not yet been described, the problem of a combined time-dependent longitudinal-torsional wave propagation has been investigated numerically by Hollweg, Jackson & Galloway (1982) and Ferriz-Mas, Schüssler & Anton (1989).

An attractive feature of the method of characteristics is that it automatically leads to the wave propagation speeds, and allows one to derive relations between the amplitudes of the various fluctuating physical quantities in the waves. Similar to the well known relations

$$\frac{v}{c_S} = \frac{\rho'}{\rho} = \frac{p'}{\gamma p} = \frac{2}{\gamma - 1} \frac{c_S'}{c_S} = \frac{1}{\gamma - 1} \frac{T'}{T} \tag{70}$$

for acoustic waves, Eqs. (45) and (56) allow us to derive the amplitude relations for longitudinal tube waves:

$$\frac{v}{c_S} = \frac{c_S}{c_T} \frac{\rho'}{\rho} = \frac{c_S}{c_T} \frac{p'}{\gamma p} = \frac{2}{\gamma - 1} \frac{c_S}{c_T} \frac{c_S'}{c_S} = \frac{1}{\gamma - 1} \frac{c_S}{c_T} \frac{T'}{T} = \frac{c_A^2}{c_S c_T} \frac{B'}{B} \, . \tag{71}$$

Here a prime indicates perturbations, considered small.

6 Magnetohydrodynamic Wave Generation

6.1 Analytical methods based on Lighthill's approach

The generation of different types of Alfvén waves and magnetoacoustic waves in the solar atmosphere has been studied primarily by using analytical methods based on the theory of sound generation by Lighthill (1952). In terrestrial applications for acoustic sound generation, Lighthill's theory is in excellent agreement with observations. Kulsrud (1955) and Osterbrock (1961) extended Lighthill's theory by including magnetic field effects, and Musielak & Rosner (1987, 1988) improved it by accommodating the presence of stratification and an embedded uniform magnetic field in the wave generation region (see also Rosner & Musielak 1989). More recently, Collins (1989a, b, 1992) has modified this type of wave generation theory to explore the excitation of MHD waves by periodic velocity fields in diverging magnetic flux tubes. The common feature of these studies is that they look at the magnetic field in a non-local way to obtain mean generated wave fluxes.

A further advance occurred when a detailed local flux tube geometry was considered. Musielak, Rosner & Ulmschneider (1989) and Musielak et al. (1995) have investigated the interaction between turbulent motions in the solar convection zone and thin magnetic flux tubes. They have considered vertically oriented magnetic flux tubes and restricted their approach to the linear regime. For a magnetic flux tube in the solar convection zone the external pressure can be written as

$$p_{ext} = p_e + p_{turb}, \tag{72}$$

where p_e is the external gas pressure and

$$p_{turb} \equiv \rho_e \left(v_x(\mathbf{r}, \mathbf{t})^2 + v_y(\mathbf{r}, \mathbf{t})^2 + v_z(\mathbf{r}, \mathbf{t})^2 \right) \tag{73}$$

the external turbulent pressure. Here v_x, v_y, v_z are the turbulent velocities in x, y, z-directions, functions of position $\mathbf{r}$ and time. Upon time averaging one gets

$$u_{xt} = \sqrt{\overline{v_x(\mathbf{r}, \mathbf{t})^2}}, \; u_{yt} = \sqrt{\overline{v_y(\mathbf{r}, \mathbf{t})^2}}, \; u_{zt} = \sqrt{\overline{v_z(\mathbf{r}, \mathbf{t})^2}} \,. \tag{74}$$

For the case of homogenous isotropic turbulence there is no longer a dependence on $\mathbf{r}$ due to the assumed homogeneity, and for the three spatial components isotropy implies that

$$u_t \equiv u_{xt} = u_{yt} = u_{zt}. \tag{75}$$

Here u_t is the rms velocity amplitude in one spatial direction, taken to be the same in the x, y and z-directions; it is independent of space and time. Note that here u_t is defined differently from in Musielak et al. (1995). From the above equations one has a time-averaged turbulent pressure

$$\overline{p_{turb}} = 3\rho_e u_t^2 \,, \tag{76}$$

and a fluctuating turbulent pressure

$$p'_{turb} = 3\rho_e u_t'^2 = 3\rho_e(v_x^2 - u_t^2). \tag{77}$$

Here one has used the fact that the velocity fluctuations in the three spatial directions are uncorrelated. For the generation of longitudinal tube waves external pressure fluctuations have to be translated into fluctuations inside the tube. From the horizontal pressure balance (34) one has for the internal pressure perturbations

$$p' + \frac{B_0 B'}{\mu_0} = p'_{turb}. \tag{78}$$

The Lighthill approach starts with an inhomogeneous wave equation

$$\left[\frac{\partial^2}{\partial t^2} - c_T^2 \frac{\partial^2}{\partial z^2} + \Omega_T^2\right] p_1(z,t) = S_t(z,t), \tag{79}$$

where $p_1 = p'/\sqrt{\rho_o B_o}$, B_0 the undisturbed field strength of the tube and Ω_T the tube (Defouw) frequency defined by Eq. (40). The source function is given by

$$S_t(z,t) = \frac{3\rho_e}{2\sqrt{\rho_o B_o}} \frac{c_T^2}{c_A^2} \left(\frac{\partial^2}{\partial t^2} + \Omega_{BV}^2\right) u_t'^2, \tag{80}$$

where Ω_{BV} is the Brunt-Väisälä frequency. For the time-averaged longitudinal tube wave flux one has

$$\overline{F(z,t)} = -B_o (1+\delta) p_1 \left(\frac{\partial}{\partial t}\right)^{-1} \left(\frac{\partial}{\partial z} + k_h\right) \left[1 + \Omega_{BV}^2 \left(\frac{\partial}{\partial t}\right)^{-2}\right]^{-2} p_1^*, \tag{81}$$

where $\delta = 5c_S^2/2\gamma c_A^2$, $k_h = (4 - 3\gamma)/4\gamma\Lambda_0$ for pressure scale height Λ_0, and complex conjugate p_1^* of p_1. By using Fourier transforms Eqs. (79) to (81) can be solved and the mean wave energy generation rate for a given flux tube can be written in the form (expressed in cgs units, erg cm^{-3} s^{-1} Hz^{-1}, the choice of the original authors):

$$\frac{\partial^2}{\partial z_o \partial \omega} \overline{F(z_o,\omega)}_{z_o,t_o} = \frac{\pi}{32} \left(\frac{\rho_e}{\rho_o}\right)^2 \left(\frac{V_t}{V_a}\right)^2 \frac{\rho_v \omega(1+\delta)}{V_t^3} \sqrt{\omega^2 - \Omega_{BV}^2} J_c(k_o,\omega), \tag{82}$$

where the convolution integral J_c is given by

$$J_c(k_0,\omega) = \frac{1}{2\pi^2} \int_{-\infty}^{+\infty} dr \int_{-\infty}^{+\infty} d\tau \left(2R_{xx}^2(r,\tau) + R_{zz}^2(r,\tau)\right) \exp^{-i(k_0 r - \omega\tau)}. \tag{83}$$

The averages extend over a suitably large height z_o and time t_o interval. Here the correlation tensors $R_{xx}(r,\tau)$ and $R_{zz}(r,\tau)$ can be expressed in terms of second order correlations averaged over z_o and t_o. These correlation tensors can be written in terms of the turbulent energy spectrum $E(k,\omega)$,

$$R_{xx}(r,\tau) = R_{yy}(r,\tau) =$$

$$\int_0^\infty d\omega \cos \omega\tau \int_0^\infty dk \; E(k,\omega) \left(\frac{\sin kr}{kr} + \frac{\cos kr}{k^2 r^2} - \frac{\sin kr}{k^3 r^3} \right) \qquad (84)$$

and

$$R_{zz}(r,\tau) = 2 \int_0^\infty d\omega \cos \omega\tau \int_0^\infty dk \; E(k,\omega) \left(\frac{\sin kr}{k^3 r^3} - \frac{\cos kr}{k^2 r^2} \right) \qquad (85)$$

where

$$E(k,\omega) = E(k)\Delta \left(\frac{\omega}{ku_k} \right) ; \qquad (86)$$

the mean velocity of a turbulent eddy with wave number k is given by

$$u_k = \left[\int_k^{2k} E(k')dk' \right]^{1/2} . \qquad (87)$$

The computation of the longitudinal tube wave flux thus reduces to the specification of the turbulent energy spectrum. The turbulent energy spectrum appropriate for the solar convection zone has been discussed in detail by Musielak et al. (1994). These authors argue on the basis of observations and numerical convection calculations that a realistic turbulent energy spectrum should be reasonably well described by an extended Kolmogorov spectrum, $E(k)$, and a modified Gaussian frequency factor, $\Delta(\omega/ku_k)$. The extended Kolmogorov spatial component is given by

$$E(k) = \begin{cases} 0, & 0 < k < 0.2k_t, \\ 0.758\frac{u_t^2}{k_t} \left(\frac{k}{k_t} \right), & 0.2k_t \leq k < k_t, \\ 0.758\frac{u_t^2}{k_t} \left(\frac{k}{k_t} \right)^{-5/3}, & k_t \leq k \leq k_d, \end{cases} \qquad (88)$$

where $k_t = 2\pi/\Lambda_0$, and the modified Gaussian frequency factor by

$$\Delta \left(\frac{\omega}{ku_k} \right) = \frac{4}{\sqrt{\pi}} \frac{\omega^2}{|ku_k|^3} e^{-(\frac{\omega}{ku_k})^2} . \qquad (89)$$

Using the above, Musielak et al. (1995) have computed longitudinal tube wave fluxes. Their results indicate fluxes of the order of several times 10^7 erg cm^{-2}s^{-1}, which seem relatively low to account for the observed enhanced heating in the chromospheric network. In a similar treatment for transverse tube waves, Musielak, Rosner & Ulmschneider (1997) show that the wave energy flux carried by these waves can be of the order of 10^8 erg cm^{-2}s^{-1}.

6.2 Numerical methods based on the direct perturbation of flux tubes

There are also methods of MHD wave generation which are not based on the Lighthill approach. In these methods solar magnetic flux tube models are perturbed from the outside with velocity or pressure fluctuations which in magnitude and spectral shape are consistent with observations.

To perturb the flux tube one assumes that v_x can be written as a spectrum of N partial waves

$$v_x = \sum_{n=1}^{N} u_n \sin(\omega_n t + \varphi_n) , \tag{90}$$

where $\varphi_n = 2\pi r_n$ is an arbitrary but constant phase angle, r_n a random number in the interval $[0, 1]$. The partial wave amplitude u_n is determined by the turbulent energy spectrum as follows. Time averaging v_x^2 one finds (Huang et al. 1995)

$$\overline{v_x^2} = u_t^2 = \frac{1}{2} \sum_{n=1}^{N} u_n^2 . \tag{91}$$

As is customary (e.g. Musielak et al. 1995) the turbulent energy spectrum is normalized to

$$\frac{3}{2} u_t^2 = \int_0^{\infty} d\omega \int_0^{\infty} dk \ E(k) \Delta\left(\frac{\omega}{k u_k}\right) = \int_0^{\infty} E'(\omega) d\omega . \tag{92}$$

From this one obtains

$$\frac{3}{2} u_t^2 = \frac{3}{4} \sum_{n=1}^{N} u_n^2 = \int_0^{\infty} E'(\omega) d\omega = \sum_{n=1}^{N} E'(\omega_n) \Delta\omega, \tag{93}$$

which allows to determine u_n

$$u_n = \sqrt{\frac{4}{3} E'(\omega_n) \Delta\omega} , \tag{94}$$

where

$$E'(\omega_n) = \int_0^{\infty} E(k) \Delta\left(\frac{\omega_n}{k u_k}\right) dk ; \tag{95}$$

for $E(k)$ and $\Delta(\omega/k u_k)$, the extended turbulent energy spectrum and the modified Gaussian frequency factor of Eqs. (88) and (89) are taken.

For a transverse wave generation calculation the velocity v_x can be directly applied as a boundary condition, representing the horizontal shaking velocity acting on the magnetic flux tube. A similar uncorrelated shaking arises from v_y. For values of $u_t = 1.0$ to 2.0 km s^{-1} and shaking at various heights, Huang et al. (1995) with a correction described by Ulmschneider & Musielak (1996) obtain total transverse wave fluxes (shaking in both horizontal directions) of $6 \cdot 10^9$ to $3 \cdot 10^{10}$ erg cm^{-2}s^{-1}.

In the process of longitudinal tube wave excitation, the tube is compressed symmetrically by the external turbulent pressure. This turbulent pressure consists of the time averaged term $\overline{p_{turb}}$ which augments the external gas pressure p_e and the fluctuating term which gives rise to longitudinal tube waves (cf. Eqs. (76) to (78)). For the generation of longitudinal tube waves these external pressure fluctuations have to be translated into fluctuations inside the tube. From the horizontal pressure balance one has gas pressure and magnetic field variations inside the tube (cf. Eq. (78)). Using the relations among the amplitudes for longitudinal tube waves one has

$$\frac{B'}{B_0} = \frac{c_S^2}{c_A^2} \frac{p'}{\gamma p_0} , \qquad (96)$$

where p_0 the undisturbed gas pressure in the tube. From this one obtains

$$p' = \frac{1}{2} p'_{turb} , \qquad (97)$$

which shows that the external pressure fluctuations are divided equally into an internal gas pressure fluctuation and a magnetic pressure fluctuation. As the tube wave code normally uses a velocity boundary condition one translates the gas pressure fluctuations into longitudinal velocity fluctuations $v_\parallel$ by using the amplitude relations for longitudinal tube waves, finding

$$v_\parallel = \frac{c_S^2}{c_T} \frac{p'}{\gamma p_0} = \frac{p'_{turb}}{2\rho_0 c_T} , \qquad (98)$$

with ρ_0 the undisturbed density in the tube.

Finally, the normalized flux of longitudinal tube waves is

$$\overline{F} = \overline{\frac{A}{A_0} \rho_0 v_\parallel^2 c_T} \qquad (99)$$

while that of transverse tube waves is

$$\overline{F} = -\overline{\frac{A}{A_0} \frac{B_0}{\mu_0} B' v_x} ,$$

where v_x is the transverse velocity and B' the transverse magnetic field perturbation.

Using this procedure Ulmschneider & Musielak (1996) have computed various longitudinal wave energy fluxes and find values of the order of several times 10^8 erg cm^{-2}s^{-1}. Note that due to the nonlinear treatment, which uses the partial wave synthesis after Eq. (90), a very spiky velocity perturbation results which leads to much larger fluxes both for the transverse and the longitudinal wave fluxes as compared to the linear results. We have the suspicion that due to the lack of cancellations when shaking simultaneously at several heights, the time-dependent nonlinear treatments tends to overestimate wave fluxes, while the linear analytical treatment tends to underestimate the fluxes.

Finally, we note that detailed studies of the various waves excited by footpoint motions of flux tubes and slabs have also been recently carried out; see, e.g., Murawski & Roberts (1993), Berghmans & De Bruyne (1995), Murawski et al. (1996), and Cargill, Spicer & Zalesak (1996). But so far such studies are aimed more at coronal conditions and as such lie outside the scope of the present review.

7   Concluding Remarks

It is evident from the above discussion that many aspects of flux tubes in the solar atmosphere are now understood. However, a number of important theoretical questions remain unanswered. For example, we note that the basic set of thin tube equations for the sausage mode presented earlier were derived under the assumption that there were no motions in the θ-direction. This has the effect of losing the torsional Alfvén waves given for a finite tube by equation (10). As recently stressed by Zhugzhda (1996), it is clearly important to extend our understanding so as to include the coupling of motions. Zhugzhda (1996) has considered such an extension of the thin tube equations to include v_θ terms, using the Taylor series expansion about $r = 0$ as originally employed by Roberts & Webb (1978). (The expansion approach has also been adopted by Ferriz-Mas, Schüssler & Anton (1990).) All extensions of this form are in principle capable of being tested against the known linear results for a finite radius tube, and thereby to examine how successful the approach is in describing the *dispersive* behaviour of the waves.

The standard derivation of thin tube equations for the kink mode makes use of the assumption that the kink mode displaces an equal proportion of fluid in the environment of the tube as is contained by the tube (Parker 1979; Spruit 1981). This has the effect of augmenting the gas density ρ_0 with ρ_e, so that the characteristic speed of the kink mode is c_k (Eq. (3)). However, to describe properly the motions in a nonlinear theory it is necessary to describe also the dispersive influence of the tube's surroundings, as given in linear theory by the long wavelength result (27). As discussed earlier, there have been a number of attempts (Choudhuri 1990; Cheng 1992, 1994; Fan, Fisher & McClymont 1994; Moreno-Insertis, Schüssler & Ferriz-Mas 1996; Osin et al. 1996) to extend Spruit's (1981) thin tube theory of the kink mode, to take proper account of inertial effects, but the matter is currently unresolved. Again, we note that it is a test of any approximate description that it is able to describe properly the *dispersive* influence of the tube's environment, evident in (26) and (27); this proved possible in the derivation of the Leibovich–Roberts equation for the sausage mode, but the kink mode remains to be fully treated.

We should note too that the presence of flows in the regions externally bordering isolated photospheric tubes has an effect on the waves a tube supports. Nakariakov & Roberts (1995) have shown that if an external steady flow of magnitude U_e exists outside a uniform magnetic slab, then the fast surface wave propagating against the direction of the flow becomes leaky if $U_e > c_{Se} - c_S$ and the slow body waves leak if $U_e > c_{Se} - c_T$. Such flows may be met in the photosphere: from our earlier illustration of speeds, we see that $c_{Se} - c_S = 1.6$ km s^{-1} and $c_{Se} - c_T = 2.9$ km s^{-1}, so downdraughts of 3 km s^{-1} might cause strong leakage from the tube.

In summary, then, while many aspects of flux tube dynamics are now understood, there remain a number of important fundamental questions for theoretical investigation. But above all, it is to be hoped that observational evidence of flux tube behaviour will point out the most important features of flux tube dynamics that require our greater study and understanding.

References

Appert, K., Gruber, R., Vaclavik, J. (1974): Physics Fluids **17**, 1471

Ayres, T. R., Testerman, L., Brault, J. W. (1986): ApJ **304**, 542

Barston, E. M. (1964): Ann. Phys. **29**, 282

Berghmans, D., De Bruyne, P. (1995), ApJ **453**, 495

Bodo, G., Rosner, R, Ferrari, A., Knobloch, E. (1989): ApJ **341**, 631

Bogdan, T. B. (1992): in *Sunspots: Theory and Observation*, eds, J. H. Thomas & N. O. Weiss (Kluwer, Dordrecht), NATO ASI Series, p. 345

Bogdan, T. B., Lerche, I. (1988): Quart. Applied Maths. **XLVI**, 365

Browning, P. K. (1991): J. Plasma Phys. Contr. Fusion **33**, 539

Cally, P. S. (1985): Australian J. Phys. **38**, 825

Cally, P. S. (1986): Solar Phys. **103**, 277

Campos, L. M. B. C. (1986): Rev. Mod. Physics **58**, 117

Cargill, P. J., Spicer, D. S., Zalesak, S. T. (1996): preprint

Chen, L., Hasegawa, A. (1974): Phys. Fluids **17**, 1399

Cheng, J. (1992): A&A **264**, 243

Choudhuri, A.R. (1990): A&A **239**, 335

Collins, W. (1989a): ApJ **337**, 548

Collins, W. (1989b): ApJ **343**, 499

Collins, W. (1992): ApJ **384**, 319

Davila, J. M. (1987): ApJ **317**, 514

Davila, J. M. (1991): in *Mechanisms of Chromospheric and Coronal Heating*, eds. P. Ulmschneider, E. R. Priest & R. Rosner (Springer-Verlag, Heidelberg), p. 464

Defouw, R. J. (1976): ApJ **209**, 266

Edwin, P. M. (1991): Ann. Geophys. **9**, 188

Edwin, P. M. (1992): Ann. Geophys. **10**, 631

Edwin, P. M., Roberts, B. (1983): Solar Phys. **88**, 179

Evans, D. J., Roberts, B. (1990): ApJ **348**, 346

Fan, Y., Fisher, G. H., McClymont, A. N. (1994): ApJ **436**, 907

Ferriz-Mas, A. (1988): Physics Fluids **31**, 2583

Ferriz-Mas, A., Schüssler, M., Anton, V. (1989): A&A **210**, 425

Goedbloed, J. P. (1975): Physics Fluids **15**, 1090

Goedbloed, J. P. (1983): *Lecture Notes in Magnetohydrodynamics*, Rijnhuizen Report 83-145, Assoc. Euratom - FOM, pp. 289

Goossens, M. (1991): in *Advances in Solar System Magnetohydrodynamics*, eds. E. R. Priest & A. W. Hood (CUP, Cambridge), p. 137

Goossens, M., Ruderman, M. S., Hollweg, J. V. (1995): Solar Phys. **157**, 75

Hardee, P. E. (1995): in *Waves in Astrophysics*, eds. J. H. Hunter & R. E. Wilson (New York Acad. Sciences, New York), p. 14

Herbold, G., Ulmschneider, P., Spruit, H.C., Rosner, R. (1985): A&A **145**, 157

Heyvaerts, J., Priest, E. R. (1983): A&A **117**, 220

Hollweg, J. V. (1986): in *Advances in Space Plasma Physics*, ed. B. Buti (World Scientific, Singapore), p.77

Hollweg, J. V. (1988): ApJ. **335**, 1005

Hollweg, J. V. (1990a): in *Physics of Magnetic Flux Ropes*, eds. C. T. Russell, E. R. Priest & L. C. Lee (AGU, Washington), Geophys. Mono. **58**, p. 23

Hollweg, J. V. (1990b): Computer Phys. Reports **12**, 205

Hollweg, J. V. (1991): in *Mechanisms of Chromospheric and Coronal Heating*, eds. P. Ulmschneider, E. R. Priest & R. Rosner (Springer-Verlag, Heidelberg), p 423

Hollweg, J. V., Jackson, S., Galloway, D. (1982): Solar Phys. **75**, 35

Huang, P., Musielak, Z. E., Ulmschneider, P. (1995): A&A **297**, 579

Ionson, J. (1978): ApJ **226**, 650

Ireland, J. (1997): these proceedings

Kulsrud, R. M. (1955): ApJ **121**, 461

Lee, M. A., Roberts, B. (1986): ApJ **301**, 430

Lighthill, M. J. (1952): Proc. Roy. Soc. London **A211**, 564

Merzljakov, E. G., Ruderman, M. S. (1985): Solar Phys. **95**, 51

Molotovshchikov, A. L., Ruderman, M. S. (1987): Solar Phys. **109**, 247

Moreno-Insertis, F., Ferriz-Mas, A., Schüssler, M. (1994): ApJ **422**, 652

Moreno-Insertis, F., Schüssler, M., Ferriz-Mas, A. (1996): A&A **312**, 317

Murawski, K., DeVore, C. R., Parhi, S., Goossens, M. (1996): Planet. Space Sci. **44**, 253

Murawski, K., Roberts, B. (1993): Solar Phys. **144**, 101

Musielak, Z. E., Rosner, R (1987): ApJ **315**, 371

Musielak, Z. E., Rosner, R. (1988): ApJ **329**, 376

Musielak, Z. E., Rosner, R., Ulmschneider, P.(1989): ApJ **337**, 470

Musielak, Z. E., Rosner, R., Gail, H., Ulmschneider, P. (1995): ApJ **448**, 865

Musielak, Z. E., Rosner, R., Ulmschneider, P (1997): ApJ, in press

Musielak, Z. E., Rosner, R., Stein, R., Ulmschneider, P. (1994): ApJ **423**,

474

Nakariakov, V. M., Roberts, B. (1995): Solar Phys. **159**, 213

Nakariakov, V. M., Zhugzhda, Y., Ulmschneider, P. (1996): A&A **312**, 691

Narain, U., Ulmschneider, P. (1990): Space Sci. Rev. **54**, 377

Narain, U., Ulmschneider, P. (1996): Space Sci. Rev. **75**, 453

Osin, A., Volin, S. & Ulmschneider, P. (1996): A&A, in press

Osterbrock, D. E. (1961): ApJ **134**, 347

Parker, E. N. (1974): ApJ **189**, 563

Parker, E. N. (1979): *Cosmical Magnetic Fields* (OUP, Oxford)

Parker, E. N. (1991): ApJ **376**, 355

Poedts, S., Goossens, M., Kerner, W. (1990): ApJ **360**, 279

Rae, I. C., Roberts, B. (1981): Geophys. Astrophys. Fluid Dynamics **18**, 197

Rae, I. C. & Roberts, B. (1982): ApJ **256**, 761

Rammacher, W., Ulmschneider, P. (1989): in *Solar and Stellar Granulation*,
    eds. R. J. Rutten and G. Severino (Kluwer, Dordrecht), p. 589

Roberts, B. (1980): Ann. De Physique **5**, 453

Roberts, B. (1981): in *Physics of Sunspots*, eds. L. E. Cram & J. H. Thomas
    (Sacramento Peak Observatory, Sunspot, New Mexico), p. 360

Roberts, B. (1985a): in *Solar System Magnetic Fields*, ed. E. R. Priest
    (Reidel, Dordrecht), p. 37

Roberts, B. (1985b): Physics Fluids **28**, 3280

Roberts, B. (1986): in *Small-Scale Magnetic Flux Concentrations in the
    Solar Photosphere*, eds. W. Deinzer, M. Knölker & H. Voigt (Vandenhoeck
    & Ruprecht, Göttingen), p. 169

Roberts, B. (1987): ApJ **318**, 590

Roberts, B. (1990a): in *Physics of Magnetic Flux Ropes*, eds. C. T. Russell,
    E. R. Priest & L. C. Lee (AGU, Washington), Geophys. Mono. **58**, p. 113

Roberts, B. (1990b): in *Basic Plasma Processes in the Sun*, eds. E. R. Priest
    & V. Krishan (Kluwer, Dordrecht), p. 159

Roberts, B. (1991a): Geophys. Astrophys. Fluid Dynamics **62**, 83

Roberts, B. (1991b): in *Advances in Solar System Magnetohydrodynamics*,
    eds. E.R. Priest & A.W. Hood (CUP, Cambridge), p. 105

Roberts, B. (1992a): in *Sunspots: Theory and Observation*, eds, J. H. Thomas
    & N. O. Weiss (Kluwer, Dordrecht), NATO ASI Series, p. 303

Roberts, B. (1992b): in *Electromechanical Coupling of the Solar Atmosphere*,
    eds. D. S. Spicer & P. MacNeice (Amer. Inst. Physics, New York), p. 24

Roberts, B., Mangeney, A. (1982): Mon. Not. Roy. Astron. Soc. **198**, 7p

Roberts, B., Webb, A. R. (1978): Solar Phys. **56**, 5

Roberts, B., Webb, A. R. (1979): Solar Phys. **64**, 77

Rosner, R., Musielak, Z. E. (1989): A&A **219**, L27

Ruderman, M. S. (1993): J. Plasma Phys. **49**, 271

Ryutov, D. D., Ryutova, M. P. (1976): Sov. Phys. J.E.T.P. **43**, 491

Ryutova, M. P. (1990a): in *Solar Photosphere: Structure, Convection and
    Magnetic Fields*, ed. J. O. Stenflo (Reidel, Dordrecht), IAU Symp. **138**,

p. 229

Ryutova, M. P. (1990b): in *Basic Plasma Processes in the Sun*, eds. E. R. Priest & V. Krishan (Kluwer, Dordrecht), p. 175

Ryutova, M. P., Persson, M. (1984): Physica Scripta **29**, 353

Schüssler, M. (1990): in *Solar Photosphere: Structure, Convection and Magnetic Fields*, ed. J. O. Stenflo (Reidel, Dordrecht), IAU Symp. **138**, p. 161

Schüssler, M. (1991): Geophys. Astrophys. Fluid Dynamics **62**, 271

Solanki, S. K. (1990): in *Solar Photosphere: Structure, Convection and Magnetic Fields*, ed. J. O. Stenflo (Reidel, Dordrecht), IAU Symp. **138**, p. 103

Solanki, S. K. (1993): Space Science Reviews **63**, 1

Solanki, S. K. (1997): these proceedings

Solanki, S. K., Livingston, W., Ayres, T. (1994): Sci **263**, 64

Solanki, S. K., Steiner, O. (1990): A&A **234**, 519

Spruit, H. C. (1981a): in *The Sun as a Star*, ed. S. Jordan (NASA, Washington), NASA SP-450, p. 385

Spruit, H. C. (1981b): A&A **98**, 155

Spruit, H. C. (1982): Sol. Phys. **75**, 3

Spruit, H. C., Roberts, B. (1983): Nat **304**, 401

Stenflo, J. O. (1989): A&AR. **1**, 3

Stenflo, J. O. (1994): in *Solar Magnetic Fields*, eds. M. Schüssler & W. Schmidt (Cambridge University Press, Cambridge), p. 301

Thomas, J. H. (1985): in *Theoretical Problems in High Resolution Solar Physics*, ed. H. Schmidt (Max Planck Institute, Munich), MPA 212, p.126

Thomas, J. H., Weiss, N. O. (1992): eds., *Sunspots: Theory and Observation* (Kluwer, Dordrecht), NATO ASI Series, pp. 428

Ulmschneider, P., Musielak, Z. E. (1996): A&A, in press

Ulmschneider, P., Zähringer, K. (1987): in *Cool Stars, Stellar Systems and the Sun*, eds. J. L. Linsky & R. E. Stencel (Springer, Berlin), Lecture Notes in Physics **291**, p. 63

Ulmschneider, P., Zähringer, K., Musielak, Z. E. (1991): A&A **241**, 625

Weisshaar, E. (1989): Physics Fluids **A1**, 1400

Zhelyazkov, I., Murawski, K., Goossens, M., Nenovski, P., Roberts, B. (1994): J. Plasma Phys. **51**, 291

Zhugzhda, Y. D. (1996): Phys. Plasmas **3**, 10

Zhugzhda, Y. D., Bromm, V., Ulmschneider, P. (1995): A&A **300**, 302

Observations of Energetic Ions During the Ulysses Mission

T. R. Sanderson

Space Science Department of ESA, ESTEC, Noordwijk, The Netherlands

Abstract. The Ulysses spacecraft was launched in October 1990, and, having completed its prime mission to overfly the south and north polar regions of the Sun, has now started on its second solar orbit. We review here energetic ion observations from the many regions encountered during the out-of-ecliptic phase of the mission: the low latitude traversal of the streamer belt, the passage through the south polar region, the fast latitude scan, and the passage over the north polar region.

1 Introduction

One of the surprises of the Ulysses mission has been the observation at high latitudes of intense fluxes of accelerated ions with energies ~1 to 10 MeV, beyond the latitudes where co-rotating interaction regions (CIR) are seen. Accelerated ions have been observed at low latitudes together with CIRs bounded by forward-reverse shock pairs, at mid latitudes together with CIRs having only a reverse shock, and at high latitudes where no shocks were observed (Sanderson et al. (1994, 1995), Sanderson (1995), Simnett et al. (1994)). In this paper we present a survey of the observations of these accelerated ions, taken from observations from the Low Energy Telescope (LET) of the COSPIN (Cosmic Ray and Solar Particle Instrumentation) instrument of the Ulysses spacecraft.

2 Instrumentation

The observations presented here were made with the Low Energy Telescope of the COSPIN instrument, which measures protons, alpha particles, and heavy ions over a range of energies from ~1 to 50 MeV/nucleon. Data from the 1.2 - 3.0, 1.8 - 3.8 and 8 - 19 MeV proton channels are presented here. The 1.2 - 3.0 MeV proton channel and the 1.0 - 5.0 MeV/nucleon alpha particle channel are used to obtain the proton-to-alpha ratio. A complete description of the instrument can be found in Simpson et al. (1992). Observations from the solar wind experiment (Bame et al., 1992) and the magnetometer (Balogh et al., 1992) are used here in the interpretation of these data.

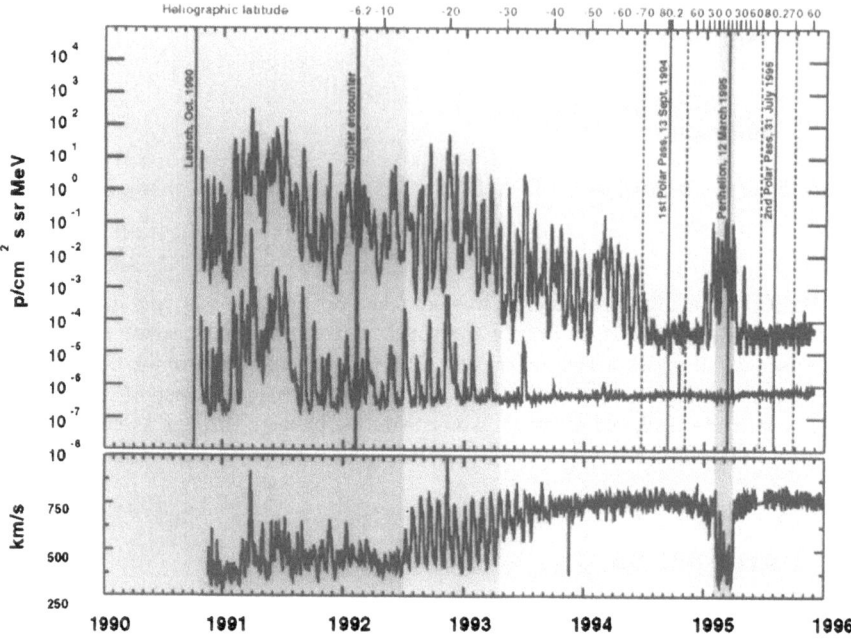

Fig. 1. Overview of the Ulysses mission from launch in October 1990 to mid 1995 as seen, in the top trace, from a low energy (1.8 - 3.8 MeV), and in the bottom trace, from a high energy ion (8 - 19 MeV) perspective. The lower trace has been multiplied by a factor of 10^{-3}. The bottom panel shows the solar wind velocity.

3   Observations

In Figure 1 we show the intensity of 1.8 - 3.8 MeV and 8 - 19 MeV protons observed during the prime mission, together with the solar wind velocity. The 8 - 19 MeV intensity has been multiplied by a factor of 10^{-3} . Vertical lines show the times of key events during the mission. Vertical dashed lines on either side of the maximum latitude show the extent of the polar passes, defined here as a latitude of $70°$. At the top of the plot we show the latitude of the spacecraft. The different regions through which the spacecraft passed, as defined by Sanderson et al. (1994, 1995) are shown shaded. These regions are as follows:

Region-1 - more or less completely immersed in slow speed solar wind flow from the streamer belt.

Region-2 - passing from the slow speed solar wind flow into the high speed flow of the polar coronal hole and back again once per rotation due to the tilt of the current sheet.

Region-3 completely immersed in high speed solar wind flow from the polar coronal hole.

In quiet times the 8 - 19 MeV proton channel (the lower trace) responds mainly to galactic cosmic rays, but during times of increased solar activity it also responds to solar particles, so can be used as a measure of solar activity. Early on in the mission, which was just after solar maximum, this channel was dominated by solar particles. Later on, in 1994 and 1995 the level of activity was so low that only one or two increases were seen. A slow increase in the background intensity of this channel can be seen due to the modulation of the galactic cosmic rays.

The 1.8 - 3.8 MeV proton channel (the upper trace) responds not only to solar particles, but also to locally accelerated particles, accelerated either by interplanetary shocks, or by CIRs. At the start of the mission, and prior to the Jupiter encounter, this channel showed the same general features as the 8 - 19 MeV channel. Later on, as the spacecraft entered Region-2, and as the number of transients decreased, regular peaks in the 1.8 - 3.8 MeV intensity occurring once per solar rotation and coinciding with the observation of CIRs were observed. These peaks continued to be observed in Region-3 (high-speed solar-wind flow) up to a latitude of around 70°S, well after the CIRs were no longer observed. Beyond this, over the southern pole the intensity remained low, and no peaks were observed. The recurrent peaks were observed again as Ulysses returned to the ecliptic plane for the fast latitude scan, entering again Region-1, the streamer belt. After leaving the streamer belt the spacecraft continued into Region-3 again as it passed over the northern polar regions.

In Figure 2 we show in more detail some of the features of Region-2. Each peak is labelled according to the numbering of the high-speed streams introduced by Bame et al. (1993), and continued by us (Sanderson et al., 1994). Here we show data for six rotations, peaks numbers 4 to 9, starting on 30 August 1992 and ending on 8 February 1993 (day 243 of 1992 to day 39 of 1993), comprising, from top to bottom, the 1.2 - 3.0 MeV proton intensity, the proton-to-alpha ratio, the solar wind velocity, and the magnetic field magnitude. The single high speed solar-wind stream reported per rotation by Bame et al. (1993) is shown in the 3rd panel from the top of Figure 2.

The streams observed in the first half of 1993 were analysed by Smith et al. (1993), who concluded that each is made up of three or four interaction regions per rotation, characterised by a large magnetic field magnitude. These structures were called a, b, c, and d by Smith et al. (1993), and are shown in the bottom panel of Figure 2. Structures a, b, and d orginated from south of the heliomagnetic equator, whilst structure c originated from north of the equator. Each structure is labelled according to the labelling of Smith et al. (1993), extending this labelling back in time from the original paper.

Most of the structures are bounded by a forward-reverse shock pair, the shocks for the dominant structure, structure-d, being denoted by the dashed vertical lines labelled F and R at the top of Figure 2. The times for these

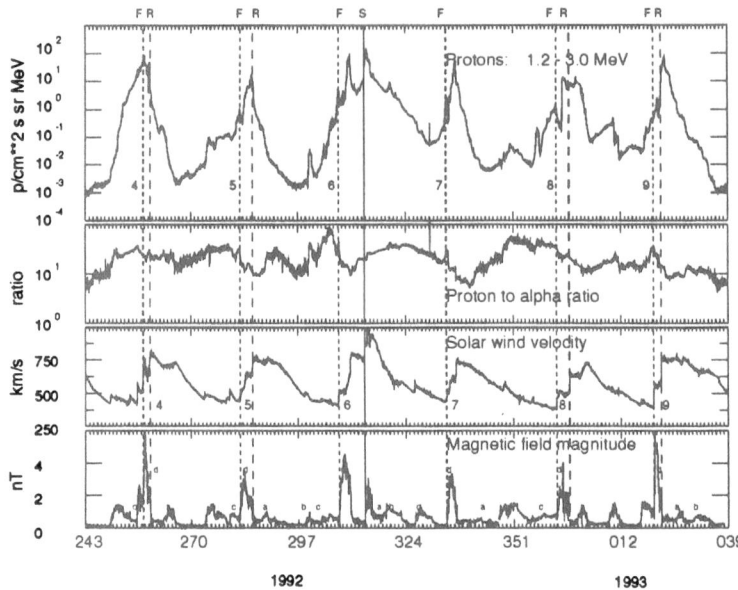

Fig. 2. Intensity of 1.2 - 3.0 Mev protons for period starting on 30 August 1992 and ending on 8 February 1993 (day 243 of 1992 to day 39 of 1993), together with the proton to alpha ratio, the solar wind velocity and the magnetic field magnitude, showing six solar rotations (4 - 9).

shocks are taken from Balogh et al. (1994). In most of the larger structures, we can also identify from the magnetic field the stream interface which marks the boundary between the fast and the slow solar-wind streams.

This period is dominated by the magnetic structure d, which during this time has the stronger shocks, i.e. larger solar wind velocity jumps, and larger magnetic field increases, than structures a, b or c. At or close to each shock of structure d we observe a spike in the proton intensity profile, the spike being larger at the reverse shocks than the forward shocks. This is typical of particle enhancements observed in co-rotating interaction regions near the ecliptic e.g., (Barnes and Simpson, 1976). The particle intensity varies between a high level of $\sim 10^2$ p/cm^{2}s sr MeV at the reverse shocks, and a low level of $\sim 10^{-3}$ p/cm^{2}s sr MeV in between the streams. The ~ 1 MeV proton enhancements fill a large fraction of the heliosphere at Ulysses as the stream rotates once , per solar rotation past the spacecraft.

The proton-to-alpha ratio rises to a peak of ~ 50 in the region around each forward shock, dropping to ~10 to 20 at the stream interface and in the enhancement associated with the reverse shock. Again, this is typical of particle enhancements in co-rotating interaction regions near the ecliptic.

The above pattern of events is disturbed during rotation number 6 by the presence of an interplanetary shock driven by a Coronal Mass Ejection (CME) (Bame et al., 1993). The shock, shown by the solid vertical line labelled S, causes the solar wind velocity at Ulysses to rise to ~1000 km/s, and is responsible for the intensity spike seen at the time of passage of the shock.

In Figure 3 we show data from Region-3 (completely immersed in high-speed flow) for 1993 and 1994. In the top panel we show the intensity as observed in the 1.8 - 3.8 MeV and the 8 - 19 MeV proton channel, the intensity of the 8 - 19 MeV channel having been multiplied by a factor of 10^{-3}. In the next panel we show the proton to alpha ratio, whilst in the bottom panel we show the spectral index measured at around 1 MeV.

Regular peaks in the intensity are observed at ~25 to 26 day intervals, the intensity at the peak falling with increasing latitude. The spacecraft entered Region-3 at the end of April 1993 (25 April, 1993, day 115). Forward-reverse shocks associated with the CIRs were still seen in Region-3 up to a latitude of ~35°S, reverse shocks up to ~42°S, but thereafter no co-rotating shocks were seen (Gosling et al., 1995a). However, the regular recurrent peaks in the particle intensity were seen well poleward of this latitude, their intensity slowly decreasing with increasing latitude.

Transient events were also observed. These can be seen as short increases in the 8 - 19 MeV intensity in Figure 3, and in the intensity of the recurrent 1.8 - 3.8 MeV proton intensity peak. The intensity at the peak then continues to decay slowly thereafter. Examples of such transient events are seen in Figure 3 at peak 15, 18, in between 23 and 24, and in between 24 and 25. (The transients in between 24 and 25, and 25 and 26 are the subject of a paper by Bothmer et al., (1995)).

In Figure 4 we show some of Region-3 in more detail. This plot, from 19 June 1993 until 28 November 1993 (day 170 to day 332 of 1993) covers the rotation numbers 15 to 20. At the start of the plot, (rotation 15) a forward-reverse shock pair was observed (Gosling et al., 1993). This is the last rotation where a forward-reverse shock pair is observed. Here the structure a has taken over from structure d in importance.

The proton intensity at each peak falls slowly with each subsequent rotation from $\sim 10^1$ to $\sim 10^{-1}$ p/cm^{2}s sr MeV over the six rotations during this period. The intensity at the peaks continues to fall after the end of the period shown in Figure 4. This can be seen in Figure 1 and 2, which shows that the accelerated protons are still being observed up to a latitude of ~70°S.

Since the dominant interaction region is now not filling the heliosphere at Ulysses for such a large fraction of a solar rotation, secondary peaks from the other interaction regions are now resolvable in between the main peaks, e.g. between peaks 15 and 16, 16 and 17, and 18 and 19.

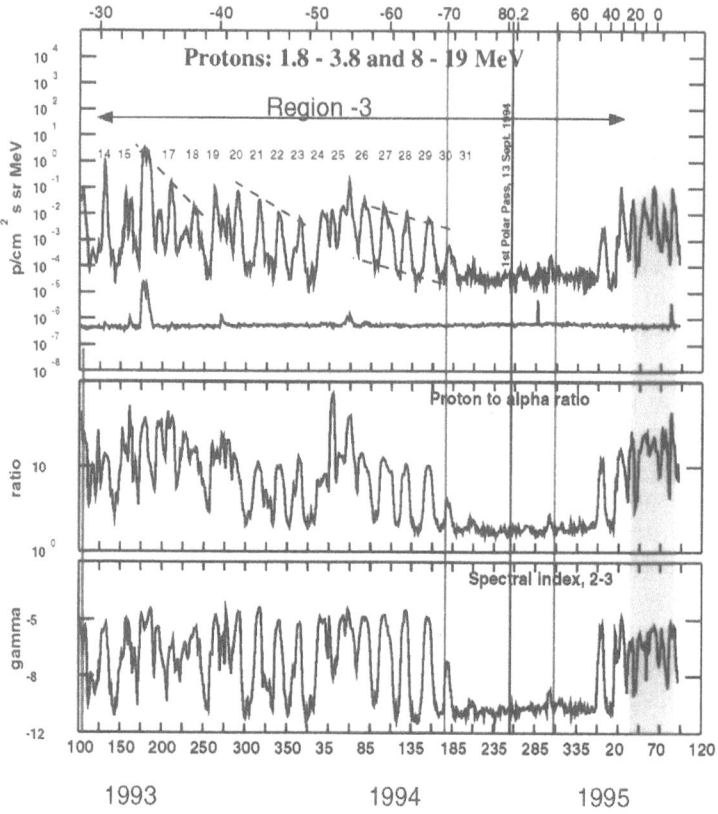

Fig. 3. Intensity of 1.8 - 3.8 Mev and 8 - 19 MeV protons observed in Region-3, together with the proton to alpha ratio at ∼1 MeV and the proton spectral index also at ∼1 MeV. The 1.8 - 3.8 MeV proton trace has been multiplied by a factor of 10^{-3}. The spacecraft was completely immersed in high-speed solar wind flow from the polar coronal hole in Region-3.

During rotations 15 and 16, the proton to alpha ratio exhibits the same variations as at lower latitudes but thereafter the ratio never rises above around 10-20. This is the same value as observed close to the reverse shocks of the forward-reverse shock pairs. The reverse shocks, which are observed up to a latitude of ∼45°S, are propagating southward from the coronal streamer belt into the polar coronal hole, according to Gosling et al. (1993).

Referring back now to Figure 3, we see that after the rise in intensity due

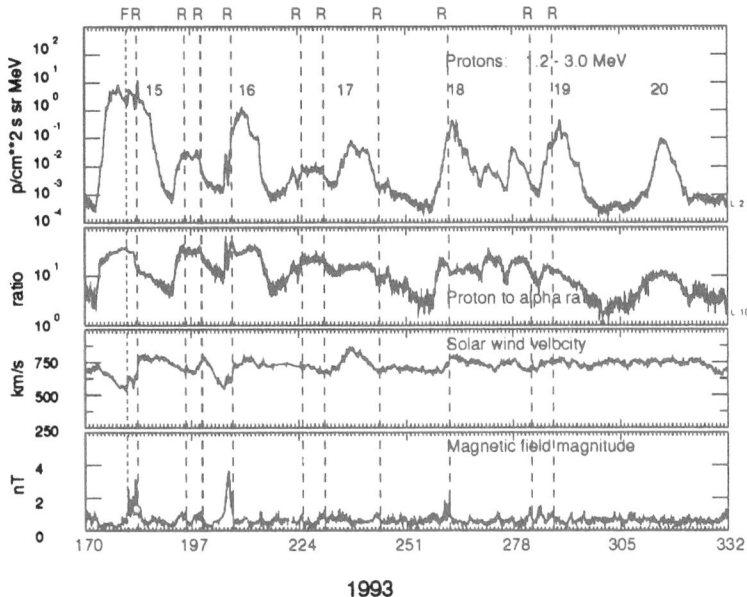

Fig. 4. Intensity of 1.2 - 3.0 Mev protons for the period starting on 19 June 1993 and ending on 28 November 1993 (day 170 to day 332 of 1993), together with the proton to alpha ratio, the solar wind velocity and the magnetic field magnitude, showing data from six solar rotations (15 to 20) observed in the high-speed solar-wind flow from the polar coronal hole.

to the passage of the transients, the intensity at the recurrent peaks drops yet again with time and with increasing latitude, as shown by the dashed lines. This behaviour suggests that the energetic solar particles from each large solar event responsible for a transient fill up the heliosphere, providing the seed particles for the recurrent CIRs to accelerate. The energetic particles slowly leak out of the heliosphere, which then gives rise to a slow decay of the intensity of the peaks. During the rise to higher latitudes, we can see the effect of the reduction in the level of solar activity as the solar cycle has progressed as a drop in the number of transient events observed at Ulysses. This is clearly evident in the 8 - 19 MeV channel shown in Figure 1 and Figure 3.

The peak proton intensity slowly drops with increasing latitude, falling to around the cosmic-ray background level at a latitude of ~70°S (day 180, 29 June 1994). Thereafter, until the highest latitude reached of 80.2°S, and then back down to ∼43°S (day 365, 31 December 1994) the intensity stays almost constant at around this level with only small peaks and variations being observed. One such minor peak was observed at ∼74°S (day 205, 24 July 1994). Another was seen at 72°S during the descent (day 305, 1 November 1994).

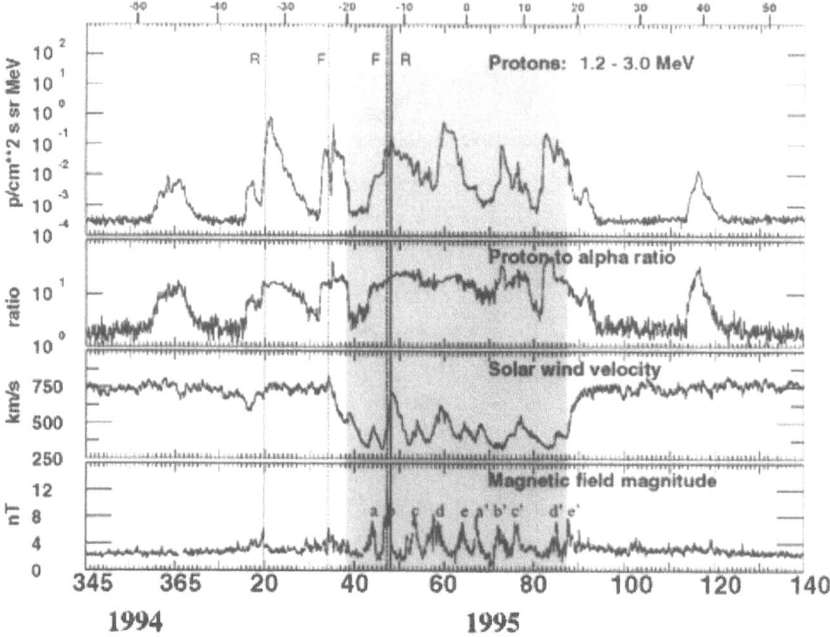

Fig. 5. Summary plot of the low-latitude pass. The top panel shows the 1.2 - 3.0 MeV ion intensity, the next panel the proton to alpha ratio, the next panel the solar wind velocity, and the bottom panel the magnetic field magnitude. The lightly shaded area is where the spacecraft is completely immersed in high speed solar wind flow, Region-3, whilst the darker shaded area is where the spacecraft is immersed in low-speed solar wind flow, Region-1.

The low-latitude pass is shown in more detail in Figure 5. Here we show, from top to bottom, the 1.2 - 3.0 MeV proton intensity, the ∼1 MeV/nucleon proton to alpha ratio, the solar wind velocity, and the magnetic field magnitude. Vertical lines show the times of preliminary identifications of forward (dashes) and reverse shocks (solid lines) (Forsyth et al., 1996).

As discussed above, the first major peak during the descent from the maximum southerly latitude was observed at the end of 1994, at around 45°S whilst the spacecraft was still well within Region-3. This peak was not associated with either a forward or reverse shock at the spacecraft, and had characteristics similar to increases without shocks observed in the high-latitude parts of Region-3 during the ascending phase (Sanderson et al., 1994). A second, much larger increase was observed at around 30°S, again whilst the spacecraft was in Region-3, but this time the increase was associated with a reverse shock. This increase had characteristics similar to increases associated with reverse shocks reported by Sanderson et al. (1994). Yet another peak was observed just before entry into Region-1, this peak being associated with a CME reported by Gosling et al. (1995b).

At around this time the spacecraft passed from Region-3 to Region-1 without ever really encountering conditions equivalent to Region-2. During ascent from Jupiter to the southern polar latitudes, the spacecraft remained in Region-2, passing back and forth from high-speed to low-speed solar wind flow every solar rotation, for approximately 11 solar rotations. During the descent from the southern polar region, the spacecraft passed from a latitude of ~30°S to ~10°S (the expected extent of Region-2 during the descent) in around 20 days, somewhat less than one solar rotation, making it difficult to identify the passage through Region-2. Region-1 was entered at a latitude of around 20°S, when the solar wind speed started to fall. However, a short excursion back into high-speed flow at ~12°S and again at ~5°S, reminiscent of Region-2, makes it again difficult to uniquely define this region.

The spacecraft passed through this region in around 55 days, during which time two major intensity increases per solar rotation were observed. Four or five corotating interaction regions, indicated by the magnetic field enhancements labelled a to e have been identified, which re-appear one solar rotation later (a' to e') (Smith et al., 1995). The particle intensity increases occurred at the same times as the interaction regions b and d, the largest being associated with interaction region d.

The first time interaction region b was observed (day 48) it was was bounded by a forward-reverse shock pair. The increase seen at this time had the same characteristics as increases seen previously, together with forward-reverse shocks pairs, i.e., an increase at the forward shock, and an even larger increase at the reverse shock (Sanderson et al., 1994). Minor increases were observed at the same time as the other interaction regions, a, c, and e. Other major increases were observed due to transient events, such as the increase seen on 22/23 March (day 81/82) which was associated with a high proton to alpha ratio typical of transient events.

A general decrease in the intensity as a function of time, both at the peaks (compare the peak b with b', and d with d'), and in-between the peaks, was seen during the low latitude pass. This is similar to the trends observed earlier on in the mission, where a large increase in particle intensity observed in the CIR was observed each time a major solar event occurred. In this case the decrease cannot be a latitude effect, since the decrease continued on as the spacecraft passed from the southern hemisphere to the northern hemisphere.

The spacecraft left Region-1 at a latitude of ~18°N (again without ever really passing through Region-2). No increases were seen in Region-3 during the first 20 days. One isolated increase was then seen around day 115 at 40°N (associated with an M7 flare which also gave rise to a proton enhancement at 1 AU (Solar Geophysical Data, 1995)). Thereafter no more increases were seen as the spacecraft ascended to the high northern latitudes. (The first major peak in intensity during the descent to low latitudes was not seen until Ulysses had reached a latitude of 43°S, day 365, 31 December 1994, considerably lower in latitude than the last major peak seen at 70°S during the ascending part of the orbit, but still inside Region-3.)

The spacecraft left Region-3 (continually immersed in the solar wind flow from the polar coronal hole) and entered Region-1 at around ~22°S (day 35, 4 February 1995) when the first indication of a slow solar wind was seen, at which point the spacecraft was more or less continually immersed in the slow speed solar wind flow of the streamer belt once again.

The latitudes at which the spacecraft encountered the boundary of the region during the descending phase in 1995, namely ~22°S is somewhat less than the corresponding latitude (29°S) in 1993 during the ascending phase of the orbit. This can be explained by the decrease in the tilt angle of the current sheet as a consequence of the decline in solar activity as solar minimum is approached. Ulysses crossed the current sheet in 1993 for the last time at ~30°S during the ascending phase, and in 1995 for the first time at a latitude of ~10°S during the descending phase of the orbit.

4   Discussion

Several theories have been proposed to explain either the regular increases in energetic ion intensity associated with the CIRs, and also the regular depression of the cosmic-ray intensity seen every solar rotation. Simnett and Roelof (1995) proposed that an acceleration site beyond Ulysses is the source of the recurrent intensity variations. A three-dimensional structure of the CIRs, based on the model of Pizzo (1994), with the acceleration site at 14 AU was found to be sufficient to explain the delay of the recurrent electron events with respect to the proton events reported by them.

Kota and Jokipii (1995) used a 3-dimensional simulation of the heliosphere with a corotating solar wind configuration, with a solar wind profile changing from slow speed in the ecliptic to high speed at high latitudes, a tilted flat current sheet, and cross-field diffusion at high-latitudes. Their model simulated the modulation of the GeV particles, and showed that the modulation in cosmic ray intensity can persist up to latitudes considerably higher than the maximum extension of the heliospheric current sheet. Their model also simulated the acceleration of lower energy (~4 MeV) particles by including a co-rotating particle source at low latitudes (at the shocks), and compression regions at higher latitudes. Reasonable agreements with the observations of the ~1MeV protons were found, the 26-day variations persisting up to higher latitudes than the source.

Fisk (1996) presented a simple model based on the differential rotation of the footprints of heliospheric magnetic field lines in the photosphere, and the subsequent non-radial expansion of the same field lines with solar wind from rigidly rotating coronal holes. This model suggests that this can give rise to excursions in latitude of the heliospheric magnetic field of more than 40 degrees of latitude, and can show how Ulysses at high latitudes can be connected to CIRs at low latitudes and thereby be connected to the acceleration site of the energetic particles.

5   Summary and Conclusions

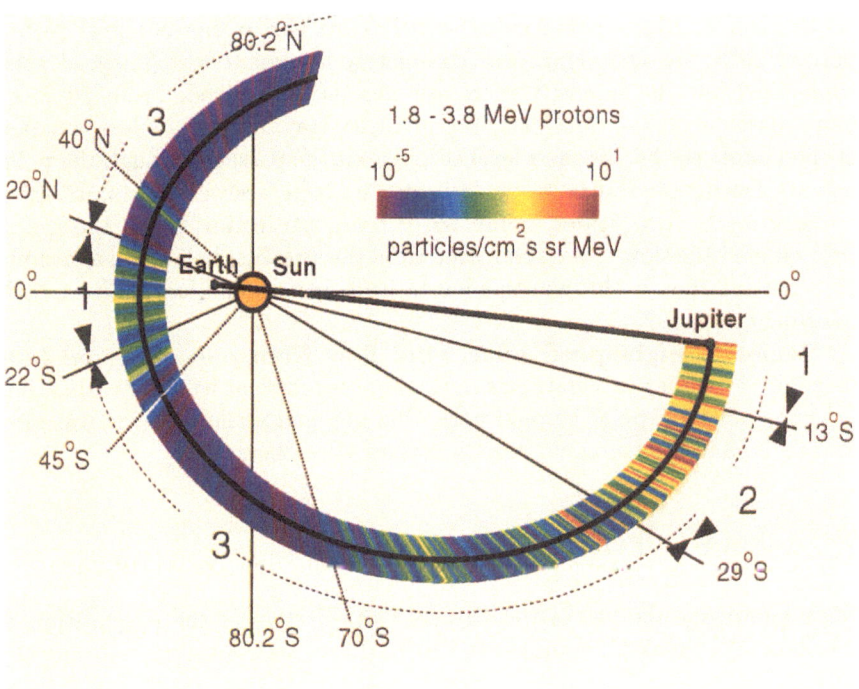

Fig. 6. Orbit of the Ulysses spacecraft, with superimposed upon it the colour-encoded intensity of the 1.8 - 3.8 MeV protons. The numbers correspond to the the different regions identified in the text.

We have presented here a summary of observations of energetic ions associated with CIRs during the Ulysses prime mission. Figure 6 is a summary of the Ulysses mission from an energetic ion perspective, showing the space-craft trajectory, with superimposed upon it the colour-coded intensity of 1.8 - 3.8 MeV protons. We summarise here, in chronological order, the different regions through which the spacecraft passed after reaching Jupiter, as follows:

Region-1, slow-speed solar wind from the streamer belt - whilst at latitudes below 13°S (Region-1), the spacecraft was most of the time immersed in slow speed solar wind flow from the streamer belt, but was already encountering irregular increases in proton intensity due to co-rotating interaction regions.

Region-2, slow-speed solar wind from the streamer belt and high-speed solar wind from the polar coronal hole - beyond 13°S the spacecraft moved in and out of high speed solar wind flow from the polar coronal hole once per solar rotation (Bame et al., 1993), and observed a much more regular pattern of intensity increases due to the CIRs (Smith et al., 1993).

Region-3, high-speed solar wind from the polar coronal hole - beyond 29°S, the spacecraft was completely immersed in high-speed solar wind flow, but the intensity increases were still observed. From 70°S, no more major increases were seen, the intensity staying more or less constant at the cosmic ray background level throughout the passage through the polar regions. During the return to low latitudes, no increases were seen until 45°S.

Region-1, slow-speed solar wind from streamer belt - the spacecraft entered the slow-speed solar wind from the streamer belt again at around 20°S passing rapidly through this low latitude region, and leaving it at 18°N (Gosling et al., 1995a).

Region-3, high-speed solar wind flow from polar coronal hole -beyond 18°N, the spacecraft was in the high-speed solar wind flow, this time from the northern polar coronal hole. One more transient increase was seen at ~40°N, and thereafter no more increases were observed.

6   Acknowledgements

We acknowledge the use of the Ulysses Data System in the preparation of this data. We thank M. Szumlas and S. T. Ho for assistance in analysing the data.

References

Balogh, A., et al., (1992): The magnetic field investigations on the Ulysses mission: Instrumentation and preliminary scientific results, Astron. Astrophys. Suppl. Ser., **92**, 221.

Bame, S. J., et al., (1992): The Ulysses solar wind plasma experiment, Astron. Astrophys. Suppl. Ser., **92**, 237.

Bame, S. J., et al., (1993): Ulysses observations of a recurrent high speed solar wind stream and the heliomagnetic streamer belt, Geophys. Res. Letters, **20**, 2323.

Barnes, C. W., and J. A. Simpson, 1976, Evidence for interplanetary acceleration of nucleons in corotating interaction regions, Astrophys. J., **210**, L91.

Bothmer, V., et al., 1995, The Ulysses south polar pass: transient fluxes of energetic ions, Geophys. Res. Letters, **23**, 3369.

Forysth, R.J., et al., 1996, The heliospheric magnetic field at solar minimum: Ulysses observations from pole to pole, Astron. Astrophys., **326**, 287.

Fisk, L.A., 1996 Motion of the footpoints of heliospheric magnetic field lines at the Sun: implications for recurrent energetic particle events at high latitudes, J. Geophys. Res., **101**, 15547.

Gosling, J. T., et al.,1993 Latitudinal variation of solar wind corotating stream interaction regions: Ulysses, Geophys. Res. Letters, **20**, 2789.

Gosling, J. T., et al., 1996a Solar wind corotating stream interaction regions out of the ecliptic plane: Ulysses, Space Sci. Rev., **72**, 99.

Gosling, J. T., et al., 1996b, The band of solar wind variability at low heliographic latitudes from near solar activity minimum: Plasma results from the Ulysses rapid latitude scan, Geophys. Res. Letters, in press.

Kota, J., and J. R. Jokipii, 1995, Corotating variations of cosmic rays near the south heliospheric pole, Science, **268**, 1024.

Sanderson, T. R., et al., 1994, Ulysses high-latitude observations of ions accelerated by co-rotating interaction regions, Geophys. Res. Letters, **21**, 1113.

Sanderson, T. R., et al., 1995, The Ulysses South Polar Pass: Energetic ion observations, Geophys. Res. Letters, **22**, 3357.

Simnett, G. M., et al., 1994, Co-rotating particle enhancements out of the ecliptic plane, Geophys. Res. Letters, **21**, 1561.

Simnett, G. M and E. C. Roelof, 1995, Reverse shock acceleration of electrons and protons at mid-heliolatitudes from 5.3 - 3.8 AU, Space Sci. Rev., **72**, 303.

Simpson, J. A. et al., 1992, The Ulysses cosmic ray and solar particle investigation, Astron. Astrophys. Suppl. Ser., **92**, 365.

Smith, E. J., et al., 1993, Disappearance of the heliospheric sector structure at Ulysses, Geophys. Res. Letters, **20**, 2327.

Smith, E. J., et al., 1995, Results of the Ulysses fast latitude scan: magnetic field observations, Geophys. Res. Letters, **22**, 3325.

Solar Wind and Interstellar Medium Coupling

David Burgess

Astronomy Unit, Queen Mary and Westfield College, London E1 4NS,
United Kingdom

Abstract. An overview is given of the current state of theory and modelling for the interaction between the solar wind and the interstellar medium (ISM). The final frontiers of the solar wind, as it pushes itself into the ISM, have been an object of speculation and analysis for many years. Observational evidence from the solar wind and heliospheric energetic particles, and measurements of the very local interstellar medium (VLISM), have led to a consensus view of an inner termination shock, a heliopause, and, probably, a heliospheric bow shock. Within this straightforward description lie several issues of basic principle, for example the acceleration of interstellar pick-up ions and their relationship to the anomalous cosmic ray component, as well as more speculative issues such as temporal variations due to solar wind and VLISM inhomogeneities. The exciting prospect of the *in situ* detection of the heliospheric boundaries will be discussed.

1 Introduction

The interaction of the solar wind and interstellar medium (ISM) represents one of the most interesting juxtapositions of solar and astrophysical ideas and concepts. The solar physicists believe they understand the solar wind with its particular blend of fine scale turbulence and large scale structure; on the other hand, the properties and behaviour of the ISM, are in the domain of astrophysics. In astrophysical terms the interaction serves as an archetypal model of two flowing plasmas, from different sources, colliding, interacting, exchanging momentum and energy, and coming into equilibrium by constructing the small scale boundary layers which are central to our view of the plasma universe. In solar terms the zone of interaction is important since it truly represents the furthest reaches of the Sun's plasma influence, and thus is the endpoint of a richly diverse sequence of plasma processes from the solar centre, through the various internal layers, to the corona, and thence to the outflowing solar wind.

The interaction is also important in terms of our knowledge of physical processes in the plasma universe, since it presents us with a stimulating mix of ingredients: the interaction of a supersonic plasma with a partially ionized plasma; the pick-up of newly ionized material in a flowing plasma; the acceleration of particles at shocks. Understanding the varied aspects of the interaction, and the implications of all possible observational constraints, gives us a unique insight into the properties of the ISM in the solar neighbourhood. Furthermore, (and concentrating the mind) we are in the unique position of having spacecraft sufficiently far from the Sun to be on the verge of providing *in situ* observations of the interstellar medium. Voyagers 1 and 2 are leaving the solar system at about 3 AU per year; Voyager 1 is currently (1996) at nearly 65 AU (9.7×10^9 km) from the Sun; both spacecraft are anticipated to operate until at least 2015 when Voyager 2 will be about 130 AU from the Sun.

The topics which we discuss here have been elaborated in many specialized reviews to which references will be found in the relevant sections. For more general reviews the following are recommended: Holzer (1989), Suess (1990), Baranov (1990), Fahr and Fichtner (1991). Jokipii and McDonald (1995) is also recommended as an introduction to the subject.

2 Boundaries and Global Morphology

We are familiar from the Earth's magnetosphere with the various boundaries that arise when the solar wind impinges on a system with its own magnetic field. Regions are segregated according to the source of the magnetic field there, with a distinct boundary (the magnetopause) between them. In gross terms the magnetopause is also the position of pressure equilibrium between the solar wind and magnetospheric plasmas. In order for the supersonic solar wind to be deflected around the magnetosphere it undergoes a transition to subsonic flow at a bow shock. The region of shocked solar wind between the bow shock and the magnetopause is called the magnetosheath. Momentum transfer between the solar wind and the magnetosphere, together with re-connection processes, leads to the formation of an extended geomagnetic tail which strongly influences what happens inside the magnetosphere. Within the magnetosphere boundaries arise associated with different physical processes, e.g., plasmapause, plasmasheet, ionosphere, etc.

Our present model (Fig. 1) of the interaction of the solar wind with the interstellar medium (hereafter "SW–ISM interaction") is analogous to a magnetosphere-solar wind interaction, but with some important differences. The solar wind is highly supersonic having a typical speed of 700 kms^{-1}, but the ISM motion, relative to the sun, is only 25 kms^{-1} (see below). Thus, if the solar wind flow is to be deflected as it encounters the ISM it must make the transition to subsonic flow. This is the same argument for the presence of a bow shock upstream of a magnetosphere: Information (e.g., pressure gradients to effect the deflection around the obstacle) about the obstacle has to be transmitted upstream, and this can only be achieved if the flow is subsonic. But in the heliospheric case the solar wind flow is completely within the ISM, so there is a termination shock, rather than a bow shock. A simple, but messy, analogy to a hydraulic jump is suggested by Axford and Suess (1994) in Fig. 2. The termination shock is in the solar wind and marks the end of the solar wind's supersonic flow, it then carries on subsonically, until it reaches the location of pressure balance between the solar wind and the

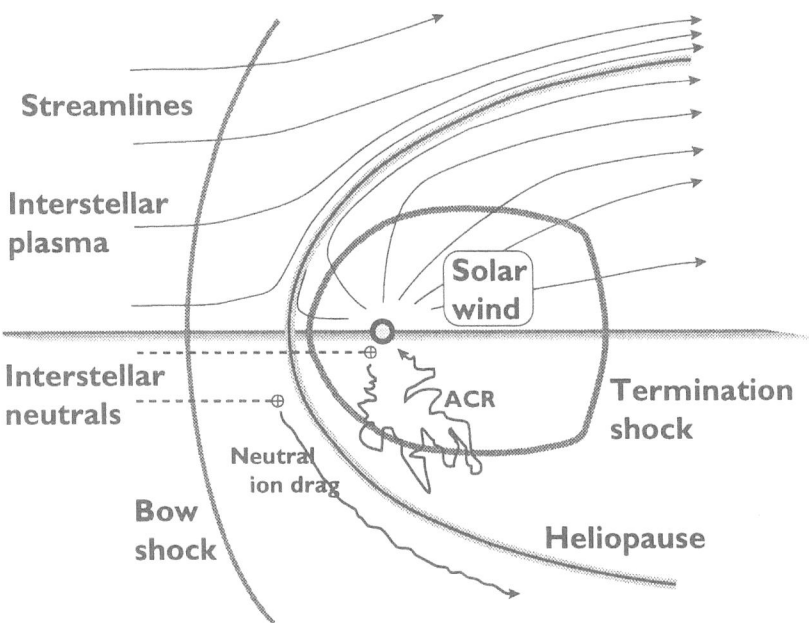

Fig. 1. Schematic of heliospheric configuration with boundaries in the case of a supersonic ISM flow. The top half of the diagram shows plasma streamlines, and the bottom half illustrates two processes discussed in the text involving interstellar neutrals.

ISM. This marks the boundary of influence of the solar plasma and magnetic field. By analogy with the magnetopause, this is called the heliopause, and is the outermost boundary of the heliosphere. In gas dynamic terms this is the contact discontinuity separating two plasmas from different sources. In plasma terms the boundary will be as complicated as the magnetopause, although best viewed as a tangential or rotational discontinuity, depending on circumstances. Because of the relative motion of the Sun and ISM there is a natural asymmetry in the interaction, and transfer of momentum at the heliopause will divert the solar wind flow downstream, stretching out the magnetic field to form a heliotail, analogous to the geomagnetic tail. What happens in the ISM outside of the heliopause depends crucially on whether the ISM is sub- or supersonic.

If it is supersonic then we can immediately invoke the magnetospheric model and infer a heliospheric bow shock in the ISM which effects a transition in the flow from supersonic to subsonic, so that it can be deflected around the heliospheric obstacle; this is sometimes referred to as the "two-shock" model. Again by analogy, the region between the supersonic ISM flow and the heliopause is called the heliosheath. The region between the termination shock and the heliopause may be thought of as a boundary layer, although it is, conceptually also a sheath flow of some sort. (Note that sometimes the

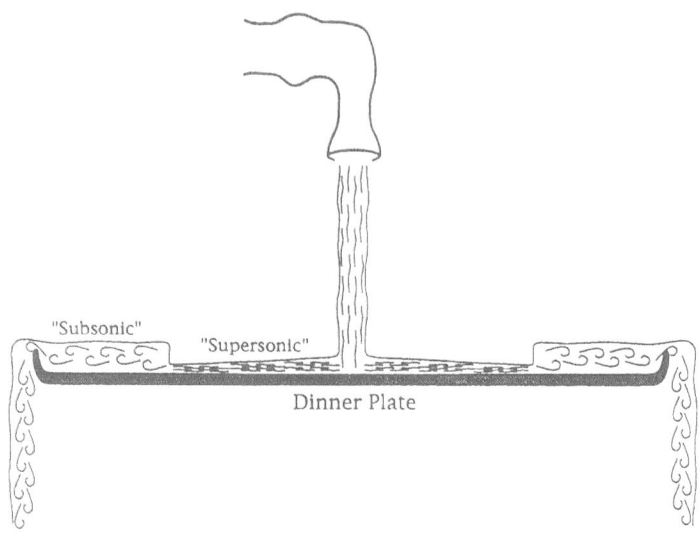

Fig. 2. Analogy for the solar wind termination shock, using water flowing over a dinner plate. The termination shock is represented by the jump between "supersonic" and "subsonic" flow. (Fig. 1 from Axford and Suess, 1994)

term "heliosheath" is used for the region beyond the termination shock up to the heliopause or even the heliospheric bow shock.)

On the other hand, if the motion of the heliosphere relative to the ISM is subsonic, then there will simply be a bow wave in front of the heliopause as the ISM diverts around it. Current indications are that this is not the case, so the subsonic ISM flow case will not be discussed here. One should be warned that this might just be current prejudice.

So far we have described the basic global phenomenology on the basis of the various boundaries that are expected. Of course there are complications: The lack of symmetry caused by solar motion through the ISM means that the termination shock will not be spherical, but rather bullet shaped. This has implications for which of the spacecraft in the outer heliosphere will first encounter the termination shock; Voyager 2 is most nearly heading "upstream." The role of the various contributions to the pressure balance at the heliopause (thermal, magnetic, galactic cosmic rays, etc.) are not exactly known, or always all included in modelling. For the two-shock model, the acceleration of the ISM in the flanks of the heliosheath means that it may become supersonic again. We know that the interplanetary magnetic field (IMF) has its characteristic spiral more tightly wound in the ecliptic plane than at high heliographic latitudes, and that the high latitude solar wind flow is very different from that in the ecliptic plane. These basic considerations imply that the three dimensional structure in the solar wind should be reflected in its interaction with the ISM. A review of three-dimensional effects in the heliospheric configuration is given by Fahr and Fichtner (1991). But, the greatest complication in our picture of the SW–ISM interaction is the fact that the ISM is actually a mixture of ionized and neutral material. (There is also a dust component which we mention later.) The two phases of the ISM are loosely coupled via collisions, but of course the neutral component is not directly coupled to the magnetic field. This introduces a set of new phenomena to investigate and understand, and which are discussed below.

All the above factors serve to complicate the physics underlying the interaction process. However, when it comes to making quantitative models for the outer heliosphere the crucial issue is that we simply do not know very accurately the parameters of the undisturbed VLISM. Of course, observations are improving, and the different observational techniques give similar results. Nevertheless, it is wise to remind ourselves of our ignorance.

The schematic of Fig. 1 is motivated and supported by two-dimensional gas dynamic modelling (Steinolfsen et al., 1994; Pauls et al., 1995). Figure 3 shows the results of such modelling work. We illustrate the two-shock model (i.e., with heliospheric bow shock) since this is, at the moment, thought to be the most likely configuration.

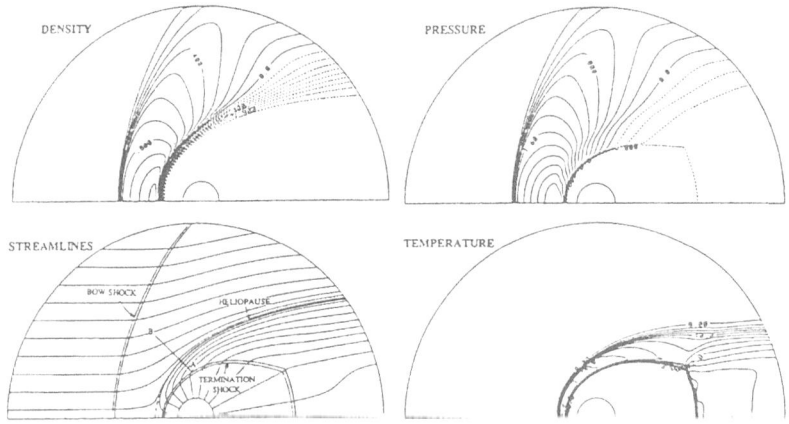

Fig. 3. Results of 2D hydrodynamic model of two-shock heliosphere. (Fig. 1 from Steinolfsen et al., 1994.)

3   The ISM in Our Neighbourhood

3.1   Nomenclature & Numbers

Before we discuss the physical processes operating in the SW–ISM interaction, we shall briefly summarize our knowledge of the ISM in the solar neighbourhood. First, since we verge on to astrophysics here, some nomenclature has to be introduced. The local interstellar medium (LISM) usually refers to scale lengths of about 100 pc. But in terms of discussing the heliospheric interaction this is far too large, so one refers to the Very Local Interstellar Medium (VLISM) being on the scale of about 0.01pc or 2000 AU. Actually, even this is larger than the currently accepted size of about 100 AU for the heliosphere. The term Local Fluff (LF) is sometimes used when referring to the structure (inhomogeneities) in the VLISM. When emphasizing the relative motion of the sun in the VLISM the term local interstellar wind (LISW) is used, and then the terms "upwind" and "downwind" are defined for the heliocentric velocity system where the Sun is at rest.

The properties of the VLISM, as currently known, are summarised in Table 1. We have not quoted experimental error bounds, constructed a range based on different observations, or tried to make an exhaustive list of references. Thus the figures should be regarded with care since they have the status of "typical best estimate." On the other hand there are usually additional techniques to estimate these quantities, and in general there is order of magnitude agreement between different methods. We note the fact that the strength of the magnetic field in the VLISM is very poorly known, which is unfortunate given that the results of modelling can depend sensitively on the assumed field strength. The measurement quoted in the table is taken over a scale of about 100 pc, so the value for the VLISM may be very different. The same argument applies to some extent to earlier estimates for the electron density based on dispersion of pulsar emissions. Secondly, observations of neutral hydrogen in the heliosphere (i.e., within the heliopause and termination shock) show a flow velocity of $21 \, \mathrm{kms^{-1}}$ and a temperature of 8000 K (HST observations by Clarke et al., 1995). The difference between the neutral helium flow speed, as measured directly, and that of neutral hydrogen is believed to be caused by charge exchange coupling between the plasma and the neutral hydrogen component at the heliopause (as discussed later).

3.2   How do we determine the parameters of the VLISM?

The LISM can be studied remotely with both low energy X-ray data and spectroscopic observation of nearby stars (see Egger et al., 1996, for review). X-ray data in the energy range 0.1-0.4 keV (ROSAT) shows absorption by the LISM indicating that the Sun is surrounded by a "bubble" of hot gas, with a radius of about 100pc, density n_h 0.005 cm^{-3} and temperature $T_h \approx 10^6$ K. This is termed the "local bubble" and is, presumably, a supernova remnant. Using observations of nearby stars (within about 150pc) it is possible to identify and measure absorption features due to the LISM (with assumptions about the normal elemental abundances etc.). These observations confirm the existence of the local bubble, but also provide indications of a very local feature, possibly a small cloud which is relatively cool and dense. It is this latter feature which defines the VLISM; similar features are seen in other interstellar clouds.

Table 1. Properties of the Very Local Interstellar Medium

Quantity	Value	References
Flow speed	$26\,\mathrm{kms^{-1}}$	Neutral helium (Ulysses)
		(Witte et al., 1993)
		UV spectroscopy
		(Lallement and Bertin, 1992)
Temperature	$7000\,\mathrm{K}$	Neutral helium (Ulysses)
(plasma and neutrals)		(Witte et al., 1993)
		UV spectroscopy
		(Lallement and Bertin, 1992)
Density (hydrogen)	$0.077\,\mathrm{cm^{-3}}$	Interstellar pickup ions
		(Gloeckler et al., 1993)
Density (helium)	$0.013\,\mathrm{cm^{-3}}$	Interstellar pickup ions
		(Gloeckler et al., 1993)
Electron density	$0.03\,\mathrm{cm^{-3}}$	H/He abundance ratio
		(Gloeckler et al., 1993, Frisch, 1995)
Magnetic field strength	$0.1\text{--}0.5\,\mathrm{nT}$	Faraday rotation
Magnetic field direction	*unknown*	

In recent years our study of the ISM has been revolutionized by what are effectively *in situ* observations, although one admits the drawbacks that such observations are in fact very local, and are influenced by the interaction of the ISM with the Sun and heliosphere. Thus for all these methods it is necessary to model the interaction and to extrapolate the observed parameters back to the undisturbed LISW. The extrapolation is usually done back to, or beyond, the termination shock, rather than to some point "at infinity."

The methods of detection can be summarized as follows:

Direction detection of interstellar neutral helium. Using a novel detection method an experiment on Ulysses (which is currently in polar orbit around the Sun) is able to detect neutral helium arrival directions (Witte et al., 1993). Assuming that the ISM distribution is Maxwellian, that the only force acting is gravity, a given ionization rate, and the neglect of charge exchange, it is possible to derive a relatively accurate velocity and temperature, and also the density (although this is more sensitive to instrumental effects).

Direct detection of interstellar dust. The Ulysses spacecraft carries a dust detector which can record the arrival speed and direction of dust particles. Beyond 5 AU from the Sun the first unambiguous detection of interstellar dust particles was made, with an observed flux of 1.5×10^{-4} particles $s^{-1}m^{-2}$ and a corresponding mass flux of $5 \times 10^{-17} gs^{-1}m^{-2}$ (Grün et al., 1994). The local detection of interstellar dust is important astrophysically, but in terms of the SW–ISM coupling dust is not thought to play an important role. Although the observed lack of small dust grains is probably due to interaction between the solar wind plasma and charged dust grains.

Interplanetary Glow (IPG). Some of the first observations of interstellar gas in the inner solar system were made by observing solar hydrogen Lyman α and helium $\lambda 584$ radiation fluorescing on neutral hydrogen and helium. The focusing effect of the Sun's gravity produces anisotropies in the IPG, from which the upstream direction as well as the velocity and temperature of the VLISM neutrals can be obtained Observations of this type have been repeated and refined over the last two decades (see Frisch, 1995, and Lallement et al., 1995, for reviews) The long base line for these measurements means that variations due to the solar cycle and other sources can be discerned.

Interstellar pickup ions. Interstellar neutrals, since they do not experience any force from the interplanetary magnetic and electric fields, might be expected to pass through the solar system only affected by the Sun's gravity and solar photon pressure. However, there are three processes which operate to ionize them: UV photoionization by solar photons, charge exchange with solar wind protons, and electron impact ionization. The latter is the least significant, but still has to be taken into account when interpreting observations. These processes produce singly ionized particles. Newly born ions are "picked up" by the solar wind flow, as explained below. The first observation of interstellar He^+ was made at 1AU (Möbius et al., 1985), and since the launch of Ulysses much new data on interstellar pickup ions has been obtained with the SWICS instrument, including, importantly, the detection of interstellar H^+ (Gloeckler et al, 1993). Isenberg (1995) provides a review of interstellar pickup ions, observations and modelling.

Anomalous cosmic ray (ACR) component. From data of energetic particles measured by space-borne instruments it was discovered, some 25 years ago, that there were anomalous flux increases in the low energy spectra of helium (<50 MeV/nucleon) and oxygen and nitrogen (<20 MeV/nucleon), as well as of other species. This component of the energetic particle distribution was also anomalous in terms of composition, with an over-abundance of species with large first ionization potential. This anomalous cosmic ray (ACR) component is believed to originate from interstellar neutral particles which penetrate into the inner heliosphere where they are picked up by the solar wind flow after ionization. Eventually they encounter the solar wind termination shock, where they undergo shock acceleration to energies of 10–100 MeV/nucleon. Being mainly singly ionized they suffer less solar modulation than galactic or solar cosmic rays of similar velocity. (It is the solar wind with its outward flow, containing irregularities which scatter particles, which screens the inner heliosphere from low energy galactic cosmic rays.) Thus they are able to diffuse back into the inner heliosphere to form the ACR component. The propagation and acceleration of the ACR component is a challenging problem, as discussed below. Klecker (1995) gives a review of ACR observational and theoretical results. In principle, measurements of the ACR component could give much information about the VLISM, but present models need to be improved before we can use them to extrapolate back to the undisturbed VLISM.

Magnetospheric measurements of ACR component. It is actually beyond the current spacecraft experiments to determine directly that the ACR component is only singly charged. But a number of methods have been used to infer the ACR component ionic charge, usually depending on propagation models of energetic particles in the heliosphere. A more immediate method is to measure the particles on spacecraft in low altitude Earth orbit. The Earth's magnetic field is then effectively used as a magnetic spectrometer. This was the aim of the SAMPEX mission, which has confirmed, with high accuracy, that all oxygen, nitrogen, and neon ions in the 8–28 Mev/nucleon energy range were singly ionized (Klecker et al., 1995). On the issue of ionization state, there is good evidence that the majority of the ACR component is singly ionized, but there is probably also some small contribution from higher charge states, by, for example, double charge exchange. Finally, we just mention the interesting fact that there has recently been discovered a radiation belt of trapped energetic particles which originate from the ACR component (Biswas, 1996).

4   Physical Processes

4.1   Interstellar Pick-up Ions

What happens when an interstellar neutral particle is ionized? If the ionization process is charge exchange, then a slow neutral becomes a slow ion, and a fast solar wind proton become a fast neutral, which can go onto to further interactions but which is usually neglected in modelling work. If the process is photo-ionization, then an ion and an electron are produced. Usually the liberated electron is ignored, because its contribution to mass loading is small, and its velocity lies within, albeit on the wings of, the solar wind electron thermal distribution. The new born ions have most effect in terms of mass loading and plasma interaction.

A newly born ion reacts immediately to the solar wind electric and magnetic fields. In the solar wind frame the motion consists of gyration around the magnetic field, together with a parallel component of motion. Considering an ensemble of such particles they are rapidly isotropized (in the solar wind frame) by ambient or self-generated fluctuations. After isotropization the ions are convecting with the solar wind. However, since the originally neutral particles more very slowly with respect to the Sun, the newly born ions have a large speed in the solar wind frame (almost equal to the solar wind speed). Thus after isotropization they form a distinct energetic population with a characteristic distribution, namely a spherical shell in velocity space with radius V_{SW} (the solar wind speed). Ions picked up closer to the Sun lose energy as they are convected by the expanding solar wind, and so shift to lower velocities. The resulting distribution at any point is then a superposition of velocity shells corresponding to pickup at different distances from the Sun. The distribution will thus form a filled in spherical sphere. Other processes (e.g., second order Fermi acceleration) will also accelerate some of the pickup ions, but this will be slower that the adiabatic cooling, so there is a sharp cutoff in the pickup distribution for $v > V_{SW}$. Since in a spacecraft frame such a distribution has a range of velocity from zero to $2V_{SW}$, the energy spectrum will have a characteristic upper energy cutoff corresponding to four times the solar wind flow energy. This is illustrated in Fig. 4 (Möbius et al., 1985) which shows the first reported observation of interstellar pickup He$^+$ at 1AU.

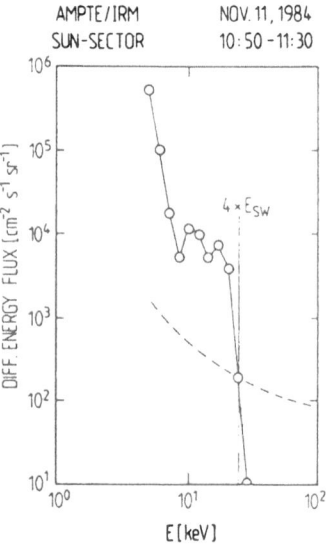

Fig. 4. Energy spectrum of interstellar pickup He$^+$ at 1AU (Fig. 2 from Möbius et al., 1985.)

The creation of pickup ions poses an interesting problem in plasma physics, since the expectation is that, since they form a non-Maxwellian distribution, they will generate waves via instabilities. Then the coupling of pickup ions to the flow (i.e., isotropization) would be mediated by these waves. Of course, since the solar wind is a turbulent medium, pre-existing fluctuations might accomplish that task faster than any self-generated contribution. The instabilities associated with pickup ions are generally varieties of the electromagnetic beam instability, and they have been much studied in the context of cometary pickup ions (e.g., Gary et al., 1989). There have been some reports of enhancements of wave activity (in the theoretically predicted frequency range) in the solar wind which have been associated with waves generated by pickup ions (e.g., Intriligator et al, 1996), but these enhancements are infrequent, and further work is necessary to disentangle the effects of Doppler shifts, wave production, and propagation, as well as correlations with local pickup ion characteristics (Isenberg, 1995).

4.2   Acceleration of ACR Component

The origin of the ACR ions is well accepted, but the actual sites and mechanisms for their acceleration are still not completely clear. The acceleration region is clearly in the outer heliosphere, because of the observed positive radial gradient out to 40AU and a modulation which is correlated with the galactic cosmic ray modulation. The clearest candidate is the solar wind termination shock, but the properties of the shock vary according to heliographic latitude. The spiral pattern of the interplanetary magnetic field (IMF) in the ecliptic plane is increasingly tightly wound further from the Sun, so that the shock normal angle of the termination shock there will be roughly perpendicular. On the other hand, at the solar poles the IMF is not so tightly wound, and one expects the shock normal angle to be more parallel. The shock normal angle θ_{Bn}, measured between the shock normal and the upstream magnetic field direction, plays a vital role in determining the shock structure and also the type of acceleration mechanism that might be operating. A major review of shock acceleration mechanisms is given by Jones and Ellison (1991). We do not have space to discuss the structure of the termination shock, but the possibility that it will resemble a cosmic ray mediated shock (where the ACR component provides the bulk of the cosmic ray pressure) is discussed by Zank et al. (1995).

When θ_{Bn} is small (the so-called quasi-parallel regime) the magnetic field threads the shock so that particle motion across the shock is relatively easy and energetic particles can scatter back and forth across the shock. If the scattering centres, which cause the energetic particles to repeatedly return to the shock, are embedded in the flow, then they have a relative motion due to the deceleration at the shock, and hence the energetic particles suffer an effective compression. This compression results in the fluid doing work on the particles and the particles gain energy. This process is first order Fermi acceleration, or diffusive acceleration. By the same fact of magnetic configuration, merely suprathermal particles can also stream into the upstream region ahead of the shock, and there lead to large amplitude turbulence which influences, indeed mediates, the whole shock transition.

When θ_{Bn} is near 90° (the quasi-perpendicular regime) it is relatively difficult for particles to escape upstream of the shock, because particle motion along the field line keeps particles in the vicinity of the shock, so that their gyration tends to return particles to the shock. Quasi-perpendicular shock structure is relatively laminar, with a definite small-scale discontinuity which resembles a gas dynamic shock. Acceleration at quasi-perpendicular shocks operates precisely by the shock's ability to retain the particle in the vicinity of the jump in the magnetic field. Because of the ∇B drift, particles will drift (in the correct sense) along the motional electric field (which is most effective in the quasi-perpendicular geometry) and thereby gain energy. This explains the name "shock drift acceleration" for this process. This acceleration continues until the particle has sufficient energy to leave the shock, either into the upstream or downstream region. Because of the geometry of the interaction the particles escaping upstream tend to be field aligned, whereas the downstream distribution tends to have a perpendicular anisotropy.

In fact, the distinction between shock drift acceleration and first order Fermi is artificial, and at any real shock particles may experience a combination. A few points should be borne in mind: Firstly, in both cases the standard explanations require some initial distribution of energetic particles. It then becomes vital to identify any additional processes operating at lower energies which take thermal particles and "inject" them into the acceleration mechanism. This turns out to be one of the important issues for the acceleration of the ACR component. Secondly, first order Fermi acceleration in some ways is a universal mechanism since it does not rely on any details of the shock (other than the compression across it) and produces a power law energy spectrum. On the other hand, because it is a diffusive process it requires a relatively long time scale to produce high energies. Thirdly, both acceleration mechanisms will have a maximum energy which they can produce, which is associated with the finite size of the shock system.

Returning now to the issue of ACR acceleration, we note that first order Fermi acceleration will tend to operate at the termination shock at the solar poles, whereas shock drift acceleration will be more effective at the termination shock in the ecliptic plane. It is possible to produce models of the diffusive acceleration process at the poles (e.g., Kota and Jokipii, 1983) and these include the effects of diffusion, convection, adiabatic deceleration and drifts along the termination shock. One important result of the drift models is the prediction of a 22 year cycle for the modulation and latitudinal gradient (the latter is observed to change sign depending on the polarity of the solar cycle). This is explained by considering the drift paths of positive ions, and the consequent time they spend near the solar poles.

However, any acceleration mechanism for the ACR component cannot be discussed in isolation from loss mechanisms. For the ACR ions (e.g., O^+) the main loss mechanisms are charge exchange with neutrals in the VLISM (strange irony!) and adiabatic deceleration. And when compared with the long time scale associated with diffusive acceleration at the quasi-parallel shock at high heliolatitudes (e.g., Klecker, 1995), this has focussed attention on acceleration at the quasi-perpendicular termination shock. In principle diffusive acceleration at the quasi-perpendicular shock (which as we have implied above is not the easiest of mechanisms to arrange) has the required short time scale, but the problem has been to show that such shocks can take the initial shell-like distribution of interstellar pickup ions and accelerate them sufficiently to inject them into a diffusive-like process.

The essential problem is that because of the tight winding of the IMF spiral pattern one expects the termination shock to be closely perpendicular, and in this case it is very difficult to show that interstellar pickup ions will be accelerated. There have been a number of approaches to this problem using plasma shock simulations. Liewer et al. (1995) used a self-consistent, one-dimensional hybrid simulation with a kinetic description for the interstellar pickup ions and thermal ions, and a fluid description for the electrons. They modelled a strong shock with between 10–20% of the upstream protons in the form of pickup ions in a shell distribution. The shocks they modelled had θ_{Bn} in the range 40–50° which is lower than might be expected from our arguments above, but they argue, from observations, that the magnetic field direction in the outer heliosphere is highly variable, so that θ_{Bn} will be in the range where they find acceleration for some major fraction of the time. The acceleration process they found was a combination of first order Fermi and shock drift.

Giacalone et al. (1994), motivated by the fact that rapid acceleration requires highly perpendicular shocks, argued that perpendicular diffusion (i.e., scattering of particles on to adjacent field lines) could keep particles close to the shock so that efficient acceleration could be achieved. However, such perpendicular diffusion requires three dimensional shock simulations over long time scales, and these are not yet computationally possible. Accordingly, they introduced, into a 1D hybrid simulation *ad hoc* scattering of the pickup ions which was intended to mimic the presence of perpendicular diffusion. In this situation they found rapid preferential acceleration of the pickup ions, and the production of a distribution suitable for further acceleration up to ACR energies. The effect of this *ad hoc* scattering on the particle distribution functions upstream and downstream of the simulated shock is shown in Fig. 5.

There remain unanswered questions about the acceleration of the ACR component. The modelling that has been done remains incomplete or ambiguous, and further studies are required. Such work is difficult because of uncertainty about the termination shock parameters and environment. For example, ACR observations discussed later indicate that the termination shock might be relatively weak. In this case would it be easier, or more difficult, to

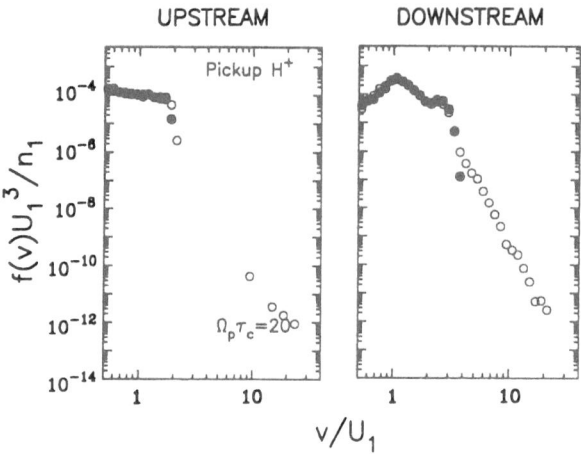

Fig. 5. Upstream and downstream velocity distribution functions for pickup protons in a hybrid shock simulation, for the case of no extra scattering (solid circles), and with extra scattering (open circles). (Fig. 1 from Giacalone et al., 1994.)

inject, and hence accelerate, the ACR component there?

4.3   Charge Exchange Drag

Since there are spacecraft in the outer heliosphere the obvious questions are: What are the distances to the various interaction boundaries (termination shock, heliopause, etc.) from the Sun? When will the spacecraft cross these boundaries?

The simplest argument for the position of the heliopause is based on pressure balance between the ram pressure of the solar wind and the sum of the various pressure components (thermal, magnetic etc.) of the VLISM. Even this simple calculation is plagued by our lack of knowledge about the magnetic field strength in the VLISM. However, using reasonable values leads to distances which are greater than those indicated by observations (some of which are discussed later). The role of interstellar pickup ions in the inner heliosphere has already been discussed, but ionization of interstellar neutrals, on both sides of the heliopause, plays a important role in the overall interaction. In the distant VLISM it is assumed that the neutral and ionized components move together, weakly coupled as they are by photoionization and charge exchange (their mutual mean free path is about 100 AU). However, closer to the heliopause the ionized material must divert around the obstacle, but the neutral component will tend to travel on, into the heliopause. The large relative velocity between the neutral and ionized components means that charge exchange becomes an important coupling mechanism, and the effect is that the neutrals impel the ionized component towards the heliopause, and, by similar arguments, decelerate the solar wind material inside the heliopause. The net effect is to increase the ram pressure of the VLISM on the heliopause, and thus decrease the distance of the heliopause from the Sun and the size of the heliosphere.

The effects of charge exchange drag were first shown using kinetic models (e.g., Baranov and Malama, 1993), but are perhaps most clearly demonstrated in gas dynamic modelling, under the simplifying assumption that the ISM neutral component can be treated as a fluid, so that a multifluid problem is simulated where there is some coupling (i.e., drag) between the H^+ component and the neutrals. The steady state, for the case of a supersonic LISW, has been investigated by Pauls et al. (1995). They demonstrated a drastic reduction in the distances to the various heliospheric boundaries when neutral H was introduced self-consistently, as suggested by earlier authors. They also noted that the mass loading of the solar wind reduced its velocity at the termination shock. Another effect of the neutral-plasma coupling is an enhancement of neutral hydrogen just upstream of the heliopause. This neutral density excess is sometimes called the "hydrogen wall," and it serves to filter the incoming interstellar neutrals from the inner heliosphere. Understanding the distribution of the ISM neutral component is also vital in interpreting IPG observations, and hence the derived properties of the ISM.

Liewer et al. (1996) using similar time dependent numerical simulations, have shown another interesting effect due to neutral drag. The pressure balance at the heliopause is basically between the fast, but underdense, solar wind plasma and the slow, but denser, ISM plasma. But the force due to neutral-ion drag is proportional to the plasma density (amongst other things), and in the sense of the neutral flow. Thus, this drag force is directed in the opposite sense to the density gradient (∇n). This situation is analogous to that of the Rayleigh-Taylor instability, and, indeed, the simulations of Liewer et al. (1996) show a similar instability of the heliopause at the upwind side of the heliosphere. Various authors have made earlier suggestions about instabilities, MHD as well as gasdynamic, at the heliopause. In Liewer et al.'s work the oscillations were found to have time scales of the order of a hundred years, and amplitudes of tens of AU, although the position of the termination shock varied by only about 10 AU. This work is only gasdynamic in nature, so that if the heliopause width was sufficiently small, then the heliopause, and any possible instabilities would have to be described by MHD, and might have very different properties and growth rates from the gasdynamic case. This topic, however, leads us on to a new perspective on the question: Does the heliosphere have a fixed size?

4.4   Variability

In everything we have discussed so far there has been the implicit assumption of steady state. But we know this is wrong. The solar wind is constantly changing, and, in particular, in the equatorial regions of the sun the speed varies between fast and slow states. This in turn leads to corotating inter-action regions (CIR), and so-called global merged interaction regions in the outer heliosphere. Solar activity also makes its contributions to the changing solar wind: flares induce transient shocks which propagate out into the he-liosphere. Figure 6 (Belcher et al., 1993) shows one day averages of the solar wind proton density as a function of radius, which shows significant day to day variations about the mean. Changes in the solar wind ram pressure will induce changes in the position of the solar wind termination shock, and it has been estimated that the shock will respond at speeds between 20–100 kms^{-1} (depending on the time scale of the variations). However, the reaction of the termination shock to changes will depend on the pressure contribution from the ACR component, which will be relatively steady. If the ACR pressure is important, then the variation in the termination shock position will be reduced.

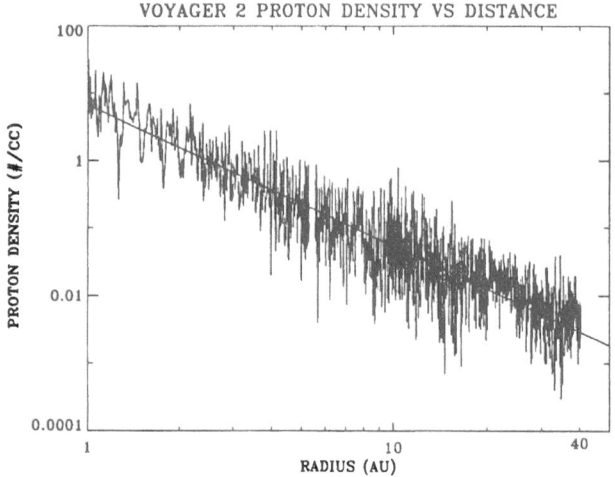

Fig. 6. One day averages of the solar wind proton density versus distance from the Sun. The straight line is a least squares fit. (Fig. 2 from Belcher et al., 1993.)

The response of the termination shock to solar wind transients has been studied by Story and Zank (1995) using one-dimensional gasdynamic mod-elling. The interaction of shocks, discontinuities and pressure pulses with a shock has been much studied by earlier authors. The work of Story and Zank is useful since it illustrates the complexity that can arise in the interaction with even simple assumptions about the solar wind changes. In their simula-tions the termination shock was set according to the shock jump conditions, and perturbations in the solar wind were introduced, and then convected self-consistently towards the termination shock. For example, Fig. 7, shows a time sequence for the density profile of the interaction of a forward-reverse shock pair (as might be expected to bound a CIR) and the termination shock. During the interaction the density profile becomes very complicated with several discontinuities, density spikes and shocks, as well as the termination shock itself, which undergoes rapid motion. Story and Zank make the obvious point that if a spacecraft encounters the termination shock during such a collision, then the actual transition would be difficult to determine observationally. The same point could also be made if one anticipates major wave activity associated with the ACR acceleration mechanism. Even at the Earth's magnetosphere, our current ability to identify boundaries has been built on a large data set. Our knowledge of cometary plasma boundaries, resting on a handful of encounters, is sketchy and still controversial. So it is natural to wonder what we will make of a single point measurement of the termination shock, which has unknown structure, and poorly known controlling physics.

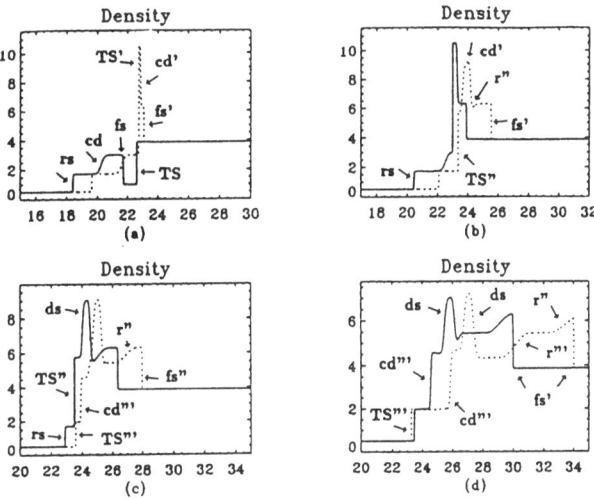

Fig. 7. Density sequence in time for the collision of a forward-reverse shock pair and a Mach 10 termination shock. In each frame two times are shown in the order solid then dotted line. (Fig. 11 from Story and Zank, 1995.)

On longer time scales the solar wind properties are modulated by the 11 year solar cycle and so we might expect consequent variations in the heliospheric boundary positions. Finally, and most unknowably, the VLISM itself has variations within it. The Sun has moved about 80AU since the first observations of the VLISM were made, and inhomogeneities on the same scale have been observed in other ISM clouds. We only have to, again, reflect on the Earth's magnetosphere and the intricate magnetic coupling between it and the solar wind, to see what complications are possible. Could there be the equivalent of geomagnetic substorms controlled by variations in the LISW magnetic field? Given that we cannot measure the VLISM magnetic field will this avenue always lead to speculation? Is this where space physics blends to astrophysics?

5   Detecting the Final Frontiers

If we take the issue of variability at face value, then we might be tempted to give up all hope of determining accurately the positions of the heliospheric boundaries. The required modelling would be sensitive to the time variations of the VLISM parameters, but at any one time those parameters can only be obtained using models for the heliospheric boundary positions and the interactions there. The situation is probably not as hopeless as this scenario suggests, although the use of *in situ*, one point, measurements to detect the heliospheric boundaries might now appear much more difficult than at first imagined. This problem (i.e., discerning boundaries in a medium constantly in change, with the boundaries constantly shifting position) is cruel, since the major motivation for the recent studies of the SW–ISM coupling has been the location of spacecraft in the outer heliosphere, and their imminent (in astrophysical terms) senescence and obsolescence.

The need to continue operating the outer heliospheric spacecraft is reinforced by three recent observations, using different techniques, which have all estimated the location of the termination shock in the range 60-100 AU. These observations have been low frequency radio emissions, Lyman α IPG observations, and radial gradients in the ACR component.

Voyager 1 and 2 and Pioneer 10 continue to supply data on the ACR component in the outer heliosphere. Using their different locations in heliographic coordinates it is possible to interpret their observations using models of the ACR production and propagation. Using data from 1987, and a model of the propagation of the ACRs along the interplanetary current sheet where the radial ACR gradient was modulated by the current sheet tilt angle, Cummings et al. (1994) estimated that the termination shock was at 67 ± 5 AU. Further analysis of the energetic particle data have shown periods (e.g., 1992) which appear not to be in equilibrium. On the other hand data from the 1993–1994 period have been modelled using a spherically symmetric equilibrium model of ACR propagation and the termination shock position has been estimated at 85 ± 5 AU (Stone et al., 1996). This latter work, suited to solar wind conditions at solar minimum, when the outer heliosphere is not dominated by global merged interaction regions, has also produced an estimate for the termination shock strength of $s = 2.63 \pm 0.14$, the density compression ratio at the shock. This is an interesting value since it is much less than the strong shock limit of $s = 4$, and so is at variance with much modelling work which assumes a high Mach number termination shock. It remains to be seen whether this will be important for explaining the ACR acceleration, and by how much current conclusions have to be modified to take account of solar cycle variations.

The Voyager and Pioneer spacecraft carry instruments to measure the hydrogen Lyman α IPG. These instruments have detected "excess" emission in the LISW upstream direction, which has been explained by a density gradient in the neutral hydrogen associated with the termination shock (the "hydrogen wall" described above). Using this data, Hall et al. (1993) have estimated, based on the kinetic Monte-Carlo model of Baranov and Malama (1993) that the termination shock lies at 75–105 AU in the upstream direction.

By observing the hydrogen Lyman α emission line profile it is possible to measure the temperature and velocity dispersion of the flow. In principle the neutral interstellar wind in the inner heliosphere contains the velocity distribution of the LISM, as well as, potentially, information about the modification of the flow at the heliospheric boundaries. The actual measurements include the convolved effects of solar gravity and radiation pressure near the Sun. Recent observations using the Hubble Space Telescope (Clarke et al., 1995) report that the line profiles are consistent with an "undisturbed" (undisturbed by near solar effects that is) flow of 21 km s^{-1} and a temperature of 8000 K. Thus the neutral hydrogen velocity is less than both the directly observed neutral helium velocity as measured on Ulysses (Witte et al., 1993), the results of IPG measurements of solar helium lines, and the results of spectroscopic observations of nearby stars (e.g., Lallement and Bertin, 1992, Bertin et al., 1993). This velocity discrepancy is believed to be due charge exchange drag on the neutral hydrogen at the heliospheric boundaries. Helium, on the other hand, because of a different cross-section for this coupling process, flows into the inner heliosphere more or less undisturbed. The point of this lengthy introduction is that, in principle, it might be possible to extract information from the Lyman α emission line profiles about the nature and position of the heliospheric boundaries, however trustworthy results might be sensitive to the chosen model. It seems more likely that this technique will return important results about the solar wind anisotropies, i.e., differences with heliocentric latitude, a review is given by Lallement et al., (1995).

The most unusual, and unexpected, observations which might elucidate the positions of the heliospheric boundaries have been made by wave instruments on Voyager 1 and 2. Exceptional radio emissions in the frequency range 2 to 3 kHz, and generally weak, have been recorded in the outer heliosphere (e.g., Kurth and Gurnett, 1991). Only a handful of events have been reported for the period 1983–1991, but they have generated many theories for their explanation. Because they are propagating electromagnetic waves, they will only be observed where the local plasma frequency is less than the emission frequency, implying that the emission source lies further from the Sun than about 10 AU. Various possible sources, such as planetary and stellar, have been ruled out, and it is accepted that the radiation is probably generated in the outer heliosphere. Most theories for the radiation mechanism rely on a one similar to type III solar radio bursts: Langmuir waves produced by an electron beam coalesce to produce electromagnetic radiation.

In order to explain the transitory nature of the radio events it is suggested that the emissions are triggered by a disturbance propagating outward from the sun. Gurnett et al. (1993) showed that the strongest radio emissions occurred about 400 days after large Forbush decreases in the cosmic ray intensity at Earth (a proxy for a strong solar wind disturbance). This idea was investigated more quantitatively by Steinolfsen and Gurnett (1995) using 2D gasdynamic modelling adjusted so that the distances to the termination shock and heliopause were consistent with the observations of the time delay between Forbush decrease and start of 2–3 kHz radio emission. This work estimated the termination shock position at 92 AU. However, it is necessary to return to the emission mechanism to decide whence the emission is actually coming. If the production of radio waves were due to the mechanism outlined above, then it would produce radiation at twice the local electron plasma frequency, $2f_{pe}$, and possibly also at f_{pe}. Given the radial dependency of the solar wind density, $f_{pe} \sim 200\,\mathrm{Hz}$ at the termination shock if placed at 100 AU. Even trying to take account of the expected jump at the shock, this implies frequencies lower than observed. Therefore, unless the shock jump is very much greater than the MHD strong shock limit (as some have suggested might be true because of the ACR pressure contribution), it seems one is to conclude that the emission comes from the heliopause, or within the heliosheath. The question of explaining the low frequency heliospheric emissions remains an area of active research (e.g., Cairns et al., 1993; Macek, 1996).

In an overview of this type some subjects are necessarily omitted or relegated. More could be said about the role of the ACR component in the termination shock's structure, waves generated by pickup ions, alternative theories and controversies – more could be said about all these, and more.

Explaining how the solar wind interacts and couples with the local interstellar medium has turned out to be a fascinating subject area, with many new and unexpected facets. Motivating much of the work has been the prospect of *in situ* observations by spacecraft beyond the termination shock, possibly even the heliopause. So much remains unknown about the VLISM. We can only hope that before they fade away those distant spacecraft can provide us with a tantalising glimpse from beyond the solar wind.

References

Axford W. I., Suess S. T. (1994): Spacecraft to explore the outer heliosphere. Eos **75**, 193–258

Baranov V. B. (1990): Gas-dynamics of the solar-wind interaction with the interstellar-medium. Space Sci. Rev. **52**, 89–120

Baranov V. B., Malama Y. G. (1993): Model of the solar-wind interaction with the local interstellar-medium - Numerical-solution of self-consistent problem. J. Geophys. Res. **98**, 15157–15163

Belcher J. W., Lazarus A. J., Mcnutt R. L., Gordon G. S. (1993): Solar-wind conditions in the outer heliosphere and the distance to the termination shock. J. Geophys. Res. **98**, 15177–15183

Bertin P., Lallement R., Ferlet R., Vidalmadjar A. (1993): Detection of the local interstellar cloud from high-resolution spectroscopy of nearby stars - Inferences on the heliospheric interface. J. Geophys. Res. **98**, 15193–15197

Biswas S. (1996): The new radiation belt of the earth from trapped anomalous cosmic- rays. Space Sci. Rev. **75**, 423–451

Cairns I. H., Kurth W. S., Gurnett D. A. (1993): Foreshock theories for the outer heliospheric radio emissions. Adv. Space Res. **13**, 205–208

Clarke J. T., Lallement R., Bertaux J. L., Quemerais E. (1995): HST GRHS observations of the interplanetary medium downwind and in the inner solar-system. ApJ **448**, 893–904

Cummings A. C., Stone E. C., Webber W. R. (1994): Distance to the solar-wind termination shock and the source flux of anomalous cosmic-rays during 1986-1988. J. Geophys. Res. **99**, 11547–11552

Egger R. J., Freyberg M. J., Morfill G. E. (1996): The local interstellar-medium. Space Sci. Rev. **75**, 511–536

Fahr H. J., Fichtner H. (1991): Physical reasons and consequences of a 3-dimensionally structured heliosphere. Space Sci. Rev. **58**, 193–258

Frisch P. C. (1995): Characteristics of nearby interstellar matter. Space Sci. Rev. **72**, 499–592

Gary S. P., Akimoto K., Winske D. (1989): Computer-simulations of cometary-ion ion instabilities and wave growth. J. Geophys. Res. **94**, 3513–3525

Giacalone J., Jokipii J. R., Kota J. (1994): Ion injection and acceleration at quasi-perpendicular shocks. J. Geophys. Res. **99**, 19351–19358

Gloeckler G., Geiss J., Balsiger H., Fisk L. A., Galvin A. B., Ipavich F. M., Ogilvie K. W., Vonsteiger R., Wilken B. (1993): Detection of interstellar pick-up hydrogen in the solar-system. Sci **261**, 70–73

Grun E., Gustafson B., Mann I., Baguhl M., Morfill G. E., Staubach P., Taylor A., Zook H. A. (1994): Interstellar dust in the heliosphere. A&A **286**, 915–924

Gurnett D. A., Kurth W. S., Allendorf S. C., Poynter R. L. (1993): Radio-emission from the heliopause triggered by an interplanetary shock. Sci **262**, 199–203

Hall D. T., Shemansky D. E., Judge D. L., Gangopadhyay P., Gruntman M. A. (1993): Heliospheric hydrogen beyond 15 AU - Evidence for a termination shock. J. Geophys. Res. **98**, 15185–15192

Holzer T. E. (1989): Interaction between the solar-wind and the interstellar-medium. ARA&A **27**, 199–234

Intriligator D. S., Siscoe G. L., Miller W. D. (1996): Interstellar pickup H+ ions at 8.3-AU - Pioneer-10 plasma and magnetic-field analyses. Geophys. Res. Lett. **23**, 2181–2184

Isenberg P. A. (1995): Interstellar pickup ions - Not just theory anymore. Rev. Geophys. **33**, 623–627

Jokipii J. R., Mcdonald F. B. (1995): Quest for the limits of the heliosphere. Sci. Am. **272**, 58–63

Jones F. C., Ellison D. C. (1991): The plasma physics of shock acceleration. Space Sci. Rev. **58**, 259–346

Klecker B. (1995): The anomalous component of cosmic-rays in the 3-d heliosphere. Space Sci. Rev. **72**, 419–430

Klecker B., Mcnab M. C., Blake J. B., Hamilton D. C., Hovestadt D., Kastle H., Looper M. D., Mason G. M., Mazur J. E., Scholer M. (1995): Charge-state of anomalous cosmic-ray nitrogen, oxygen, and neon - SAMPEX observations. ApJ **442**, L 69–L 72

Kota J., Jokipii J. R. (1983): Effects of drift on the transport of cosmic-rays .6. a 3-dimensional model including diffusion. ApJ **265**, 573–581

Kurth W. S., Gurnett D. A. (1991): New observations of the low-frequency interplanetary radio emissions. Geophys. Res. Lett. **18**, 1801–1804

Lallement R., Bertin P. (1992): Northern-hemisphere observations of nearby interstellar gas - Possible detection of the local cloud. A&A **266**, 479–485

Lallement R., Kyrola E., Summanen T. (1995): Interstellar gas in the heliosphere and the solar-wind anisotropies. Space Sci. Rev. **72**, 455–466

Liewer P. C., Rath S., Goldstein B. E. (1995): Hybrid simulations of interstellar pickup ion-acceleration at the solar-wind termination shock. J. Geophys. Res. **100**, 19809–19818

Liewer P. C., Karmesin S. R., Brackbill J. U. (1996): Hydrodynamic instability of the heliopause driven by plasma-neutral charge-exchange interactions. J. Geophys. Res. **101**, 17119–17127

Macek W. M. (1996): Emission mechanism for low-frequency radiation in the outer heliosphere. Space Sci. Rev. **76**, 231–250

Mobius E., Hovestadt D., Klecker B., Scholer M., Gloeckler G., Ipavich F. M. (1985): Direct observation of He+ pick-up ions of interstellar origin in the solar-wind. Nat **318**, 426–429

Pauls H. L., Zank G. P., Williams L. L. (1995): Interaction of the solar-wind with the local interstellar-medium. J. Geophys. Res. **100**, 21595–21604

Steinolfson R. S., Gurnett D. A. (1995): Distances to the termination shock and heliopause from a simulation analysis of the 1992–93 heliospheric radio-emission event. Geophys. Res. Lett. **22**, 651–654

Steinolfson R. S., Pizzo V. J., Holzer T. (1994): Gasdynamic models of the solar-wind interstellar-medium interaction. Geophys. Res. Lett. **21**, 245–248

Stone E. C., Cummings A. C., Webber W. R. (1996): The distance to the solar-wind termination shock in 1993 and 1994 from observations of anomalous cosmic-rays. J. Geophys. Res. **101**, 11017–11025

Story T. R., Zank G. P. (1995): The response of a gasdynamic termination shock to interplanetary disturbances. J. Geophys. Res. **100**, 9489–9501

Suess S. T. (1990): The heliopause. Rev. Geophys. **28**, 97–115

Witte M., Rosenbauer H., Banaszkiewicz M., Fahr H. (1993): The Ulysses neutral gas experiment - Determination of the velocity and temperature of the interstellar neutral helium. Adv. Space Res. **13**, 121–130

Zank G. P., Cairns I. H., Webb G. M. (1995): The termination shock - Physical processes. Adv. Space Res. **15**, 453–462

Flows Through the Magnetically Structured Solar Atmosphere

Brigitte Schmieder[1,2]

[1] Observatoire de Paris, Section Meudon, 92195 Meudon Cedex Principal, France
[2] Universitetet i Oslo, P.B. 1029 Blindern, 0315 Oslo, Norway

Abstract. The upper solar atmosphere until three solar radii is controlled by the magnetic field. The plasma β (P_{kin}/P_{magn}) is decreasing with increasing altitude in the low corona but is already less than unity in the middle chromosphere. The dynamics of the chromosphere and the transition region are driven by magnetic effects, even in the "quiet" Sun. We will review some promising research programs concerning steady flows and oscillatory phenomena in the small scale structures, mainly based on new two-dimensional spectral diagnostics. The filamentary behaviour of the solar atmosphere may be the key to understanding basic problems like coronal heating. The importance of the chromospheric network, plages and penumbra is emphasized by the fact that they represent footpoints of coronal loops. Large-scale coronal structures (streamers, jets, post-flare loops, arcades) are linked to photospheric and chromospheric events and give a three-dimensional view of the atmosphere.

1 Introduction

The outer solar atmosphere is governed by different ingredients or forces that we have to take into account, in order to understand the flows through the atmosphere. Mass motions below the photosphere drive the magnetic field which gives rise to the solar cycle (Raadu 1983). The photosphere is dominated by convection. The magnetic field lines as they emerge, due to magnetic buoyancy, are pushed and concentrated into flux elements of a few hundred km diameter, as a consequence of small-scale convective motions seen as granules and supergranules. This concentration of flux tubes leads to the formation of the chromospheric network, pores, faculae and sunspots. Photospheric motions and magnetic buoyancy may be the origin of the continued modification of the coronal magnetic field and the dynamical events observed, i.e. on the small scale: reconnection, on the large scale: expansion.

In the chromosphere β ($\beta = 8\pi P/B^2$) is already less than unity and the magnetic pressure is dominates. Plasma flows generally follow magnetic structures (Schmieder 1992a) for good physical reason. In the corona until the solar wind region, β is lower than 1, if we consider a fully-ionized plasma $\beta = 16\pi n_e kT/B^2$ with $P_{kin} = n_e kT_e + n_i kT_i$ and $T_e = T_i = T$, $P_{kin} = 2n_e kT$, and the magnetic field controls the structure of the atmosphere. Flux tubes play an important role in the upward transport of energy and mass into the

Table 1. Summary of the dynamic events observed through the solar atmosphere according to their scale: short, medium and large

B longitudinal	Magnetic Structures	Chromosphere-Transition region (Spectra)	V km/s	Corona (Proper motions)
concentration of the field parasitic polarity	Open field lines long loops	network mottles/spicules surges	3-20 15-40 50-100	(absorption lines) jets and X ray loops
		explosive events	250	
bipolar region	Closed field lines	Spot Evershed effect Active region	1.8-5	dark area bright/dark loops expansion
emerging/ cancelling flux quadrupolar region	Compact loop	Arch filament System Flares: 2 ribbon flare evaporation red shift (Hα line)	10-15 150 40	loops Xray Bright Point loop/cusp (Ca XIX)
magnetic channel (B=0)	Arcade Dipole	Quiet prominence Eruption Post-flare loops	0.5 - 10 100 - 400 150 1-10	streamer/helmet CME loop system/cusp expansion

corona. In the photosphere horizontal flows are generally larger than vertical ones: for example, cork motions are seen with $V_{hor} = 300\text{-}500$ m s^{-1} and $V_{ver} = 50\text{-}100$ m s^{-1} (Yi and Molowny-Horas 1995). The vertical flows in sunspots are even an order of magnitude less than the horizontal flows (1.8-2 km s^{-1}) (Shine et al 1994). In upper levels the amplitudes of both vertical and horizontal flows are increasing. Yi and Molowny-Horas (1995) measured in the chromosphere $V_{hor} = 1\text{-}1.5$ km s^{-1}, and $V_{ver} = 1\text{-}2$ km s^{-1}. This confirms the validity of the continuity equation (ρ V = const.) in flux tubes.

The transport of angular momentum between the poles and the equator cannot be investigated without taking into account the large-scale motions of the convective zone. Rotation and convection are obviously connected. The coupling of differential rotation and convection in the ionized solar interior is basically the mechanism of solar activity. Solar rotation (14 degrees/day at the equator and 11.5 degrees/day at a latitude of 60°) can be measured from tracers, i.e. sunspots, filaments and photospheric magnetic fields. Therefore, to account for this differential rotation, large-scale meridional circulation has to be invoked (Nesmes-Ribes et al 1993). This would indicate some shear motions at the low boundary of the convection zone. Such shear could be of great importance for the creation of twisting emerging flux tubes which have been observed in active regions (van Driel-Gesztelyi and Leka 1994, Leka et al. 1996, Emonet and Moreno-Insertis 1996).

With all these basic mechanisms in mind, we can review the different flows which are observed through the atmosphere. In Table I a summary of the different structures is presented where flows are identified, classified according to their scale and the type of field lines involved. In the chromosphere and in the transition region Dopplershifts can be derived from spectral observations. In the transition region the lines are optically thin; that is not the case for the chromospheric lines, which need special diagnostics in order to be interpreted. In the first section we shall present some spectroscopic diagnostics. In the second section we will present **small-scale dynamic structures**. Concentration of magnetic field over the solar surface produces what is commonly termed the "network" and organizes the structures anchored at the basis of the chromosphere and expanding in the corona, *i.e.* mottles, spicules. These structures follow open field lines or long loops. Some of them are bent and form the canopies. With the emergence of new flux, we observe instabilities which lead to strong, dynamic events such as surges, jets and explosive events.

The **medium-scale phenomena** involve bipolar and quadrupolar regions which take into account emerging or cancelling flux. The field lines are mainly closed. The dynamics in and over sunspots, i.e. the Evershed effect in the photosphere and the inverse effect in the chromosphere and in the transition region, are quasi-static phenomena. Over active regions expansion of loops is commonly observed (Uchida et al 1992). Again in this particular configuration any emerging flux leads to instabilities which can be considered as X-ray bright points and compact or eruptive flares. We will not develop this point here because the Evershed effect is discussed by Solanki (this issue) and all the flare aspects are reviewed in the following chapters. We concentrate in the last section on the third class of phenomena associated with magnetic channels and concerning **large-scale events**. Prominences are formed along magnetic channels with a typical shear of 25°, under a helmet streamer. Frequently eruptions are observed, leading to Coronal Mass Ejections (CME) in the corona. After the eruption an arcade may be formed above the filament channel and in this case the filament itself is reformed. In two-ribbon flares, the reconnection site appears as a hot cusp with cool/hot loops, called post-flare loops (Forbes and Acton 1996), below the cusp. The matter is highly dynamic inside the cool loops (Wiik et al 1996). In the corona bulk flow is measured by proper motions. Spectral data have up to now been only sparsely available but now with SOHO this situation will be remedied. The main instruments which have obtained coronal spectral data are the Bragg Crystal Spectrometer (BCS) aboard the Solar Maximum Mission (SMM) and Yohkoh, and the two UV spectrographs aboard SOHO: the Coronal Diagnostic Spectrometer (CDS) and the Solar Ultraviolet Measurements of Emitted Radiation (SUMER).

2   Spectroscopic Diagnostics

Chromospheric structures such as spicules, filaments, mottles and surges are observed in a few lines formed around the temperature plateau ($5000 < T <$ 10 000K), such as Ca II H and K and Hα in the visible; He lines in the IR; and the Lα wing, the Lyman continuum and the Mg II h and k lines in the UV (Vernazza et al 1981).

2.1   Non-LTE model

The interpretation of chromospheric line profiles requires appropriate techniques to extract the values of physical quantities because of radiative transfer coupling in the atmosphere. To understand the energetics and physical make-up of the chromosphere we have to take into account atomic structure, spectral energy and spatial distribution in a non-LTE approach. The coupling between dynamics and radiative transfer is difficult to solve in a non-LTE atmosphere. Even though there are some attempts to realize this goal, most of the models are static and one-dimensional. Some efforts have been made to take into account the inhomogeneities and a finite geometry. A grid of 140 non-LTE models has been computed for prominence-like structures located at a height of 10000km and represented by a vertically-standing 1D slab of thickness D (Gouttebroze et al 1993). We will come back to this model in its application to chromospheric structures., i.e spicules and mottles.

2.2   Velocity diagnostics

The models combining chromospheric velocity diagnostics and radiative transfer are more or less empirical iterative methods. Mein et al (1985) have proposed a technique of non-linear profile inversion for the Ca II resonance line. They found that large gradients of high velocities (5-10 km s^{-1}) are associated with temperature enhancements, for example in bright points. If there is a discontinuity in the gradient of the physical parameters, cloud model techniques are suitable. In that case the chromospheric fine structure is supposed to be an optically-thin cloud over an uniform atmosphere, but which contributes to the observed profiles. Since the first proposal of a "cloud model" by Beckers (1964), many investigations have been devoted to mottles

(Grossmann-Doerth and von Uexküll 1977, Bray 1973, Tsiropoula et al 1993) and to arch filaments (Alissandrakis et al 1990, Tsiropoula et al 1992).

The first step is to fit the observed contrast profile ($C(\lambda) = \frac{I(\lambda) - I_0(\lambda)}{I_0(\lambda)}$ where $I_0(\lambda)$ is the background intensity and $I(\lambda)$ the intensity of the cloud) with a set of four parameters assumed to be constant within the cloud: the source function, the Dopplerwidth, the Dopplershift and the optical thickness. The reference profile is observed in the quiet chromosphere and is assumed to be identical to the intensity below the structure. To improve the solution in the case of an inhomogenous background such as faculae, Mein and Mein (1988) proposed a differential cloud model (DCM1). Another development comprises a second-order differential cloud model (DCM2) which takes account the gradient of the velocity inside the structure. This can be of great importance in twisted ropes or spicules. Other techniques have been developed using two clouds (Gu et al 1994, Heinzel and Schmieder 1994).

2.3   Source function

Another important assumption concerns the source function. In the optically thin case, it is more accurate to determine the source function by the reference incoming radiation than to derive it from the contrast profile. The cloud model leads to the determination of the four parameters described above. However, recent theoretical studies show that source functions for hydrogen lines are functions of the optical depth in vertical-standing 1one-dimensional slabs irradiated from both sides by isotropic radiation (Gouttebroze et al 1993). In the optically thick case, it is not realistic to assume that the source function is constant throughout the structure. These models were also applied to mottles, which were considered as slabs of thickness D=500-1000 km and with temperatures 8000-10000 K (Heinzel and Schmieder 1994). Using these models Heinzel and Schmieder tried to reproduce the observed profiles with one or two clouds and derived velocities for mottles between 5-10 km s^{-1}. They found this result for $P \geq 0.5$ dyne/cm^2. They have explained the difference between bright and dark mottles as due to the variation of the gas pressure. In fact the chromospheric structures are not really vertical struc tures and seen on the disk they can be better approximated by a horizontal one-dimensional plasma slab, strongly irradiated from the bottom. In such a model the source function is decreasing from the bottom to the top where no incident radiation is imposed.

Theoretical computations for a one-dimensional model of a horizontal slab, obtained in a non-LTE approach, were used to solve the cloud-model inversion problem of Hα profiles of an arch filament system (Mein et al 1996). The classical cloud model gives lower optical thickness than the theoretical ones. For large thicknesses the discrepancy can reach 40 % for $\tau \cong 5$. Here τ represents the optical depth. The determination of velocities is not so sensitive. A better approximation is to compute the variation of the source function in a two-dimensional geometry (Heinzel 1994). Paletou (1995) developed in a two-dimensional geometry (Heinzel 1994). Paletou (1995) developed a two-dimensional non-LTE model of prominences which is very promising for two-dimensional structures observed in the chromosphere. The knowledge of the velocity field from chromospheric lines requires an important theoretical work in radiative transfer in a non-LTE approach. In the case of faculae or flares we have to use a model of a semi-infinite atmosphere; inversion line techniques or iterative direct computations should be used. In the case of spicules, surges and mottles the cloud model methods can give reasonable solutions particularly for the determination of velocity fields.

3   Small-Scale Dynamic Phenomena

The network is principally observed in various lines like Ca II, Hα and metal lines (Bray and Loughhead 1974, Gaizauskas 1985, 1994). A two-dimensional coronal model taking into account the concentration of magnetic field at the supergranule cell boundaries has been proposed by Gabriel (1992). We focus in this section on the flows above the intercell magnetized medium. The dynamic network over the cells is discussed in this issue by Carlsson.

3.1   Spicules

Observations with high spatial resolution of the chomosphere show it to be highly inhomogeneous. At the limb, jet-like structures called spicules are observed protruding into the hot corona. Spicules have a temperature $\sim 10^4$ K and densities of $\sim 10^{-13}$ g cm^{-3}. They rise to a maximum height of about 10 000 km with a velocity of about 20 km s^{-1}. The average time for ascent is 3-5 minutes and they typically return to their original state in 20 minutes. Most of the spicular mass undoubtedly has to return to the solar surface. Observationally spicules appear to either fade from view or fall back along the same path (Beckers 1972). The potential energy of spicules is 4×10^{24} ergs and relatively low compared to a surge ($E_p = 4 \times 10^{28}$ ergs).

There are different indirect observations of the presence of cold material at high levels in the corona. Observations by the Normal Incidence X-ray Telescope (NIXT) aboard a rocket have produced an image in X-rays contaminated by white light. A dark band obscures the bright corona above the white-light limb. This implies a substantial optical depth of absorbing material extending upwards into the corona. That material could correspond to the spicules. Compared to the existing models (Vernazza et al 1981), the decrease of the hydrogen number density N_H is slow and N_H has already a relatively low value at 3000 km: 10^{11} cm^{-3} with an order of magnitude of uncertainty (Daw, DeLuca, Golub 1995).

A second approach is the interpretation of the Lyman lines. The introduction of partial redistribution for a three-level hydrogen atom has not succeeded in explaining the enhanced wing emission in the hydrogen Lα and

Lβ lines in the chromosphere. However, Heinzel (1991) proposed a hybrid chromospheric model which links one-dimensional models of the overlying inhomogeneous structures and atmosphere models such as VAL2C (Vernazza et al 1981). Fontenla et al (1988) give an overview of the main features of the solar atmosphere as observed in Lα by SMM. They interpret the highly variable central reversal of the line by the existence of a "dynamic layer " (cloud layer) overlying the limb. It can be responsible for distorting the background profile of the line and can lead to the observed asymmetric profiles. The optical depth of the cloud would be around two at the head of the Lyman continuum, and rather large at Lα line center (10^4). Another solution is that the slit covers many different shifted profiles of material, such as spicules. This leads to the problem of the filling factor of the structures which is an important parameter for deriving physical quantities. The UV spectrographs SUMER and CDS aboard SOHO with their high spatial resolution should help to make progress on the determination of the filling factor.

Many models have been proposed to explain the formation of spicules, but up to now no completely satisfactory theory exists. The thermal jet theory (Athay 1984) proposes that the energy of spicules is derived from the downward conductive energy flux coming from the hot corona which cannot be entirely dissipated through radiation. The magnetic acceleration theory suggests that the spicules are generated by the interaction of nearby regions of opposite polarity (Pikel'ner 1969). Models due to a pressure gradient pulse or due to magnetic reconnection could be approached in the same was as for surges. This model agrees with the recent observations of the Multichannel Subtractive Double Pass (MSDP) spectrograph (Tsiropoula et al 1994). However, Suematsu et al (1995) claim that they did not see the bipolar regions at the base of spicules in the form predicted by this model. The French-Italian telescope, THEMIS, with its high spatial resolution and sensitivity, should answer this question.

A very extensively-studied theory is the wave theory suggesting that the spicule is a non-linear manifestation of wave motions in a flux tube (Hollweg 1982, Sterling and Hollweg, 1988, Cheng 1992 a,b). A wave generated impulsively in the photosphere and propagating along a vertical flux tube will grow in amplitude as it moves up. Due to strong non-linear effects, shocks are formed and move the chomosphere from below into the transition region. They are identified as the spicules. Extensive models were developed based on the rebound shock train, which is excited by the back-falling material. A recent model taking into account radiative loss and heat conduction seems to explain the observations satisfactorily (Cheng 1992 b).

Another new series of models is based on a self-excited dynamo for magnetic elements of the periphery of the supergranules. The element or structure is maintained vertically through the presence of azimuthal electric currents (Lorrain and Koutchmy 1993). Hénoux and Somov (1996) show that in an initially weak magnetic field, a radial inflow of neutral atoms can generate azimuthal DC currents. The forces produced by two current systems of opposing directions can play a role in the structure and dynamics of flux tubes which can extend above the temperature minimum region.

3.2 Mottles

When viewed on the disk the mottles are gathered at the supergranulation boundaries (Bray and Loughhead 1974). Surrounding elementary pores, they are called rosettes. However Doppler measurements of the velocities fail to indicate blueshifts of magnitude seen in spicules (Grossmann-Doerth and von Uexküll 1978, Grossmann-Doerth and Schmidt 1992). From Hα spectra they measure a mean value of $\pm$ 4 km s^{-1}. On the other hand some attempts using the cloud-model method with high-resolution MSDP observations (Tsiropoula et al 1993) or using a cinematic method with tunable-filter observations (Suematsu et al 1995) derived velocities for mottles as large as those of the proper speeds (15 km s^{-1}) which is consistent with spicule upward velocities. The discrepancy between the different authors is mainly due to interpretation. Mottles should be the base of the spicules.

3.3 Downflow in the network

Between 5000 to 30000 spicules are visible on the disk according to Mouradian (1965). The great number of spicules combined with their measured velocities imply an upward mass flux two orders of magnitude larger than that of the solar wind. It would be high enough to completely fill the corona in three hours. A question arises: how does the mass flow come back to the chromosphere? Analysis of spectral line profiles formed in the transition zone displays a net red-shift (Doschek et al 1976, Bruner et al 1976, Dere et al 1984). The observed redshift is maximum for regions of temperature of 10^5 K as observed in the O IV line (Fig. 1). Klimchuck (1987) found that the redshifts in active regions tend to occur where the field is strong (B> 100G) whereas blueshifts occur in regions of weak magnetic field. There are three different models to explain the observed red shift in the transition lines by this time (Brynildsen et al 1996), but only one is related to spicules. Cheng (1992 b) gave support to a rebound shock model which relates downflows with the return of spicules. However Hansteen and Wikstol (1994) have recently shown that during the first phase of the event, in the ascending phase before the shock, the density is higher than in the descending phase. So in a time-average observation the line will reflect the motion of the higher density plasma and should be blueshifted. This is not observed so they deduce that the rebound shock model does not work. This could rule out the idea of explaining the redshift of the transition region by the return of the spicule material. Some other explanations are based on loop model or wave-pulse model. Achour et al (1996) suggest that a flow model with loops of different filling factors should be defined with a ratio larger than 2 between upflows

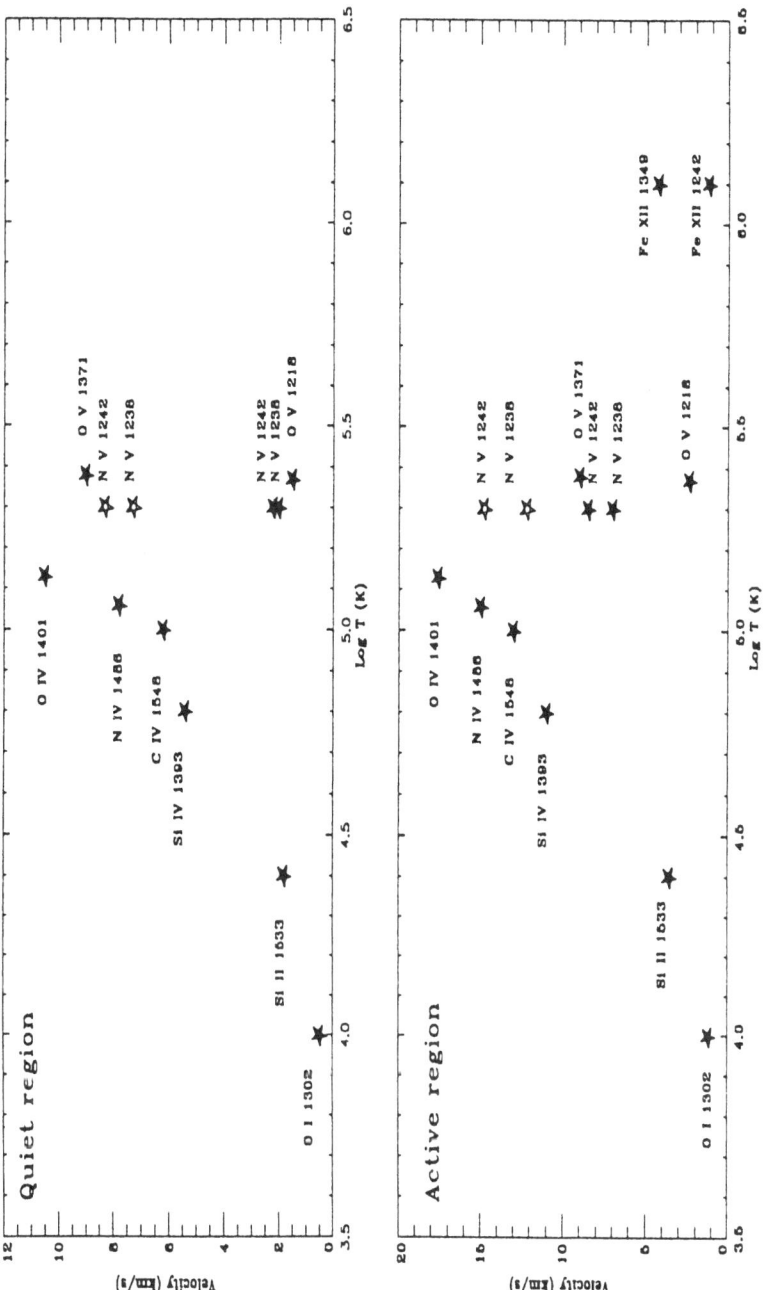

Fig. 1. Redshift for a series of transition region lines in the quiet Sun and in an active region observed with HRTS (Achour et al 1996)

and downflows. Episodic heating in the corona could create disturbances that propagate along magnetic field lines towards the chromosphere. Acoustic and Alfvén waves have been considered (Hansteen and Maltby 1992). The spectrograph SUMER on SOHO should provide observations to test these models.

3.4   Low frequency oscillations in the network

Since the work of Mein and Mein (1976), which showed broad spectra of the Ca II K3 line with apparently some significant differences in bright or dark zones, different authors have pointed out the existence of a low frequency component in chromospheric spectra of the network, compared with the 3 minute peak internetwork spectra. Fleck and Deubner (1989), Deubner and Fleck (1990) analyzed the Ca IR lines, Lites et al (1993) the Ca II H line, Lites et al (1994) and Bocchialini et al (1994) the He I line. This low frequency component has no real correlation with the photosphere below, according to phase difference and coherence spectra (Lites et al 1993). Bocchialini and Baudin (1995) show in a wavelet analysis that they consist of a wave-train characterized by a duration of 700s to 1000s. Deubner and Fleck (1990) suggest that this component is due to *magneto-gravity waves* guided from higher levels of the chromosphere down into the magnetic funnels of the chromospheric network. The 3 min resonance is better visible in the dark cell region than in the highly-filamentary network structures. Von Uexküll et al. (1989), analysing a two-dimensional sequence obtained with the MSDP , confirm that the oscillatory behaviour dominates in the chromospheric cells whereas at the boundaries they observe random motions on scales of 2-10″. They suggest that this random behaviour of Hα structures outlines the *permanent rearrangement of the magnetic field lines* pushed around by the subphotospheric flow. It is a view opposite to that of Deubner and Fleck (1990) who speculate on *downflows in the network*. The fact that the slow network oscillations are not related to the underlying atmosphere means either that they are confined in the chromosphere or that they are excited by a source not simply just below the chromospheric point but related horizontally to it. Lites et al (1994) would prefer upward magneto-acoustic wave energy transport along rosette structures to the Deubner and Fleck downflow. This five-minute oscillation could be the response of granular motions and propagation of waves along the magnetic lines of force. It may be due to an *instability* of the chromosphere itself, via a nanoflare process. Such a process is driven indirectly by the dynamics of the photosphere. These long-period disturbances have also been detected in sunspot penumbra. Their mechanism is still not understood. They could be connected to the five-minute photospheric oscillations or be connected with the quasi-perodic fluctuations in the Evershed flow observed by Shine et al (1994). Another possible interpretation is related to the pseudo-periodic time evolution of the *mottles or spicules* (Suematsu et al 1995 and Tsiropoula et al 1994). They could produce such wave-trains. However the amplitudes of the network oscillations are rather small (1-2 km s^{-1}) compared to spicule mass flows. This inconsistency may be due to the resolution of the instruments or to the orientation of the magnetic structures versus the line-of-sight. Apparently the spicules visible at the limb have a large velocity dispersion (Heristchi and Mouradian 1992) and could be responsible of the low frequency component.

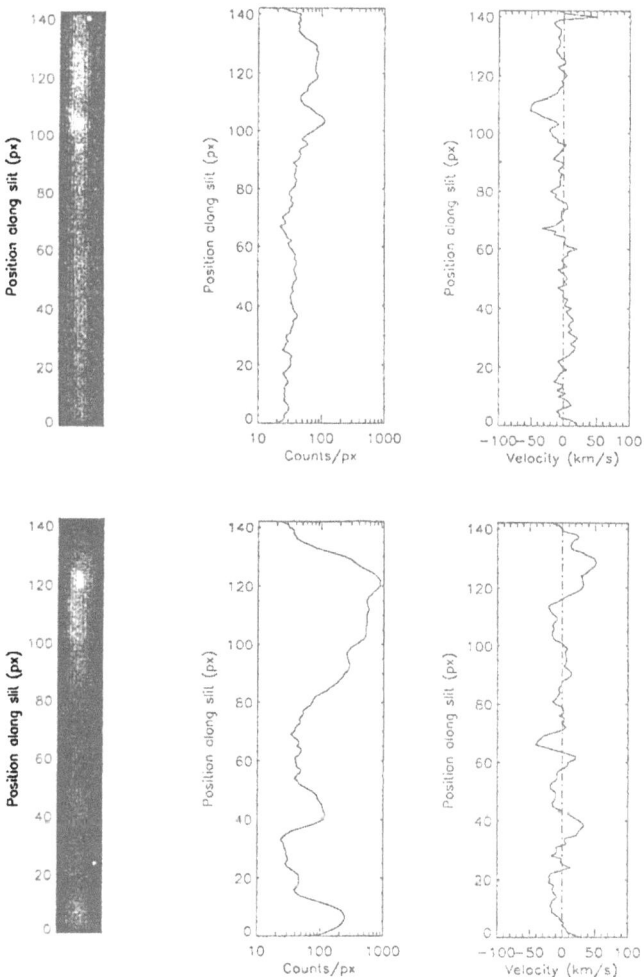

Fig. 2. Flows in the corona observed with CDS/SOHO. From left to right, spectrum, variation of the intensity in counts along the slit, variation of the velocity, top panels in Mg X 624 Å (T = 1.1 10^6 K), low panels in O V 629 Å (T= 2.4 10^6 K) (Brekke et al 1996).

3.5 Explosive events

In the transition region there are the explosive events well observed from the
High Resolution Telescope and Spectrograh (HRTS) data (i.e. Dere et al.
1989, Dere et al. 1991, Dere 1994) and with SMM (Porter et al. 1987). Coordinated observations show them to be located close to but not exactly in the
network (Porter and Dere 1991). A model of cancelling magnetic flux was proposed by Dere et al (1991). The SUMER observations confirm the existence
of the explosive events and will add much new information regarding their
detailed structure and time development. Particularly their birth rates will
now be much better determined. Preliminary estimates point to at least 5000
present on the disc at any time; this number is now of the same order of magnitude as the spicule number. As for the question as to whether the explosive
events extend to higher temperatures than 100,000 K, this was approached
by Cheng and Kjeldseth-Moe (1991), Kjeldseth-Moe and Cheng (1994) and
more recently by the use of the spectrograph CDS/SOHO (Kjeldseth-Moe et
al 1996). CDS covers a wide temperature range from 300,000 K to 2.5 MK.
So far one high velocity event has been found extending into the corona up to
2 MK. It shows a velocity dispersion exceeding 450 km s^{-1} in all lines from
He I 584 to Fe XV.

4   Large-Scale Dynamic Events

Large steady loops connecting the penumbrae of sunspots to the network
have been identified with X-ray images (NIXT and Yohkoh) (Schmieder et
al. 1996 c). Apparently the magnetic field lines joining umbrae of sunspots are
not observed because the gas pressure is too low and the plasma is not heated.
Let us remember that sunspots are dark because there are dense bushes of
flux tubes. The gas pressure inside the tube is less than outside. This is
the well-known Wilson depression at the location of the spots if we look at a
constant pressure level. In the higher parts of the tubes the pressure stays low
because there is no mass exchange with the outer atmosphere. The convection
or subphotospheric motions push gathered sets of flux tubes forming pores
or sunspots. These motions cannot produce thermal instabilities inside the
tubes and do not heat the plasma. However anchored in the penumbra, the
tubes have a higher density and can be heated to coronal temperatures and
observed by coronal instruments. With the EUV spectrograph CDS on board
SOHO, high velocities have been measured recently above active regions (Fig.
2): 50 km s^{-1} at T $\sim 10^6$ K (Mg IX 368 Å, Mg X 624 Å) and a relatively
smaller value in hotter plasma (at 2 and 2.7 10^6 K) (Brekke et al 1996). Are
these large flows in hot lines the expected signatures of magnetic reconnection
in the corona which could be at the origin of the coronal heating ? (Démoulin
and Priest 1996).

Otherwise the dynamic phenomena are directly related to solar activity.
Twisting and shearing of flux tubes may be expected to set up force-free field

currents along the field lines. On the large scale, differential rotation will induce an extended current system which increases continuously, leading to an outer expansion of magnetic field lines. The effective electrical conductivity of the solar corona is so high that the field structures resulting from flux emergence cannot immediately relax to a current-free state. The presence of currents in the corona can lead to the formation of different structures such as prominences (Raadu and Kuperus 1973); any disruptions or reconnections lead to ejections (Heyvaerts, Priest and Rust 1977), CMEs or flares.

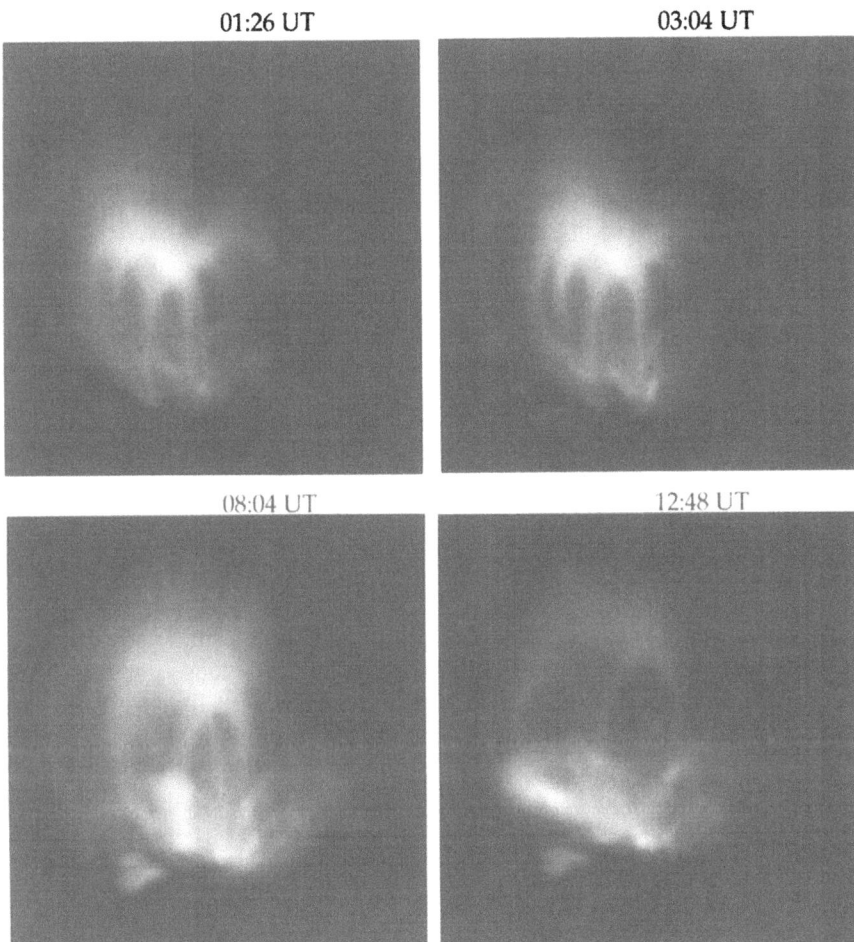

Fig. 3. Post-Flare loop images observed by Yohkoh/SXT (T = 5.6 10^6 K) extracted from the long series of observations lasting 14 hours on June 26 1992 (Schmieder et al 1995 a, 1996 b). The images represent a field-of-view of 128×128 pixels with a pixel size of $2.46'' \times 2.46''$

4.1 Prominences

Prominences are cool structures embedded in the hot corona, supported by horizontal magnetic field. The global structure is rising with an upward motion of about 0.5 km s^{-1}, while various flows are observed in the fine structures. The feet are an enigma with up-and down-flows reaching 10 km s^{-1}. These structures are relatively stable into the filament channel, visible during one to several rotations (Schmieder 1989, 1990, Tandberg-Hanssen 1994). Taking into account that matter can flow many times across the structure during its lifetime, we conclude that prominences are dynamic structures with permanent mass exchanges. The filament channel is well observed in radio at 17 GHz (Nobeyama) and metric radio sources of thermal origin appear to be associated with neutral lines of photospheric magnetic field but not necessarily with stable and large filaments (Alissandrakis and Lantos 1996). Prominences can be heated and are then observed in hot-plasma lines.

When destabilized they erupt, leading to coronal mass ejections (Hundhausen 1993). In the corona in magnetic equilibrium, the tension holds the helmet streamer overlying a prominence. A slight change can lead to the break-up of the helmet and a CME is initiated. Each event carries a total mass in the range of 10^{15} to 10^{16} g and on average there are 2 CMEs per day depending on the degree of the solar activity (Low 1996). The energy is about 10^{31} to 10^{32} ergs. That represents only 10% of the steady solar wind. The speed of the ejected matter is between 100 and 10^3 km s^{-1} with an average value of 350 km s^{-1}. It is more than the sound speed which is around 150 km s^{-1} and less than the Alfvén speed (700 km s^{-1}) in the corona. An arcade of hot loops is formed after the eruption (see Yohkoh/SXT movie, McAllister et al 1996), and the filament is reformed in the channel (Malherbe 1989).

4.2 Post-flare loops

Cool and hot flare loops appear frequently during the gradual phase of a large two-ribbon flare (Švestka 1989, Schmieder 1992b). The loops develop slowly and persist for hours. Recently a long series of observations of post-flare loops has been obtained during a coordinated campaign between Yohkoh and ground-based observatories (Schmieder et al 1995a, 1996b, Wiik et al 1996). The hot flare loops were visible in X-rays for 16 hours on June 26, 1992 (Fig. 3). The cool loops were located just below the hot loops and showed a similar arcade to the hot ones. The apparent rate of rise of the arcade was 10 to 4 km s^{-1} during the first hours and then around 1 km s^{-1}. This latter value is relatively small compared to the values obtained with SMM (Švestka et al. 1987). Along the cool loops, bubbles with downflows of 100-150 km s^{-1} were observed in Hα (Fig. 4). These flows are lower than free-fall motions, confirming previous results (Heinzel et al 1992). Two kinds of loops are observed in Figure 4, the vertical arcade of post-flare loops with downflows along the both legs, and horizontal sheared loops with flows going along the flaring loops (Moore et al 1997).

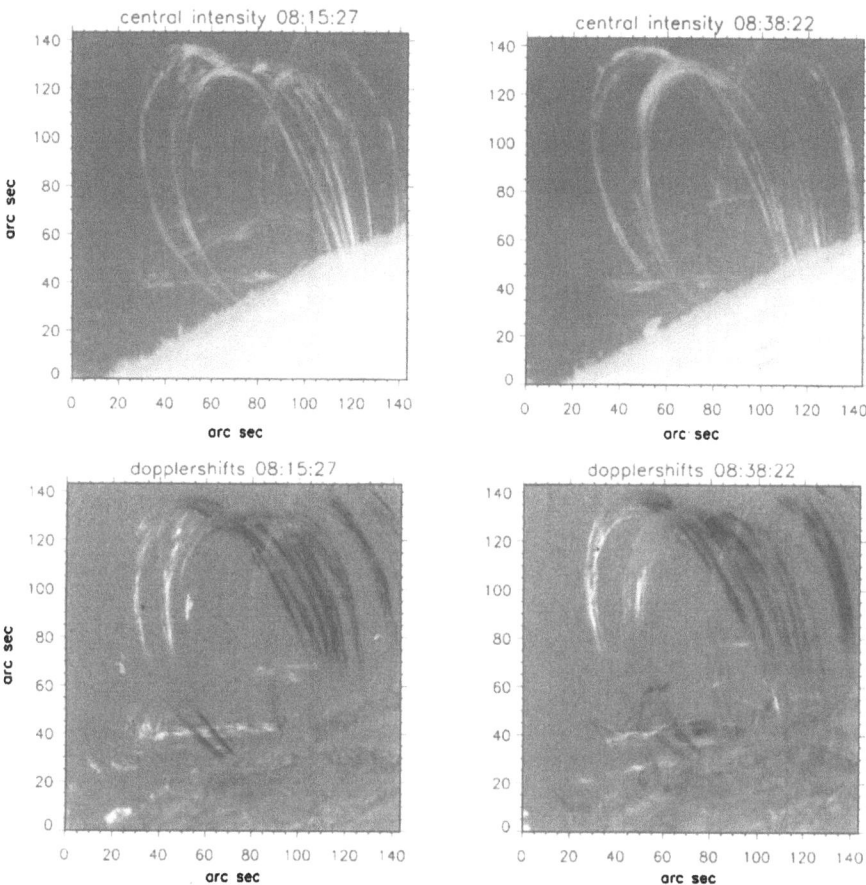

Fig. 4. Post-Flare loops observed in La Palma in Hα with the Swedish tele-scope by the Lockheed group on June 26 1992 (Wiik et al 1996). The up-per panels show intensity and the lower panels Dopplershift (black/white represent red/blue shifts). Note also the spicules at the limb and the high dynamical state of the chromosphere. (J.M.Malherbe et al 1997)

Post-flare loops are now widely believed to be the physical evidence of on-going magnetic field line reconnection. According to theoretical models (Forbes and Malherbe 1986 and Forbes and Acton 1996) reconnection keeps forming new loops at an ever-increasing altitude. The newly-formed loops are filled with hot plasma by chromospheric evaporation and can cool quickly through radiative and conductive processes to the temperature of 10^4 K.

Many questions remain unclear: How is this process maintained for such a long time? Is the evaporation process efficient enough to supply material

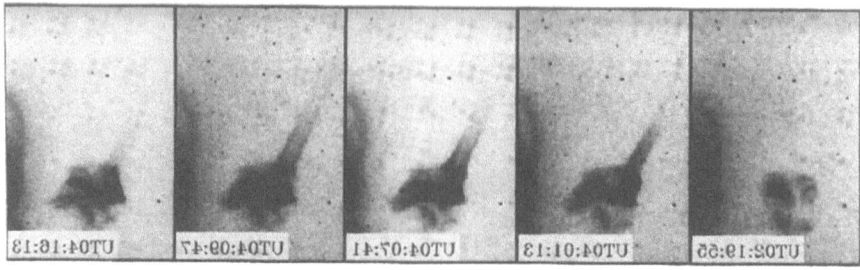

Fig. 5. Jet observed by Yohkoh/SXT (Shimojo et al 1996)

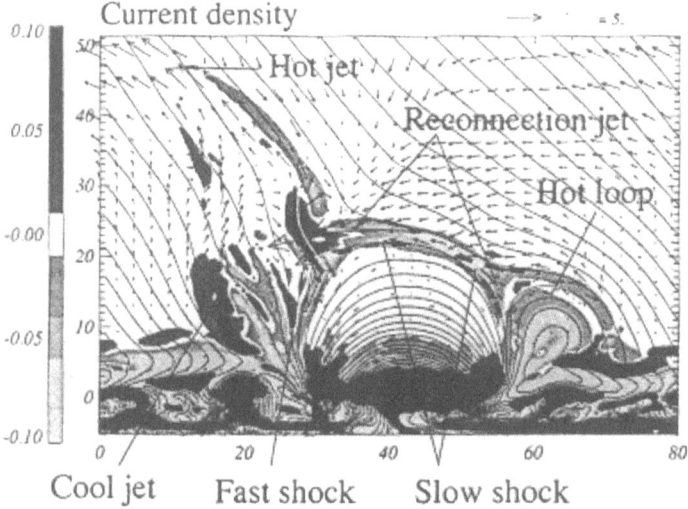

Fig. 6. Simulation of a model of jet and surge in an oblique configuration of the coronal magnetic field (Yokoyama and Shibata 1996)

during the whole life-time of the loops? How is the fine structure (bubbles) in cool loops formed ? Is there a flow of material in the hot loops?

4.3   Jets

The coronal X-ray jets, defined as transitory X-ray enhancements with an apparent motion, were first mentioned during the SMM mission and have since been well observed with Yohkoh/SXT (Fig. 5). The velocity range of the jets is 100 to 400 km s^{-1} with an average value of 200 km s^{-1}, the life-time is between a few minutes and a few hours, the length is 10^5 to 5×10^5 km, the energy involved is 10^{27} to 10^{28} erg (Fig. 6) (Shibata et al 1992 b, 1995,

Shimijo et al 1996). In many cases they are associated with small flares at the footpoints and are similar to transient loops brightenings (Shimizu 1995). These ejections of material are interpreted by magnetic reconnection models (Shibata et al. 1992 a, Yokoyama and Shibata 1996). The counterparts of the jets in the chromosphere are the surges or cool jets and some brightenings at the base of the surges (Schmieder et al 1995 b, 1996 a, Kotrč et al 1996). It is well known that surges occur when there is a parasitic polarity. This is the signature of a new flux emerging in a pre-existing magnetic field, with which a reconnection of the field lines can occur.

5   Conclusion

The list of phenomena investigated in this review is not exhaustive. Velocity fields are involved in all layers of the Sun and in most physical mechanisms. The flows are controlled by many physical processes which are often coupled:

- convection and solar rotation
- gas-pressure gradients
- pressure and magneto-gravity waves
- Lorentz forces and magnetic instabilities

At the same time the diagnostic methods are not always simple. The interpretation of spectral lines involves very often non-LTE conditions and serious radiative transfer problems. Radiative transfer should be considered in the analysis of the spectral lines and also in the radiative energy exchange through the atmosphere. Flows require a three-dimensional analysis over a large range of temperature. Reconnection processes are often proposed to explain the dynamic behaviour of the highly-filamentary atmosphere. At the small scale the energy release through reconnection could be a serious contender to explain coronal heating. On the larger scale, reconnection of magnetic field lines is a good candidate to explain spicules, jets, explosive events, surges and post-flare loops. Coordinated observations with SOHO, Yohkoh and ground-based observatories are very promising for such a study.

Acknowledgments
The author wants to thank particularly the following persons who have agreed to discuss this review, given some fruitful comments and read the manuscript: Drs. Tandberg-Hanssen, Heinzel, Maltby, Kjeldselth-Moe and Simnett.

References
Achour H., Brekke P., Kjeldseth-Moe O., Maltby, P. (1996): ApJ **453**, 945
Alissandrakis C.E., Tsiropoula G. and Mein P. (1990): A&A **230**, 200
Alissandrakis C.E., Lantos P. (1996): Solar Phys. **165**, 61
Athay R.G. (1984): ApJ **287**, 412
Beckers, J.M. (1964): Thesis in Utrecht

18   Brigitte Schmieder

Beckers J.M. (1972): Ann. Rev.Astr. Ap. **10**, 73

Bocchialini K., Vial J.C., Koutchmy S. (1994): A&A **423**, L70

Bocchialini K., Baudin, F. (1995): A&A **299**, 893

Bray R.J. (1973): Solar Phys. **29**, 317

Bray R.J., Loughhead R.E. 1974, *The solar chromosphere*, London: Chapman and Hall.

Brekke P., Kjeldseth-Moe O., Brynildsen N., Maltby P., Haugan S.V.H., Harrison R.A., Thompson W.T., Pike C.D. (1996): ApJ , in press

Bruner E.C., Chipman E.G., Lites B.W., Rottmann G.J., Shine R.A., Athay R.G., White O.R. (1976): ApJ **210**, L97

Brynildsen N., Kjeldseth-Moe O., Maltby, P. (1996): ApJ **462**, 534

Cheng C.C., Kjeldseth-Moe O.(1991): *Dynamics of solar Flares*, eds. B.Schmieder and E. Priest, 101

Cheng Quing Qi. (1992a): A&A **262**, 581

Cheng Quing Qi. (1992b): A&A **266**, 537

Daw A., DeLuca E.E., Golub L. (1995): ApJ **453**, 929

Démoulin P., Priest E. (1996): *JGR*, 100, 23, 443

Dere K.P. (1994): Space Science Rev. **70**, 21

Dere K.P., Bartoe J.D.F., Brueckner G.E. (1984): ApJ **281**, 870

Dere K.P., Bartoe J.D.F., Brueckner G.E. (1989): Solar Phys. **123**, 41

Dere K.P., Bartoe J.D.F., Brueckner G.E., Ewing J., Lung P. (1991): Journal of Geophysical Research, **96**, 9399

Deubner F.L., Fleck B. (1990): A&A **228**, 506

Doscheck G.A., Feldmann U., Bohlin J.D. (1976): ApJ **205**, L177

Emonet, T. and Moreno-Insertis, F. (1996): ApJ **458**, 783

Fleck B., Deubner F.L. (1989): A&A **224**, 245

Fontenla J., Reichmann E.J., Tandberg-Hanssen E. (1998): ApJ **329**, 464

Forbes T.G., Malherbe J.M. (1986): ApJ **302**, L67

Forbes T.G., Acton L.W.. (1996): ApJ **459**, 330

Gabriel A.H. (1992): *The Sun, a Laboratory for Astrophysics*, eds. J.Schmeltz, and J.C Brown, Kluwer

Gaizauskas V. (1985): *Chromospheric Diagnostics and Modelling*, ed. B.W.Lites, NSO conf., Sacramento Peak, 25

Gaizauskas V. (1994): *Solar Surface Magnetism*, eds. R.Rutten and C.J. Schrijver, Vol. **433**, 133

Gouttebroze P., Heinzel P., Vial J.C. (1993): *Astron. Astrophys. Suppl. series* **49**, 513

Grossmann-Doerth U., von Uexküll M.(1977): Solar Phys. **55**, 321

Grossmann-Doerth U., Schmidt W. (1992): A&A **264**, 236

Gu X.M., Lin Jun, Luan Ti, Schmieder B. (1992): A&A **259**, 649

Hansteen V., Maltby P. (1992): Comm.Astrophys., **16**, 137

Hansteen V.H., Wikstol O. (1994): A&A **290**, 995

Heinzel P. (1991): Kluwer Academic Publishers, Crivellari et al (eds)

Heinzel P. (1994): in *Numerical Methods for Multidimensional Radiative Trans-*

fer Problems, ed. R. Rannacher, S. Turek and R. Wehrse, Univ. of Heidelberg
Heinzel P., Schmieder B. (1994): A&A **282**, 939
Heinzel P., Schmieder B., Mein P. (1992): Solar Phys. **139**, 81
Hénoux J.C., Somov B.V. (1996): A&A , in press
Heristchi D., Mouradian Z. (1992): Solar Phys. **142**, 21
Heyvaerts J., Priest E., Rust, D. (1977): ApJ **216**, 123
Hollweg J.V. (1982): ApJ **257**, 345
Hundhausen A.J. (1993): J.Geophys. Res., **98**, 13177
Kjeldseth-Moe O., Cheng C.C. (1994): Space Science Rew., **70**, 85
Kjeldseth-Moe O., Brynildsen N., Brekke P., Harrison R.A. (1996): Madison
Meeting in May 1996
Klimchuck J.A. (1987): ApJ **323**, 368
Kotrč, P., Schmieder, B., Karlicky, M. and Heinzel, P (1996) *Solar Physics*,
in press
Leka, K.D., Canfield, R.C. McClymont, A.N. and van Driel-Gesztelyi, L.
(1996): *Astrophys. Journ.*, **462**, 547-560.
Lites B.W., Rutten R.J., Kalkofen W. (1993): ApJ **414**, 345
Lites B.W., Rutten R.J., Thomas J.H. (1994), in R.J.Rutten and C.J.Shriver
(ed) *Solar Surface Magnetism*, NATO ASI:C, Vol 433, Kluwer, Dordrecht
Lorrain P., Koutchmy S. (1993): A&A **269**, 526
Low B.C. (1996): Solar Phys. **167**, 217
Malherbe J.M. (1989): *Dynamics and Structure of Quiescent Prominences*,
ed. E. Priest, Kluwer, Dordrecht, 115
Malherbe J.M., Tarbell T.D., Wiik J.E., Shine R., Schmieder B., Franck Z.,
van Driel-Gesztelyi L. (1997): ApJ , in press
McAllister A.H., Kurokawa H., Shibata K., Nitta N. (1996): Solar Phys. ,
in press
Mein N., Mein P. (1976): Solar Phys. **49**, 231
Mein P., Mein N., Malherbe J.M. (1985): Proceeding of Munich conference,
MPA/LPARC, ed. Schmidt H.V., 303
Mein P., Mein N. (1988): A&A **203**, ,162
Mein N., Mein P., Heinzel P., Vial J.C., Malherbe J.M. and Staiger J. (1996):
A&A **309**, 275
Moore R.L., Schmieder B., Hathaway D.H., Tarbell T.D. (1997), Solar Phys.,
submitted
Mouradian Z. (1965): *Annales d'Astrophys.*, **28**, 805
Nesmes-Ribes E., Ferreira E.N., Mein P (1993): A&A **274**, 563
Paletou F. (1995): A&A **302**, 387
Pikel'ner S.B. (1969): AZh, **46**, 328
Porter J.G., Moore R.L., Reichmann E.J., Engvold O., Harvey K.L. (1987):
ApJ **323**, 380
Porter J.G., Dere K.P. (1991): ApJ **370**, 775
Raadu M.A. (1983): Space Sciences Review, **34**, 37
Raadu M.A., Kuperus M. (1973): Solar Phys. **28**, 77

Schmieder B (1989): *Dynamics and Structure of Quiescent Prominences*, ed. E. Priest, Kluwer, Dordrecht, 15

Schmieder B. (1990): Lecture Notes in Physics, Springer-Verlag, Tandberg-Hanssen and Ruždjak (Eds), **363**,85

Schmieder B. (1992 a): Simuris meeting, ESA SP-344,35

Schmieder B. (1992 b): Lecture Note in Physics, **399**, *Eruptive Solar Flares*, eds. Z. Švestka, B.V.Jackson and M.Machado (NY), 124

Schmieder B., Heinzel P., Wiik J.E., Lemen J., Anwar B., Kotrc P., Hiei E. (1995 a): Solar Phys. **156**, 337

Schmieder B., Shibata K., van Driel-Gesztelyi L., Freeland, S. (1995 b): Solar Phys. **156**, 245

Schmieder B., Rovira M., Simnett G.M., Fontenla J.M., Tandberg-Hanssen E. (1996 a): A&A **308**, 957

Schmieder B., Heinzel P., van Driel L., Lemen, J. (1996 b): Solar Phys. **165**, 303

Schmieder B., Démoulin P., Aulanier G., Golub L. (1996 c): ApJ **467**, 881

Shibata K., Nozawa S., Matsumoto R. (1992 a): PASJ **44**, 265

Shibata K., Ishido Y., Acton L.W. etal. (1992 b): PASJ, **44**, L173

Shibata K., Masuda S., Shimojo M., Hara H., Yokoyama T., Tsuneta S., Kosugi T., Ogawara Y. (1995): ApJ **451**, L83

Shimizu T. (1995): PASJ **47**, 251

Shimojo M., Hashimoto S., Shibata K., Hirayama T., Hudson H.S., Acton L. (1996): PASJ **48**, 123

Shine R.A., Title A.M., Tarbell T.D., Smith K., Franck Z.A. (1994): ApJ **430**, 413

Sterling A.C., Hollweg J.V. (1988): ApJ **327**, 950

Suematsu Y., Wang H., Zirin H. (1995): ApJ **450**, 411

Tandberg-Hanssen E. (1994), *The Physics of Solar Prominences*, Kluwer

Švestka Z. (1989): Solar Phys. **121**, 399

Švestka Z., Fontenla J.M., Machado M.E., Martin S.F., Neidig D.F., Poletto G (1987): Solar Phys. **108**, 237

Tsiropoula G., Georgakilas A.A., Alissandrakis C.E., Mein P. (1992): A&A **262**, 587

Tsiropoula G., Alissandrakis C.E., Schmieder B. (1993): A&A **271**, 574

Tsiropoula G., Alissandrakis C.E., Schmieder B. (1994): A&A **290**, 285

Uchida Y., McAllister A., Strong K.T., Ogawara Y., Shimizu T., Matsumoto R., Hudson, H.S. (1992): PASJ, **44**, L155

van Driel-Gesztelyi, L., and Leka, K.D. (1994): *ASP Conf. Series*, 68, 138

von Uexküll M., Kneer F., Malherbe J.M., Mein P. (1989): A&A **208**, 290

Vernazza J.E., Avrett E.H., Loeser R. (1981): ApJ **S45**, 635

Yi Z., Molowny-Horas R. (1995): A&A **295**, 205

Yokoyama T., Shibata K. (1996): PASJ **48**,353

Wiik J.E., Schmieder B., Heinzel P., Roudier, T. (1996): Solar Phys. **166**, 89

Chromospheric Dynamics — What Can Be Learnt from Numerical Simulations

Mats Carlsson[1] and Robert F. Stein[2]

[1] Institute of Theoretical Astrophysics, P.O. Box 1029 Blindern, N-0315 Oslo, Norway

[2] Department of Physics & Astronomy, Michigan State University, East Lansing, MI 48824, USA

Abstract. Observations of the solar chromosphere are often interpreted using methods derived from static modeling (e.g., the Vernazza et al. 1981 model atmospheres and work based on such models) or linear theory (e.g., phase relations). Recent numerical simulations have shown that such an analysis can be very misleading. It is found that enhanced chromospheric emission, which corresponds to an outwardly increasing semi-empirical temperature structure, can be produced by wave motions without any increase in the mean gas temperature. Thus, despite long held beliefs, the Sun may not have a classical chromosphere in magnetic field free internetwork regions. This dynamic picture is consistent with observations in CO lines and the calcium H and K bright grains. More opaque lines, on the other hand, seem to show emission all of the time. This indicates the existence of a hotter, magnetic, component that increases in importance with height.

The simulations closely match the observed behaviour of $Ca\,II\ H_{2v}$ bright grains down to the level of individual grains. The bright grains are produced by shocks near $1\,Mm$ above where the optical depth is unity at $500\,nm$ ($\tau_{500} = 1$). These shocks are primarily due to waves from the photosphere with a frequency slightly above the acoustic cutoff frequency. The concept of a fixed formation height is of little use in the chromosphere. The temperature spikes at shock fronts may produce doubly-peaked intensity contribution functions with one peak at $\tau_{\nu} = 1$ and another at the shock. The mean height of formation for lines and continua formed around $1\,Mm$ can vary greatly with time and does not necessarily correspond to the actual layers emitting the photons. When waves in the chromosphere have large amplitude, linear perturbation theory is not valid since the passage of waves changes the atmosphere fundamentally.

1 Introduction

What is the temperature structure of the solar chromosphere? Does it even exist as a quasi-static phenomenon or is it wholly dynamic in nature? These questions have to be answerered before it is meaningful to address the question of the energy balance at chromospheric heights (what is heating the chromosphere?).

Does the Sun even have a chromosphere in magnetic field free regions? This may seem to be a stupid question since we certainly observe enhanced emission over radiative equilibrium values. The question, however, is whether this means the gas kinetic temperature increases outward above a temperature minimum? At some altitude this must be the case since the existence of a million degree corona has long been known from the identification of coronal emission lines as arising from highly ionized atoms. Many one-dimensional hydrostatic solar models have been constructed that reproduce the observed continuum and line intensities in an average sense (both temporally and spatially), with a chromospheric temperature rise starting at about 500 km above the photosphere. Familiar examples are the Vernazza, Avrett & Loeser (1981, VAL) models representing regions of different continuum intensity ranging from dark cell interiors to very bright network locations. However, spatially and temporally resolved observations of Ca II (e.g., Lites et al. 1993), and recently of CO (Solanki et al. 1994, Uitenbroek et al. 1994) are consistent with a large fraction of the internetwork solar surface having no temperature rise in the first 500 km above the temperature minimum in the VAL models.

Our main source of information about the solar atmosphere comes from photons that escape. Before the photons reach us they have interacted with the matter so that the information about local conditions in the atmosphere has been convolved both in space and in time. The solar chromosphere is a dynamic radiating medium where the hydrodynamics, thermodynamics and radiation couple in a non-local and non-linear fashion. The diagnostics of the solar chromosphere from the solar spectrum is therefore a complicated task.

The outline of this paper is as follows: In Sect. 2 we discuss modeling strategies, in particular the inverse problem versus the forward problem; in Sect. 3 we discuss numerical simulations of the solar chromosphere; in Sect. 4 we discuss our results and in Sect. 5 we summarize our conclusions.

2   Modeling Strategies

We illustrate the basic concepts of the inverse problem versus the forward problem with a simple linear case:

$$g_i = \int f(T)k_i(T)dT + \delta g_i \qquad (1)$$

where g_i is the "data" function with observation errors δg_i, $f(T)$ is the source term and $k_i(T)$ is the kernel containing our model of what leads to observable g_i. In the inverse problem this equation is inverted to obtain the source term $f(T)$; in the forward approach a set of models $f(T)$ are obtained, inserted into the equation to get a series of observables g_i to be compared with observations.

The inverse approach has been much used for the interpretation of optically thin emission lines from the solar transition region and corona ("differential emission measure" analysis). In an optically thin plasma all photons that are created escape from the medium. We can write the total intensity as a function of temperature (T) and electron density (n_e) only if we assume:

1. the gas is optically thin to all radiation that can influence the level populations
2. statistical equilibrium: the Lagrangian time derivative of the population densitiy of all levels is zero
3. all perturbing particles have Maxwellian distribution functions with temperature
4. all element abundances are constant (or simple functions of T and n_e)

We can then write

$$g_i = \int_Z K_i(T(z), n_e(z)) n_e^2(z) dz = \int_{\Delta T} \int_{\Delta n_e} K_i(T, n_e) \mu(T, n_e) dT dn_e \quad (2)$$

which defines the source term $\mu(T, n_e)$ and where the kernel function is of the form

$$K_i(T, n_e) = \frac{h\nu_i}{4\pi} n_i(T, n_e) A_i / n_e^2. \quad (3)$$

The choice of this form for the kernel function is that for optically thin resonance lines the emission is roughly proportional to n_e^2 thus giving a kernel almost a function of temperature only. The source term $\mu(T, n_e)$ is measuring the fractional volume of emitting plasma differentially as a function of both temperature and density within the plasma, weighted by n_e^2.

Given the assumptions above, the kernel functions are functions of atomic rates only. If these are known with high accuracy and one can find kernels with different dependencies on temperature and electron density it is possible to solve for the source term given a set of measured integrated intensities.

In the case of the solar transition region and corona there are reasons to doubt some of the fundamental assumptions needed to cast the equations into the form of (2); rates slower than typical dynamical time-scales cause the Lagrangian time-derivative in the rate equations to be non-zero; distribution functions may be non-Maxwellians; and abundances may be non-constant. Even *within* the framework of (2), Judge et al. (1997) show in their seminal paper that the inversion problem is hopelessly ill conditioned because (1) the kernels are rather weakly dependent on n_e, (2) they have asymptotically identical dependence on n_e, (3) the kernels cannot (and may never) be calculated with an accuracy of better than 10%. In order to solve the ill-conditioned inverse problem it is necessary to impose constraints on the solution; like smoothness. What seemed to be a model-independent inversion of data has then become as model-dependent as a forward model. The disadvantage with the inverse approach is that smoothness of the solution is a constraint motivated by the solution strategy rather than by the physics in the problem. The smooth solution may be a very bad representation of the real source term.

The advantages of the forward approach are that the approximations needed to solve the problem may be made from physical rather than mathematical arguments (if free parameters without physical meaning are introduced, this advantage is lost). Since the observational data are not used in the solution, the degree of agreement with observations is some check of the forward model. There is, however, a severe problem with non-uniqueness so it is important to investigate a substantial parameter space. With physically well defined approximations the *discrepancies* between the model and observations can give much insight; especially if the discrepancies are robust against changes in model parameters.

3   Chromospheric Dynamics – Simulations

The investigation of the temperature structure in the chromosphere from the inversion of observations is not possible for several reasons: The Planck function is an almost linear function of temperature only in the long wavelength limit. In the UV there is almost an exponential dependence on temperature. The temporal mean of intensities will therefore not give information on the mean temperature but weight the high temperatures more than the low temperatures. Even semi-empirical modeling of *time resolved* observations will be misleading. The kernel in the inversion problem is the contribution function to intensity – a function that is quite non-local. The decoupling of the source function from the Planck function introduces further non-locality and non-linearity. Furthermore, because of the low densities of the chromosphere one does not expect statistical equilibrium at the instantaneous thermo-dynamical state. Slow rates, especially slow recombination rates, affect both the thermal response of the atmosphere to dynamic perturbations (see Carlsson and Stein 1992) and the radiation in the diagnostic continua. The only solution is self-consistent forward modeling.

   The H and K resonance lines of ionized calcium are the only strong lines formed in the solar chromosphere in the visible part of the spectrum. There exists a wealth of observations with clear dynamical signatures (e.g., Rutten and Uitenbroek 1991 and references therein; Harvey et al. 1992; Bocchialini et al. 1994; Von Uexkuell and Kneer 1995; Hofmann et al. 1996; Steffens et al. 1996). The diagnostic value is, however, limited without a proper understanding of dynamic line formation. An inverse method is bound to fail. The kernels of the inversion depend in a very non-local manner on the temperature, density, velocity and radiation field and the inversion becomes hopelessly ill-conditioned. Again the way to proceed is by forward modeling.

   We will here summarize results of such a self-consistent radiation-hydrodynamic treatment under solar chromospheric conditions with the important radiative transitions treated in non-LTE. More details can be found in Carlsson & Stein (1992, 1994, 1995, 1997).

   We describe our methodology in Sect. 3.1. In Sect. 3.2 we discuss the formation of chromospheric continua. We will see that concepts that work well in a quasi-static photosphere do not work in the dynamic chromosphere. In Sect. 3.3 we discuss the consequences dynamic formation of continua has on the classical semi-empirical static picture of the chromosphere. Finally, in

Sect. 3.4 we discuss the formation of line emission, in particular the formation of the H resonance line of ionized calcium and what can be inferred from the observed asymmetry of the line core emission (called Ca II H_{2V} bright grains).

3.1   Method

A schematic representation of the computational scheme is shown in Fig. 1. We solve the one-dimensional equations of mass, momentum, energy and charge conservation together with the non-LTE radiative transfer and population rate equations, implicitly on an adaptive mesh. We employ 6 level model atoms for hydrogen and singly ionized calcium. We include in detail all transitions between these levels, thus including four Lyman lines (α, β, γ, δ), three Balmer lines (α, β, γ), Paschen α, β, Bracket α, the Lyman-, Balmer-, Paschen-, Bracket- and Pfundt continua of hydrogen, the H and K resonance lines, the infrared triplet and the photoionization continua from the five lowest levels of singly ionized calcium. Other continua are treated as background continua in LTE, using the Uppsala atmospheres program (Gustafsson 1973). Microturbulence broadening was set to a constant 2 km/s throughout the atmosphere.

Our initial atmosphere is in radiative equilibrium above the convection zone (for the processes we consider) without line blanketing and extends 100 km into the convection zone, with a time constant divergence of the convective energy flux (on a column mass scale) calculated with the Uppsala code without line blanketing.

Waves are driven through the atmosphere by a piston located at the bottom of the computational domain (100 km below $\tau_{500} = 1$). The piston velocity is chosen to reproduce a 3750 second sequence of Doppler shift observations in an Fe I line at $\lambda 396.68$ nm in the wing of the Ca H-line (Lites et al. 1993).

The velocity spectrum as a function of frequency changes with height, both in amplitude and phase. The velocity amplitude of propagating waves increases with height in a stratified atmosphere to maintain a constant flux as the density decreases. Damping reduces this amplitude increase. Propagating modes also show a phase shift due to their finite phase speed. Evanescent modes are attenuated but have no change in phase as a function of height.

We calculate this change in amplitude and phase between the piston height and 260 km as a function of wave frequency and multiply the observed Doppler-shifts with the inverse of this transfer function to obtain our piston velocities. Comparing the simulated velocities at 260 km with the observed Doppler shift in the iron line provides a check of this procedure (Fig.1). We applied this procedure to the observed Doppler shifts at five different slit positions. At the top of the computational domain there is a transmitting boundary condition.

In the dynamic calculation only hydrogen and calcium are treated self-consistently in non-LTE. With the time-variation of hydrodynamic variables (density, temperature, electron density, velocity) taken from this calculation,

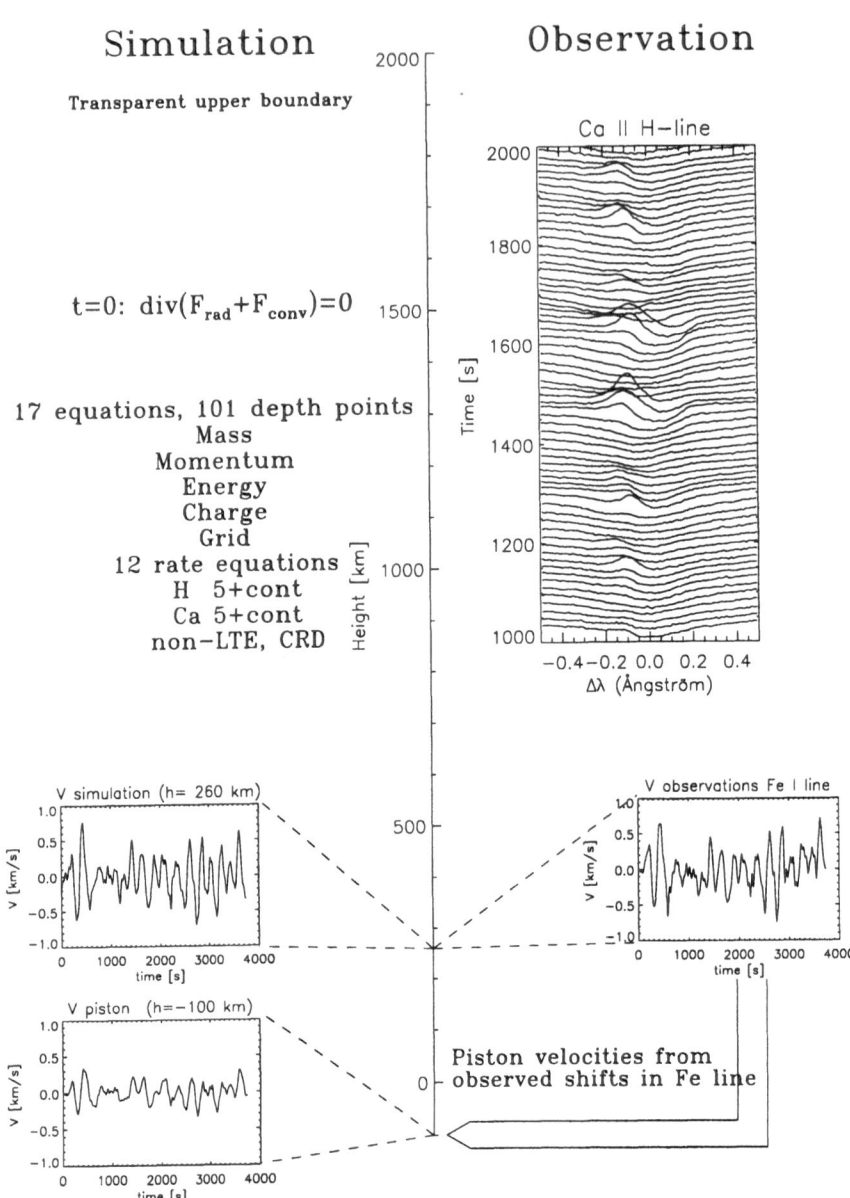

Fig. 1. A schematic representation of the computational method. Observed Doppler shifts in an iron line are transformed in amplitude and phase to get a piston velocity that gives a simulated velocity at the iron line formation height (260km) close to the observed Doppler shifts. The equations of conservation of mass, momentum, energy and charge are solved together with the non-LTE rate equations for 6 levels of hydrogen and 6 levels of ionized calcium implicitly on an adaptive mesh.

the statistical equilibrium equations were solved for an additional set of "minority species" (solution assumed not to influence the hydrodynamics or energetics of the atmosphere): neutral aluminium, magnesium, silicon and carbon.

3.2   Continua Formation

In this section we discuss the formation of the continuum radiation at four wavelengths (the dominant bound-free opacity source given in parenthesis): 207 nm (Al I), 152 nm (Si I), 110 nm (C I) and 91.2 nm (Lyman continuum). The intensities discussed below are for the wavelength just short-ward of the respective bound-free opacity edge. These continua span a range of heights of formation and ionization potentials.

Figure 2 gives the evolution in time over a short part of the full dynamic simulation of 3750 seconds. The Al I continuum with its edge at 207 nm is formed deep, around $\lg \tau_{500} = -0.66$ at a height of 92 km. The mean depth of formation is below $\tau_\nu = 1$ because the contribution function gives more weight to larger depths due to the outward temperature drop. Aluminium has a low ionization potential, so it is nearly all ionized and Al I is not the main ionization stage. This makes the opacity very temperature sensitive. The source function is slightly larger than the Planck function due to the non-LTE over-ionization. The source function at $\tau_\nu = 1$ is thus above the Planck function at that point and the emergent intensity is even higher due to the mean formation depth being below $\tau_\nu = 1$. These effects do not vary much in time and the emergent intensity, the source function (at the time varying height where $\tau_\nu = 1$) and the Planck-function thus vary in phase (Fig. 2). However, because of the temperature sensitivity of the ionization and hence opacity, an increase in temperature produces further ionization, which leads to a smaller opacity, so one sees in deeper to yet higher temperature. Thus the variations in radiation temperature, source function and Planck-function will be larger than the variations in gas temperature at the *mean* formation height; the rms of the radiation temperature in the simulation is 40 K while the rms of the gas temperature at a *fixed* height of 92 km is 17 K. The variation in radiation temperature is thus dominated by the change in formation height caused by the opacity change and *not* by the temperature change at a fixed height. The rms variation in mean formation height is 0.04 in $\lg \tau_{500}$ corresponding to 7 km.

The silicon continuum at 152 nm is formed around $\lg \tau_{500} = -4.3$ at a height of 648 km. The rms variation in the mean height of formation is 0.11 in $\lg \tau_{500}$ corresponding to 49 km. Silicon has a moderate ionization potential. Hence, near the formation depth, silicon is $\approx 80\%$ ionized. Due to the lower temperatures in the dynamic model, compared with a classical model with a chromosphere, and the temperature sensitivity of the silicon ionization, the formation height is substantially higher than in a semi-empirical model. The source function and Planck function are nearly uniform in the formation region, because it is below the region where shocks significantly modify the

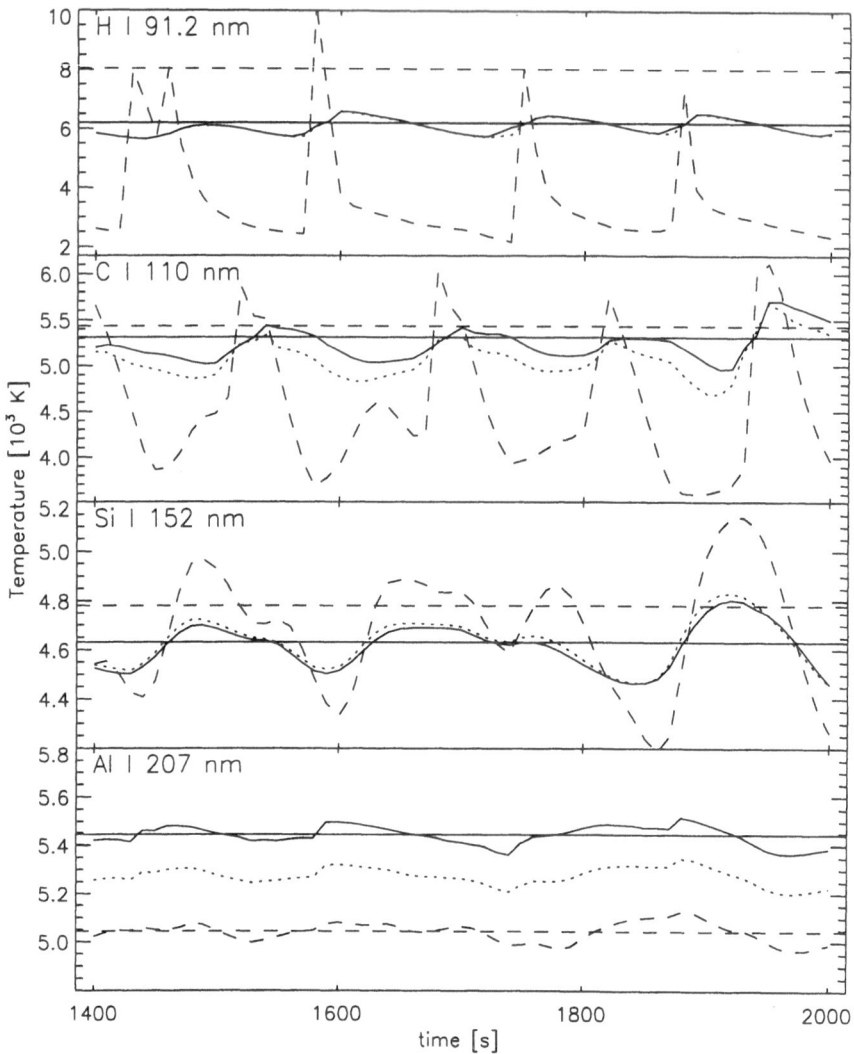

Fig. 2. The radiation temperature of the outgoing intensity (solid), the source function at $\tau_\nu = 1$ (dotted) and the temperature at $\tau_\nu = 1$ (dashed) as functions of time for a small part of the dynamical simulation for four wavelengths. The horizontal lines show the radiation temperature of the mean of the outgoing intensity (solid) and of the mean of the Planck function (dashed) with the mean taken over this part of the simulation. The means from the complete simulation will be slightly different. The non-linear averaging is clearly shown with the radiation temperature of the intensity mean being close to the maximum intensity and likewise for the Planck function. The source function varies much less than the Planck function due to the non-LTE decoupling. The radiation temperature of the outgoing intensity does not show the shock signature of a discontinuous rise for the same reason.

temperature structure (and hence the source function) and above the region of decreasing photospheric temperatures. This layer is near the region where shocks create an instantaneous temperature increase as they pass through. The intensity is produced close to $\tau_\nu = 1$ (2), and the classical contribution function there dominates the intensity formation. There may be a secondary peak in the contribution function at a shock at smaller optical depth, but such secondary maxima always make smaller contributions to intensity. As a result, fluctuations in the radiation temperature follow the thermal temperature at $\tau_\nu = 1$. However, the radiation temperature fluctuations have a smaller amplitude (rms of 86 K) than those in the gas temperature (rms of 210 K) due to the decoupling of the non-LTE source function and the Planck function.

The carbon continuum at 110 nm is formed at $\lg \tau_{500} = -5.6$, just below the region where the shocks become strong. The rms variation in the mean height of formation is 0.32 in $\lg \tau_{500}$ corresponding to 168 km. Carbon has a high ionization potential, so is nearly all neutral there. Since this is near the level of strong shock formation, the source function responds to both the high temperatures at the shock and the low temperatures in their wake. However, the source function varies much less than the Planck function, because it is controlled by radiation, not by collisions. The intensity contribution function is often bi-modal, with one peak near $\tau_\nu = 1$ and the other at the shock. The radiation temperature is generally above the gas temperature at $\tau_\nu = 1$ and does not dip as low. Only when a strong shock is formed by merging of shocks at small heights does the temperature in the wake fall low enough to make the source function follow. The radiation temperature of the emergent intensity shows a large rms variation of 165 K. The gas temperature at $\tau_\nu = 1$ shows an rms variation of 636 K.

The Lyman continuum at 91.2 nm is formed close to the top of the computational domain. The monochromatic optical depth at the top boundary is around 0.1 and the emergent intensities will be affected by the treatment of the matter leaving the computational domain and by the absence of a magnetic component in the simulations. Keeping this in mind, the continuum radiation is formed around $\lg \tau_{500} = -8$ with an rms variation of 0.27 in $\lg \tau_{500}$ corresponding to 59 km in height. The source function is almost completely decoupled from the Planck function; this gives a much smaller variation in the radiation temperature of the emergent intensity (rms of 256 K) than in the gas temperature at $\tau_\nu = 1$ (rms of 1910 K). The gas temperature shows a rapid almost discontinuous rise when the shocks pass while the radiation temperature shows no such shock signature due to the decoupling of the source function from the Planck function. The time-scale for hydrogen ionization/recombination is long at the formation height of the Lyman continuum. Recombination then takes place behind the shocks leading to a maximum radiation temperature after the maximum gas temperature at $\tau_\nu = 1$.

Continua formed in the photosphere, e.g. Al I with the edge at 207 nm, and up to ~ 0.5 Mm, e.g., Si I with the edge at 152 nm, thus have contribution functions peaked near $\tau_\nu = 1$ and have no secondary maxima at the height where shocks exist as the number of their atoms at that height is very small. The C I continuum with the edge at 110 nm is formed close to where shocks form and often exhibits a bimodal contribution function. Even though the number of atoms at shock forming heights is small, the exponential temperature sensitivity of the Planck function outweighs this factor.

Above the photosphere, the source function is so decoupled from the Planck function that variations in intensity can *not* be taken as a proxy for gas temperature variations (Fig. 2). Even in the photosphere such a one-to-one correspondence between intensity variation and gas temperature variation is not possible because of the dependency of the formation height on the opacity and therefore on the temperature.

It is important to note that there is no shock signature in the form of a saw-tooth intensity variation at any continuum wavelength. This is because the source function is non-locally dominated at the shock formation heights and because the intensity is formed over a range of heights. Statements like "we don't see any shocks in the observations" must therefore be treated with caution. The decoupling of the source function from local conditions and the width of the contribution functions to intensity makes a direct interpretation of observed intensities in terms of local temperature impossible.

It is also important to note that for these minority species (aluminium, magnesium, silicon and carbon) the rates were assumed to be instantaneous (statistical equilibrium equations solved rather than the full rate equations including the time-derivative and advection terms). Slow recombination rates would even further decouple the solution from local conditions (a decoupling in time rather than in space).

The concept of a fixed formation height is thus of limited use in the cases where: (1) the opacity is temperature sensitive; (2) the source function is decoupled from the Planck-function; (3) the formation is in the chromosphere where the atmosphere is very dynamic and the contribution function may be doubly peaked.

3.3   Semi-Empirical Model Atmosphere

In order to investigate the usefulness of a static analysis of time averaged intensities, we analyze the dynamical simulation in a way similar to the construction of the VAL models. The time averaged intensity as a function of wavelength from the simulation is taken as the quantity to be reproduced by a semi-empirical, hydrostatic, model atmosphere. For an assumed temperature structure the equations of hydrostatic equilibrium, statistical equilibrium and radiative transfer are solved and the computed intensities are compared with the time average from the dynamical atmosphere. The semi-empirical temperature structure is adjusted and the process is iterated to convergence.

Shortward of infrared and mm wavelengths only UV continua are formed above the position of the temperature minimum in the VAL models. The

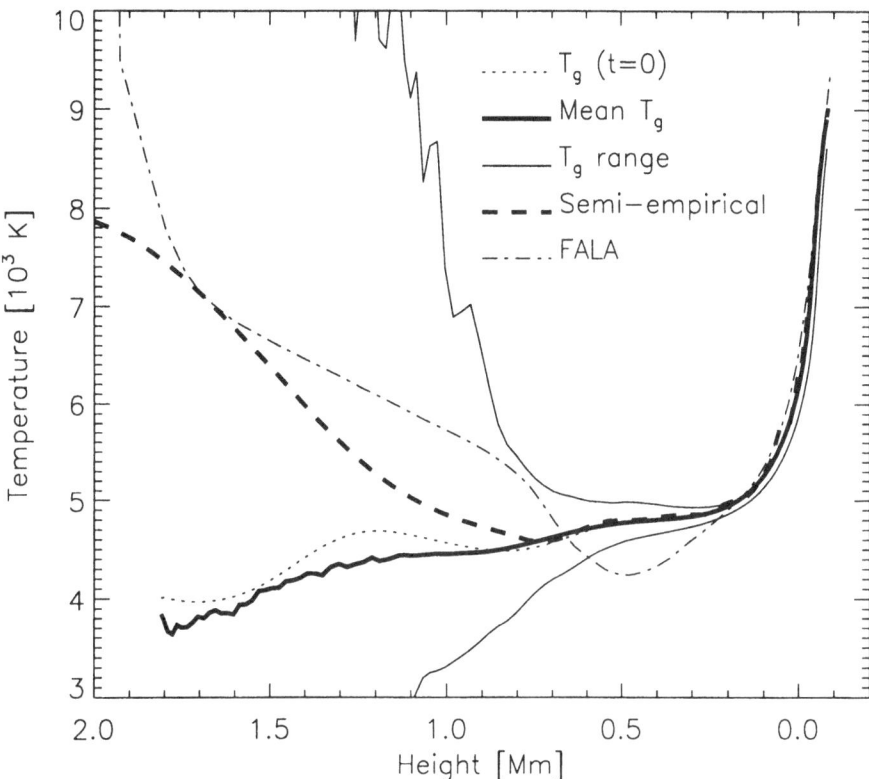

Fig. 3. Time average of the temperature in the dynamical simulation, the range of temperatures in the simulation, the semi-empirical model that gives the best fit to the time average of the intensity as a function of wavelength calculated from the dynamical simulation, the starting model for the dynamical simulation and the semi-empirical model FAL-A. The maximum temperatures are only reached in narrow shock spikes of short duration. The semi-empirical model giving the same intensities as the dynamical simulation shows a chromospheric temperature rise while the mean temperature in the simulation does not.

semi-empirical temperature structure is thus determined by the time averaged UV intensities from 150 nm (formed around the VAL temperature minimum) to the Lyman continuum (formed in the upper chromosphere in the VAL models). The final semi-empirical temperature structure is typically able to reproduce the time averaged intensity to within 20 K in radiation temperature above 100 nm and to within 100 K in the Lyman continuum below 91.2 nm.

Figure 3 shows the time average of the gas temperature as a function of height in the dynamical simulation (thick solid line) and the semi-empirical model that gives the best fit to the time average of the calculated intensities (thick dashed line). Also shown are the range of temperatures in the simulation, the starting model, and the semi-empirical model FAL-A constructed to reproduce the solar dark internetwork regions (Fontenla et al. 1993, FAL).

The striking feature of Fig. 3 is that the time average of the temperature as a function of height in the dynamical simulation shows *no chromospheric temperature rise*. Instantaneously, the viscous dissipation in the shocks together with the pressure work leads to a high temperature. This shock spike is very strong because the long time scales for hydrogen ionization/recombination initially prevents the energy from going into ionization energy (Carlsson and Stein 1992). Further back into the post-shock region hydrogen is ionized which lowers the temperature. Even further behind the shock, recombination occurs releasing the energy in the form of radiative cooling. Integrated over time at a given height, the viscous dissipation is balanced by the radiative cooling and there is no increase in temperature. There is, however, a radiation increase compared with the initial radiative equilibrium atmosphere due to the radiative cooling. To reproduce this intensity increase the best match semi-empirical model needs a chromospheric temperature increase.

A semi-empirical analysis of the dynamic simulation is thus completely misleading. Due to the non-linear averaging any analysis of average intensities is misleading whenever there is a large temporal variation of the intensity. From observations in the Ca II H and K resonance lines (see next section), as well as from recent SUMER observations, it is clear that this is certainly the case in the solar chromosphere.

3.4   Ca II Grain Formation

We made simulations with piston velocities chosen to reproduce the measured Doppler shifts in the iron line at 396.68 nm at five different slit positions in the observations by Lites et al. (1993). Fig. 4 compares the computed behaviour of the H line with these observations for two different slit positions. At the beginning of the simulation there are startup effects clearly visible for slit position 110 up to about 1000s. After this startup period there is a general correspondence between the simulations and the observations down to the level of individual grains. The simulations show the characteristic behaviour of H_{2V} grains: a brightening in the wings that propagates toward line centre and creates a bright grain when reaching a wavelength of about -0.1 Å. The dark line core moves slowly to the red and then abruptly to the blue just after the appearance of the grain. Many individual grains are well reproduced: at slit position 106 the strongest observed grains are at t=1855, 1990, 2560, 3000 and 3525s. These correspond to grains in the simulations at t=1880, 2020, 2600, 3050 and 3550s. At slit position 110 the strongest observed grains are at t=1460, 1605, 1815 and 3015s. These correspond to grains in the simulations at t=1460, 1630, 1800 and 3040s. Other grains are stronger in the simulations than in the observations.

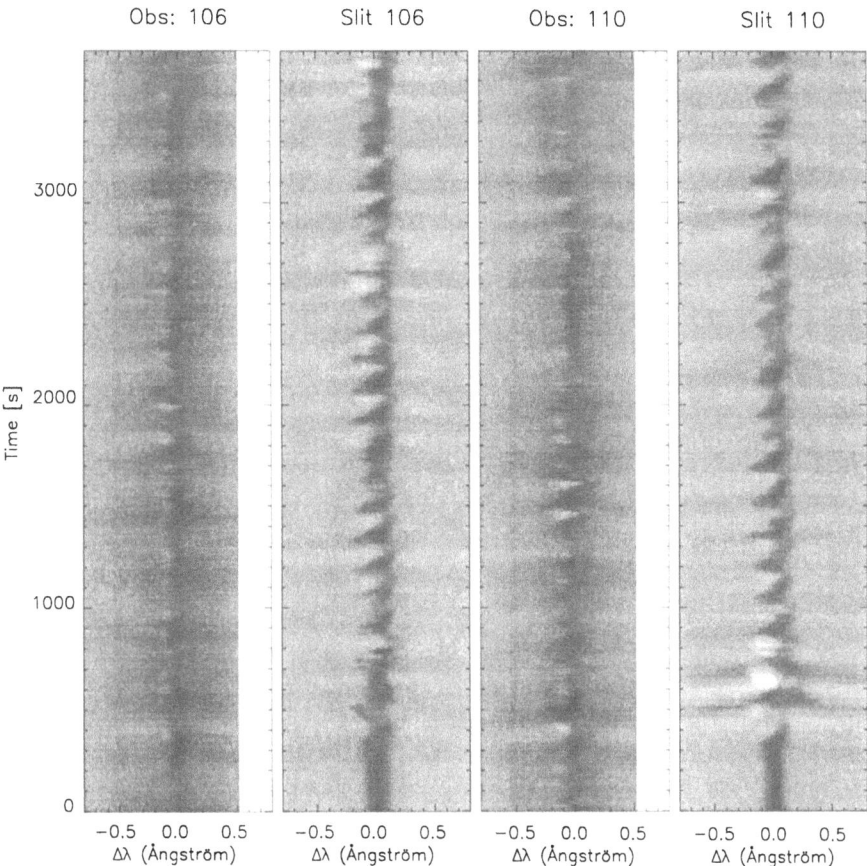

Fig. 4. The computed Ca II H line intensity as a function of wavelength and time compared with observations for slit positions 106 and 110. To mimic effects of scattered light and seeing the simulations have been convolved with a Gaussian with FWHM of 5 km/s, scattered light of 1% of the continuum intensity has been added and the sequence have been shifted in time with a random function sharply peaked at no shift. There is close correspondence between the simulated and observed grains after the initial startup period.

These two slit positions are only two arc-seconds apart (1.5 Mm) and some grains are visible at both slit positions (e.g., around t=3000s). Many grains are, however, only visible at one slit position, e.g., the grains at t=1855, 1990, 3525 and 3665s at slit position 106 and at t=1460, 1605, 1815 and 3600s at slit position 110. These grains are well reproduced in the corresponding simulation and are weak or absent in the simulation for the other slit position. This shows that the size of the grains is determined already by the velocity field at 260 km height.

There are also several significant differences which give us important information on where the theory and the observations need to be improved. The occurrence time of the simulated grains is different than the observed grains by up to a few tens of seconds. This seems to be due to imperfections in the transfer function, so that the velocities do not match the observed iron line Doppler shifts sufficiently accurately. The grains produced primarily by high frequency waves are stronger in the simulation than in the observations. This may be due to the 1D nature of the calculation, so that these high frequency waves to not spread horizontally; it may be due to a resonance that occurs in the calculation; or it may be due to excess power in the iron line observations at high frequencies because of image motion and seeing effects. The simulated grains are stronger than the observed grains, even after smoothing to represent seeing effects. This may also be due to the 1D plane wave nature of the calculation, or to the neglect of partial redistribution, or to neglect of important coolants, or to seeing. Finally, the simulated line cores are darker than the observations. This may be due to effects of scattered light on the observations, or to the existence of a hotter magnetic region in the line core forming layers which lie above the bright grain forming layers by several hundred kilometres.

Where in the atmosphere is the intensity formed at the time of occurence of the bright grains? What is the cause of the asymmetry of the Ca II line profile? What in the velocity field causes the bright grains? The advantage with a forward model is that these questions can be answered for the grains formed in the simulation. By studying the simulation in detail (Carlsson & Stein 1997) it is clear that the grains are caused by shocks 1-1.3 Mm above $\tau_{500} = 1$. At the shock there is a high temperature spike. In the mid-chromosphere the source function is still sufficiently coupled to the Planck function that this high temperature also corresponds to a maximum in the source function. If the shock formation is higher up the coupling between the Planck function and the source function is much weaker leading to no source function increase. At lower heights the waves have not shocked yet and the temperature and source function increase is smaller. This is seen in Fig. 5 where the temperature and source function as functions of height are displayed.

The asymmetry of the line profile is caused by a large velocity gradient across the shock. Material motion Doppler shifts the frequency where atoms emit and absorb radiation. The shock typically propagates into downfalling matter. At the location of the temperature and source function maxima the matter is propagating upwards and the maximum emission is Doppler shifted to the blue. There is little matter above that point with a similar velocity capable of absorbing the radiation and we get a local maximum in the line profile to the blue of line centre. At the corresponding frequency on the red side of line centre there are very few atoms that can emit photons because of the material motion Doppler shift to the blue. In addition, many atoms in the downfalling matter above the shock can absorb radiation, thus giving

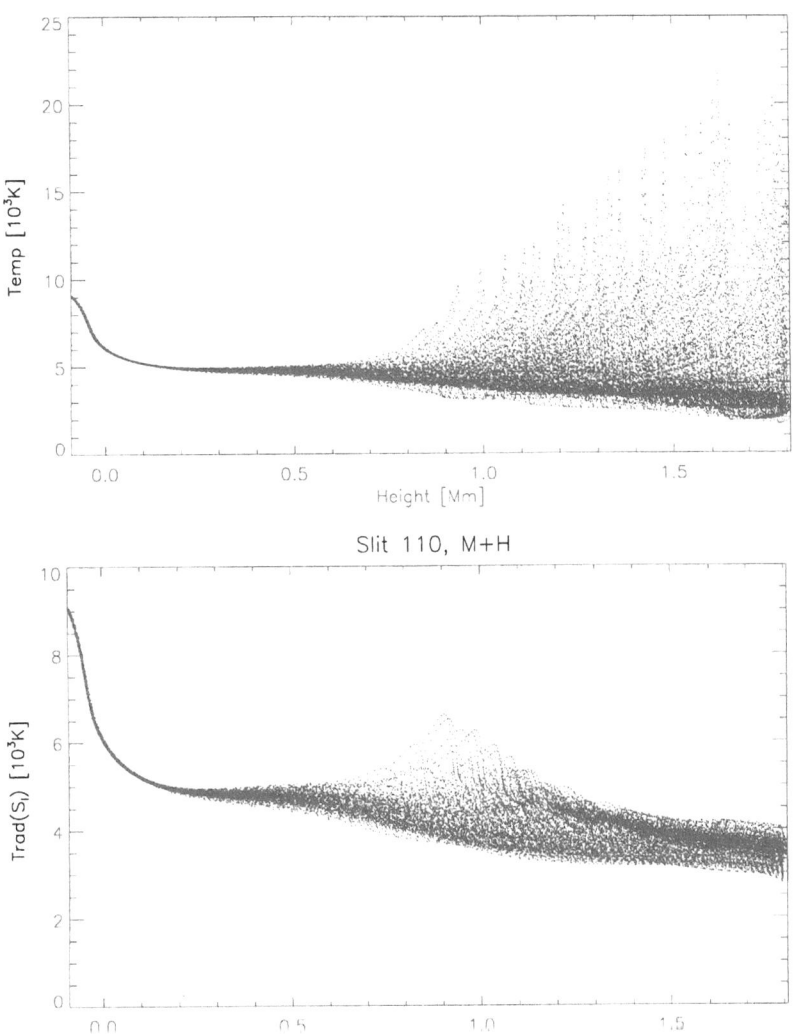

Slit 110, M+H

Fig. 5. Temperature (upper panel) and the radiation temperature of the source function $T_{rad}(S_1)$ (lower panel) as functions of height at all time-steps shown as scatter-plots. The decoupling of the source function from the Planck function at about 0.9 Mm is clearly seen. Note that the temperature scales are different in the two panels.

rise to higher optical depth, further decreasing the intensity on the red side.

What type of waves are responsible for the shocks that cause the grains? Explanations in the literature range from very high frequency waves (period

on the order of 30s) through the process of shock overtaking (Rammacher &
Ulmschneider 1992), three minute waves (Carlsson & Stein 1992), five minute
oscillations (Kalkofen et al. 1992) to impulse excitation (Kalkofen 1996). To
investigate the possible driving mechanism in the simulation we made a series
of simulations with the piston velocities filtered at different frequencies. The
results are shown in Fig. 6. The full piston spectrum run looks very similar
to the medium plus high frequency and the medium plus low frequency cases
after the initial startup. Studies of the wavelet transform of velocities and
the filtered velocities as function of height and time (see Carlsson & Stein
1997) confirm that it is the waves with frequencies slightly above the acoustic
cut-off frequency that produce most grains.

4   Discussion

The observations in the Ca II resonance lines clearly show the absence of a
quasi-static chromospheric temperature rise; most of the time in the internetwork *there is no emission* in the line core (Lites et al. 1993). Ca II is the
dominant ionization stage of calcium in a very wide temperature range under solar chromospheric conditions; in the dynamic simulations practically
all calcium is in the form of Ca II all the time even though the temperature may vary from 2500 K to 25 000 K. Therefore, the optical depth scale in
the resonance lines only depends on the column mass and is independent of
temperature fluctuations. Since the source function has a large contribution
from the local temperature, the absence of core emission implies there is no
temperature rise up to at least a height of 1 Mm most of the time.

The resonance lines of Mg II form slightly higher than those of Ca II due to
the larger magnesium abundance. Spatially resolved balloon observations of
these lines by Lemaire and Skumanich (1973) and Staath and Lemaire (1995)
show them to be in emission over the whole disk. SUMER observations of
emission lines from neutral species also show emission all the time (even
though the amount of emission varies greatly with time). We ascribe this
to the presence of a magnetic component. At higher altitudes the magnetic
fields can no longer be contained by the sharply decreasing plasma pressure
and this magnetic component will increase in importance for the total emission with increasing height. Spatially resolved CO observations (Solanki et al.
1994 and Uitenbroek et al. 1994) are compatible with the absence of a quasi-
static temperature rise in internetwork regions. The shocks in the dynamical
simulation form at about 1 Mm and will probably not be visible in CO lines
formed deeper in.

A comparison with continuum observations is not straightforward because line-blanketing is strong in the UV but is neglected in the dynamical
simulations. Furthermore, there is a lack of *spatially resolved* spectra with an
absolute calibration in this wavelength regime (see Brekke and Kjeldseth-Moe
1994). The time averaged spectrum from our dynamical simulations agrees

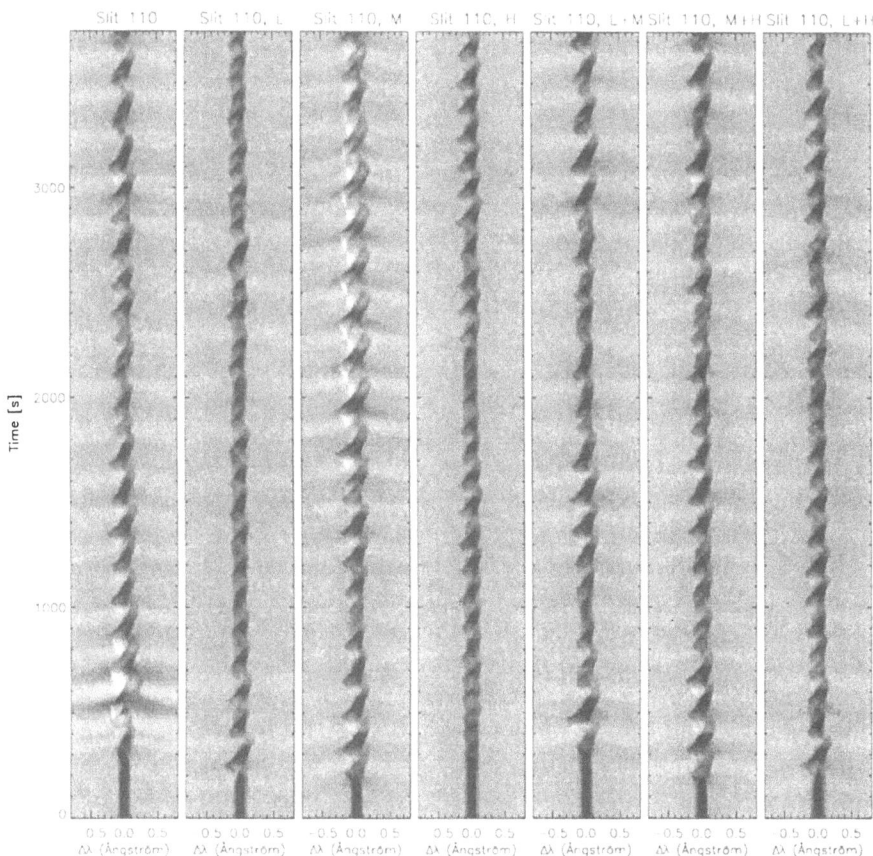

Fig. 6. Intensity as a function of wavelength displacement from line centre and time for 7 calculations with different frequency ranges of the piston
velocity, all for slit position 110. The full piston velocity spectrum case is
compared with those for only low frequencies ($\nu < 4.7$ mHz), medium frequencies ($4.7 < \nu < 7.1$ mHz), high frequencies ($\nu7.1$ mHz), low plus medium
($\nu < 7.1$ mHz), medium plus high ($\nu4.7$ mHz) and low plus high ($\nu < 4.7$ and
$\nu7.1$ mHz) respectively from left to right. The bright grain pattern is seen
as a brightening in the wings that shift toward the line core with advancing
time, followed by a brightening on the blue side of the line core. The full
piston spectrum run looks very similar to the medium plus high frequency
and the medium plus low frequency cases after initial startup.

rather well with the observations within these uncertainties. Time-resolved
images at 160 nm have been obtained with the HRTS instrument (Cook et al.
1983). These observations show a small variation in the radiation temperature with a peak-to-peak variation of up to 60–80 K, corresponding to a rms

variation of 21–28 K (Hoekzema 1994). The simulations show a rms variation of 39 K, a somewhat larger variation than in the observations.

5   Conclusions

There is no theoretical or observational evidence for a temperature rise in the magnetic field free internetwork lower chromosphere. The observed enhanced emission can be produced by temporally varying waves that generate short intervals of high temperatures, without any outward increase in average temperature. Because of the exponential dependence of the Planck function on temperature in the ultra-violet, these short intervals of high temperature dominate the time averaged intensity, even though decoupling of the source function from the Planck function tends to reduce this sensitivity. Hence, the radiation temperature represents the peaks in gas temperature rather than its mean value. The extra energy that is radiated away comes primarily from the energy dissipated by the wave motions, which goes directly into radiation without passing through a mediating state of enhanced mean thermal energy.

One should also be aware that significant differences exist between hydrostatic model atmospheres and the average state of a dynamic atmosphere. The presence of waves changes the mean state of the atmosphere, so that procedures that work well in the photosphere may fail badly in the chromosphere. The temperature spikes at shock fronts produce doubly-peaked intensity contribution functions with one peak at $\tau_\nu = 1$ and another at the shock. Thus the mean height of formation for lines and continua formed around 1 Mm can vary greatly with time and does not necessarily correspond to the actual layers emitting the photons. Therefore, static formation heights and contribution functions cannot be used for analyzing observations of chromospheric continua and lines from an inherently time-dependent atmosphere. When waves in the chromosphere have large amplitude, linear perturbation theory is not valid since the passage of waves changes the atmosphere fundamentally.

Spatially and temporally resolved observations of Ca II H & K lines in internetwork regions show no emission most of the time. Hence, there can be no general chromospheric temperature rise or emission would always be present. Also, high spatial resolution observations of CO emission off the limb show no evidence for a temperature rise in the low chromosphere.

Thus, despite long held beliefs, the Sun may not have a chromosphere in the internetwork regions, at least not one with an outward increasing temperature.

The bright grains are produced by shocks near 1 Mm above $\tau_{500} = 1$. Shocks in the mid chromosphere produce a large source function (and therefore high emissivity) because the density is high enough for collisions to couple the CaII populations to the local conditions. The asymmetry of the line profile is due to velocity gradients near 1 Mm. Material motion Doppler shifts the frequency where atoms emit and absorb photons, so the maximum opacity is located at – and the absorption profile is symmetric about – the local fluid velocity, which is shifted to the blue behind shocks. The optical depth depends on the velocity structure higher up. Shocks propagate generally into downflowing material, so there is little matter above to absorb the blue Doppler shifted radiation. The corresponding red peak is absent because of small opacity at the source function maximum and large optical depth due to overlying material. The brightness of the violet peak depends on the height of shock formation. The lower the shock, the higher the density and the larger the source function. The position in wavelength of the bright violet peak depends on the bulk velocity at the shock peak and the width of the atomic absorption profile (described with the microturbulence fudge parameter).

The bright grains are produced primarily by waves near and slightly above the acoustic cutoff frequency. The precise time and strength of a grain depends on the interference between these waves at the acoustic cutoff frequency and higher frequency waves. When waves near the acoustic cutoff frequency are weak, then higher frequency waves produce grains. The "five-minute" trapped p-mode oscillations are not the source of the grains, although they can modify the behaviour of higher frequency waves. The wave pattern that exists at the solar surface is due to the interference of many trapped and propagating modes, so that the grain pattern has a stochastic nature.

The grain pattern varies with the input velocity field and the grain appearance is very dependent on the shock strength and shock height. The simulations thus confirm the diagnostic potential of the Ca II H and K resonance lines. To make a proper interpretation of the observations it is, however, necessary to perform self-consistent radiation hydrodynamic simulations similar to the ones discussed here.

Acknowledgements

Lites and Rutten are thanked for making available their observations. This work was supported by a grant from the Norwegian Research Council and by grant NAGW-1695 from NASA. The computations were supported by a grant from the Norwegian Research Council, tungregneutvalget.

References

Bocchialini, K., Vial, J.-C., and Koutchmy, S. (1994): Dynamical properties of the chromosphere in and out of the solar magnetic network. ApJ **423**, L67–L70

Brekke, P. and Kjeldseth-Moe, O. (1994): New Radiometric Values of the Solar UV Continuum Radiation 1500–1700 Å. ApJ pp. L55–L58

Carlsson, M. and Stein, R. F. (1992): Non-LTE Radiating Acoustic Shocks and CaII K2V Bright Points. ApJ **397**, L59

Carlsson, M. and Stein, R. F. (1994): Non-LTE Radiation Shock Dynamics in the Solar Chromosphere., Proc. Mini-Workshop on Chromospheric Dynamics, Institute of Theoretical Astrophysics, Oslo, 47–77

Carlsson, M. and Stein, R. F. (1995): Does a Nonmagnetic Solar Chromosphere Exist?. ApJ **440**, L29–L32

Carlsson, M. and Stein, R. F. (1997): Formation of Calcium H and K Bright Grains. ApJ (submitted)

Cook, J. W., Brueckner, G. E., and Bartoe, J.-D. F. (1983): High-resolution telescope and spectrograph observations of solar fine structure in the 1600 Å region. ApJ **270**, L89–L93

Fontenla, J. M., Avrett, E. H., and Loeser, R. (1993): ApJ **406**, 319–345

Gustafsson, B. (1973): Uppsala Astr. Obs. Ann. **5**, No. 6

Harvey, J., Jefferies, S., Pomerantz, M., and Duvall, T., J. (1992): Global Observations of Chromospheric Oscillations. BAAS **180**, 1705

Hoekzema, N. (1994): On CI Jets and 160 nm Internetwork Bright-points., Proc. Mini-Workshop on Chromospheric Dynamics, Institute of Theoretical Astrophysics, Oslo, 111–120

Hofmann, J., Steffens, S., and Deubner, F. L. (1996): K-grains as a three-dimensional phenomenon. II. Phase analysis of the spatio-temporal pattern. A&A **308**, 192–198

Judge, P. G., Hubeny, V., and Brown, J. C. (1997): Fundamental Limitations of Emission Line Spectra as Diagnostics of Plasma Temperature and Density Structure. ApJ (in press)

Kalkofen, W. (1996): Chromospheric Oscillations in K 2v Bright Points. ApJ **468**, L69–L72

Kalkofen, W., Rossi, P., Bodo, G. and Massaglia, M. (1992): The 3 min Oscillations in Chromospheric Bright Points. in M.S.Giampapa, J.A. Bookbinder (eds.), Seventh Cambridge Workshop on Cool Stars, Stellar Systems and the Sun, Astron. Soc. Pac. Conf. Series, **26**, 543–545

Lemaire, P. and Skumanich, A. (1973): A&A **22**, 61

Lites, B. W., Rutten, R. J., and Kalkofen, W. (1993): Dynamics of the Solar Chromosphere. I. Long-Period Network Oscillations. ApJ **414**, 345–356

Rammacher, W., Ulmschneider, P. (1992): Acoustic waves in the solar atmosphere IX. Three minute pulsations driven by shock overtaking. A&A **253**, 586–600

Rutten, R. J. and Uitenbroek, H. (1991): Ca II H_{2V} and K_{2V} Cell Grains. Solar Phys. **134**, 15–71

Solanki, S. K., Livingston, W., and Ayres, T. (1994): New Light on the Heart of Darkness of the Solar Chromosphere. Science **263**, 64–66

Staath, E. and Lemaire, P. (1995): High resolution profiles of the Mg II h and Mg II k lines. A&A **295**, 517

Steffens, S., Hofmann, J., and Deubner, F. L. (1996): K-grains as a three-dimensional phenomenon. I. Statistics and spatial evolution. A&A **307**, 288

Uitenbroek, H., Noyes, R. W., and Rabin, D. (1994): Imaging spectroscopy of the solar CO lines at 4.67 μm. ApJ **432**, L67–L70

Vernazza, J. E., Avrett, E. H., and Loeser, R. (1981): Structure of the Chromosphere. III. Models of the EUV Brightness Components of the Quiet Sun. ApJS **45**, 635–725

Von Uexkuell, M. and Kneer, F. (1995): Oscillations of the Sun's chromosphere. VII. K grains revisited.. A&A **294**, 252–259

Topologically Forced Reconnection

Åke Nordlund[1] and Klaus Galsgaard[2]

[1] Theoretical Astrophysics Center & Niels Bohr Institute for Astronomy, Physics, and Geophysics, Juliane Maries Vej 30, 2100 Copenhagen Ø, Denmark
[2] University of St. Andrews, The Department of Mathematics, St Andrews, FIFE, KY16 9SS, Scotland

Abstract. A magnetically dominated plasma driven by braiding motions on boundaries at which magnetic field lines are anchored is forced to dissipate the work being done upon it, no matter how small the electrical resistivity may be. Recent numerical experiments have clarified the mechanisms through which balance between the boundary work and the dissipation in the interior is achieved. The results largely confirm Parker's (1972) idea of "topological dissipation"; dissipation is achieved through the formation of a hierarchy of electrical current sheets. Current sheets form as a result of the topological interlocking of individual strands of magnetic field. The average level of dissipation is well described by a scaling law that is independent of the electrical resistivity.

1 Introduction

Magnetic fields are ubiquitous in astrophysical objects; circumstances where there is no magnetic field present are exceptions. Indeed, much of the non-thermal activity that is observed in astrophysical systems is probably related to the presence of magnetic fields.

The Sun provides a convenient, nearby laboratory, where a range of processes may be observed with a spatial and time resolution sufficient to challenge our theoretical ideas. In particular, the Sun presents us with a duality that must be generic to many equivalent circumstances in the Universe; on the one hand a dynamo process that generates a magnetic field in the high-density region below the surface of the Sun, and on the other hand the low-density corona above the surface, where magnetic dissipation, even if it is only a fraction of the total magnetic dissipation, may completely dominate the local energetics.

Gravity is not only the cause of the mass density stratification, it is also the source (or at least the mediator) of mechanical energy input; in the Sun as well as in accretion disks and jets, the motions involved (e.g., differential rotation and convection) are intimately related to gravitation. In addition, gravity causes the buoyancy that tends to bring magnetic fields up from the high-beta dynamo region to the low-beta dissipation region, thus creating and maintaining topological connections between the high- and low-beta regions. It is through these topological connections, that energy is transferred from the dense region to the low-beta region.

Thus, even though the magnetic field itself is unaffected by gravity, gravity plays a crucial role here. Analogous arguments may be made for accretion disks and for jets related to active galactic nuclei; a magnetic field that pervades the system, be it in a very turbulent or a more systematic fashion, is able to act as an agent that allows energy to be extracted from dense regions and to be deposited in less dense regions.

Both in connection with the generation of magnetic fields by astrophysical dynamos, the diffusion of magnetic fields by a turbulent medium, and the dissipation of magnetic energy in a magnetically dominated and highly conducting plasma, one faces theoretical problems related to the extremely large magnetic Reynolds numbers R_m that characterize astrophysical plasmas ($R_m = UL/\eta$, where U, L, and η are characteristic magnitudes of velocity, length, and electric resistivity).

The issues may appear to be: 1) Does dynamo efficiency and magnetic diffusivity dwindle in the limit of large R_m, or do "fast dynamos" exist? 2) What is the asymptotic behavior of magnetic dissipation in the limit of large R_m; i.e., does "fast reconnection" exist? Astrophysical reality shows us, however, that instead of asking "does it work?" we must ask "why and how does it work"; for without turbulent diffusion of magnetic fields, the time scale of the solar dynamo could not be as short as 22 years, without a fast dynamo process, the solar dynamo could not keep up with the diffusion, and without fast reconnection, the large scale connectivity of the solar corona could never readjust as fast as it does.

Numerical experiments can provide further clues to the answers to these questions, but here we face a complementary problem; how can we say anything about the asymptotic behavior for large R_m from numerical experiments that are limited to much smaller values than those that characterize the real world systems? In the subsequent sections, these questions are further discussed, in the light of recent numerical 3-D experiments with boundary driven magnetic dissipation (Galsgaard and Nordlund 1994, 1996, 1997ab).

2   Magnetic Dissipation

The magnetic dissipation problem may be paraphrased as

Given some circumstance with free magnetic energy E_B, characteristic integral scales L, velocities U, and an oh-so-large Reynolds numbers $R_m = UL/\eta$, how does nature go about burning the free energy at a rate $Q_J \sim E_B/\tau = E_B U/L$, or some fair fraction thereof, rather than at a ridiculously small rate $Q_\eta \sim E_B \eta/L^2 \sim Q_J/R_m$, or $Q_J/R_m^{\frac{1}{2}}$ or so?

The problem of "fast reconnection" in a single current sheet is a particular incarnation of the generic magnetic dissipation problem: How does the reconnection in a single current sheet scale with increasing Lundquist number

$N_L = V_A L/\eta$ (V_A is the Alfvén speed); are there geometries that allow a reconnection rate that does not decrease as some power of the Lundquist number N_L^{-p} (cf. Parker 1979, Priest and Forbes 1992, Parker 1994, Biskamp 1993)?

As emphasized by Priest and Forbes, the reconnection problem is intimately related to boundary conditions. Monolithic current sheets with vastly different reconnection rates may be set up; each corresponds to a particular combination of boundary and initial conditions.

2.1 Boundary driven magnetic dissipation

One may choose to approach the problem from the opposite side, and study what type of current sheets that develop with a given set of boundary conditions. Such an approach is indeed closely related to the astrophysical problems that motivate the interest in reconnection in the first place. The solar corona, for example, may be regarded as a semi-infinite low-beta region that is constantly being worked upon by braiding motions at a spherical boundary.

Braiding of the magnetic field causes stretching of the field lines, and hence an increase of the free magnetic energy. The increase of the free magnetic energy is closely related to increasing topological complexity (Berger 1993, Ricca and Berger 1996). Continued braiding would increase the free magnetic energy indefinitely, if reconnection of magnetic field lines did not occur. The corona above an accretion disc is in a similar situation; there, both the differential rotation and the small scale turbulence of the disc provide a continuous input of work that must be dissipated (Brandenburg et al. 1995).

Regardless of the particular geometry involved, these are examples of the general "problem of boundary driven magnetic dissipation"; how does a low-beta, high R_m plasma that is being worked upon by a boundary manage to dissipate the work? A large number of papers have addressed this, or similar problems, both from the analytical and the numerical point of view (Parker 1972, 1975, 1982, 1983abc, 1987, 1988, 1991; Sturrock and Uchida 1981; Heyvaerts and Priest 1984, 1992; van Ballegooijen 1985, 1986, 1988; Mikić et al 1989; Berger 1991; Strauss 1991; Gómez and Fontán 1992; Longcope and Sudan 1994; Einaudi & Velli, these proceedings).

Galsgaard & Nordlund (1996, 1997ab) used a 3-D MHD code that includes artificial viscous diffusivity and magnetic resistivity (Stein et al. 1994, Nordlund and Galsgaard 1996) to perform numerical simulations that cast new light on the problem.

3 Resistive 3-D MHD Experiments

The simplest experiments model a magnetically dominated (low-beta), highly conducting ($N_L \gg 1$) plasma, with periodic y,z boundaries and "perfectly conducting" rigid x-boundaries with specified motions, e.g., sinusoidal shear with randomly changing direction and phase, acting on an initial magnetic field with straight field lines, and with numerical resolutions ranging from 24^3 to 136^3. (Galsgaard and Nordlund 1996).

The main results of these experiments are:

1. Ubiquitous formation of current sheets within a few correlation times of the boundary motions.
2. The current sheets form a hierarchical structure; "smaller current sheets ride piggy-back on larger ones..."
3. On the average, the dissipation soon reaches equilibrium with the boundary work, but both the instantaneous dissipation and the instantaneous boundary work fluctuate significantly.
4. The level of work and dissipation at which the equilibrium is obtained does not depend noticeably on the resistivity.

Items 1 and 2 are illustrated in Figs. 1–3, showing isosurfaces of the electric current density in a snapshot from one of the experiments with the highest numerical resolution. The magnetic field line traces in Fig. 2 and Fig. 3 illustrate the discontinuities in magnetic field direction across the current sheets. Note the small, twisted current sheet (Fig. 2 and Fig. 3 center) that protrudes from the bottom current sheet and bridges over to another current sheet (to the left in Fig. 3). Magnetic field lines to the left on Fig. 3 pass under the "bridge", whereas magnetic field lines from the right pass over it.

The isosurface representation is able to nicely bring out the three-dimensional structure of the current sheets, but has the drawback that it only shows one particular level of the current density. In order to show weaker current sheets, the isosurface level must be lowered, but that quickly leads to cluttering of the surfaces, since the entire model is filled with a hierarchy of current sheets of varying strength. These weaker current sheets are better made visible in two-dimensional slices of the current density. Figure 4 shows such a slice, halfway between the two driving boundaries. By letting the brightness be proportional to the current density raised to some power smaller than one, one may enhance the visibility of the weaker current sheets.

One particularly remarkable result from these experiments is a demonstration of that a single, stationary boundary shearing pattern can lead to the formation of a current sheet in a finite time, as long as the resistivity is finite.

Figure 5 shows the evolution of the electrical current density in an experiment with such a stationary pattern of boundary motions; orthogonal, sinusoidal shears on the two boundaries. In the first few frames, the peak current density is increasing exponentially, and the half width of the current concentration is decreasing exponentially, until it becomes comparable to the grid size, and a numerical current sheet forms. Reconnection of field lines in the current sheet causes jets (or "jet sheets") to form along the edges of the current sheet. The jets blow "bubbles" in the surrounding medium (cf. panels

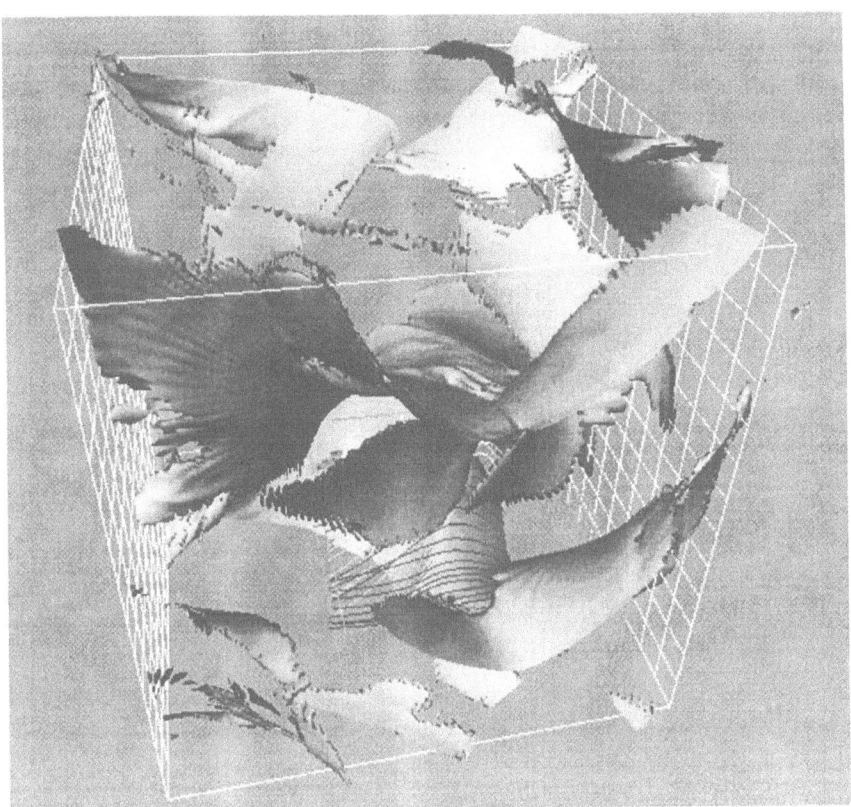

Fig. 1. Isosurfaces of electric current density, from an experiment with boundary driven magnetic dissipation (this figure is available in color at URL http://www.astro.ku.dk/~aake/papers/spm96). The driving boundaries are to the left and right. The aspect ratio is 1:1:1 and the numerical resolution is 136^3. The small box near the bottom is shown in more detail in Fig. 2.

5–7 in Fig. 5), and these subsequently interfere with the periodic images of the current sheet (panels 8–10).

The formation of the current sheet and the subsequent dynamic activity are necessitated by the formation of pairwise interlocking flux bundles; as the boundary motion proceeds, the two flux bundles in a pair are pressed against each other with steadily increasing force, and reconnection is necessary to release the interlocking.

3.1   Interlocking flux bundles

Figure 6 illustrates how the interlocking of a pair of flux bundles leads to the formation of a current sheet. The left panel shows an initially sheared

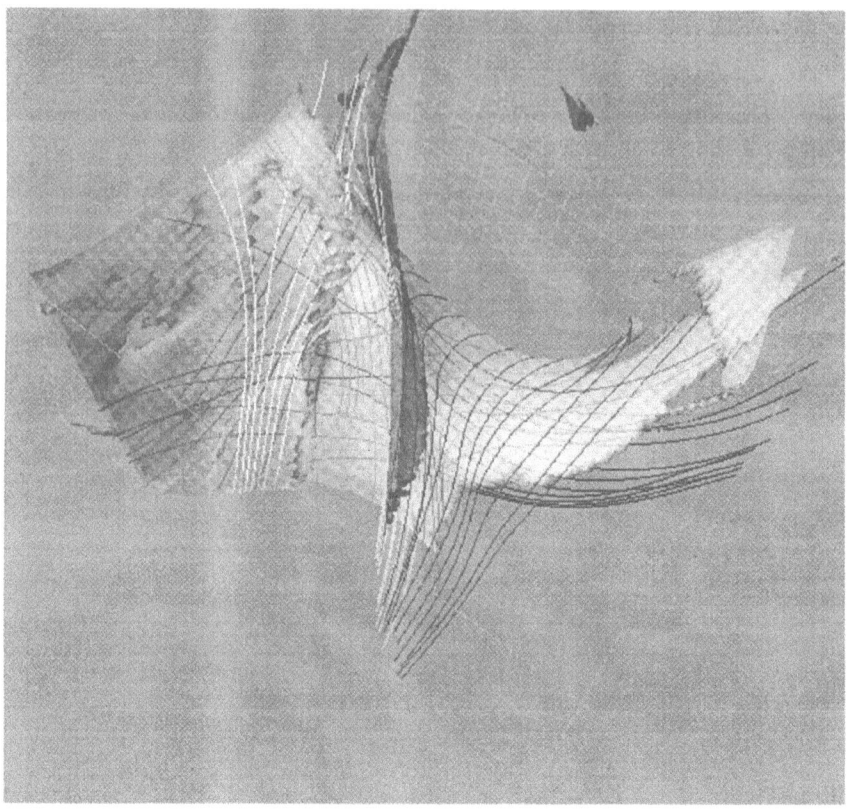

Fig. 2. A blow-up of the small box at the bottom of Fig. 1. Current sheets are shown as transparent isosurfaces, with magnetic field lines traced out on both sides of the current sheets. On the color version of this figure (see http://www.astro.ku.dk/~aake/papers/spm96), each color indicates an independent magnetic flux system, with a unique magnetic connectivity.

configuration. A subsequent orthogonal shear leads to interlocking of field lines, and the creation of a current concentration in the region between the two flux bundles (middle panel).

The right-most panel shows field lines near the current concentration, in projection onto a plane half-way between the boundaries. A force balance is maintained, where the tension force from the interlocked flux strands pushes inwards in the "entry region", towards the line of symmetry, and is balanced by a slight excess of magnetic pressure there. The excess magnetic pressure pushes field lines out along the line of symmetry, where the gradient of the magnetic pressure (weaker than in the perpendicular direction) is balanced by the tension from the weakly curved field lines in the "exit region".

For purposes of illustration, this example shows interlocking as a conse-

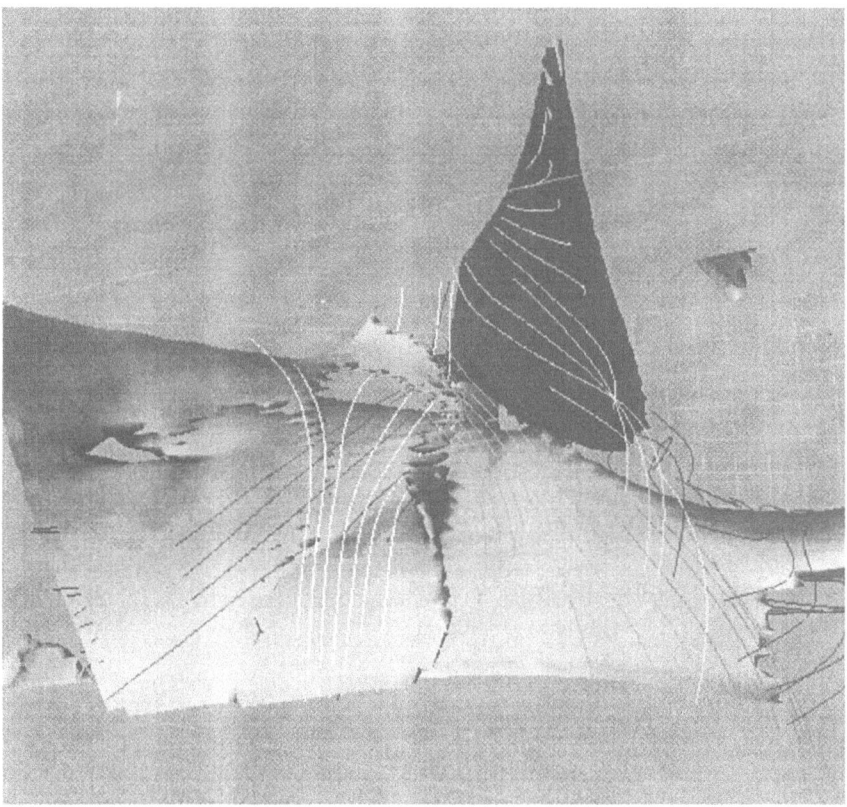

Fig. 3. A view from the lower left of the region shown in Fig. 2 (in color at URL http://www.astro.ku.dk/~aake/papers/spm96).

quence of two subsequent, orthogonal sinusoidal shear patterns, but many other simple, even stationary, boundary motion patterns (e.g., the orthogonal sinusoidal shears mentioned earlier) lead to the formation of interlocked flux strands in the same way.

3.2   Exponential growth of the current density

The exponential growth of the current density in-between the flux strands that was illustrated in Fig. 5 is a generic phenomenon that must occur whenever tension is increased on such a pair of interlocked flux bundles. The cause is the stagnation point type motion pattern (indicated by he dashed flowlines in the right-most panel of Fig. 6) that is induced as the flux bundles are pressed together.

The two interlocked strands of magnetic field become increasingly curved as the boundary motions proceed, and field lines from the two strands are

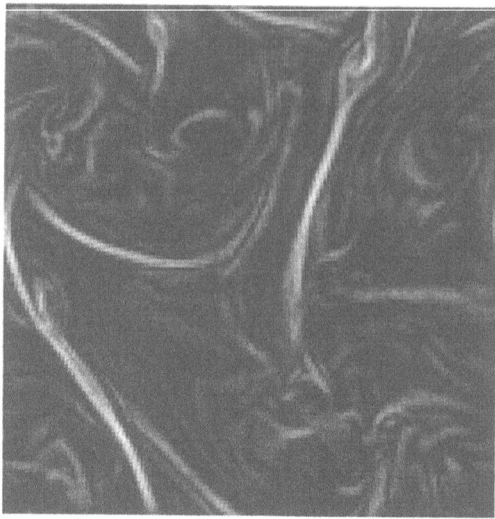

Fig. 4. Electric current density in a cross section halfway between the boundaries, for a snapshot from a run with resolution 136^3.

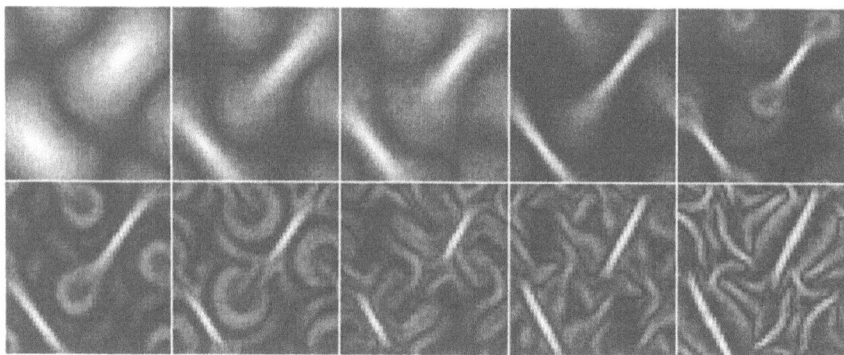

Fig. 5. Time evolution of the electric current density in a cross section half way between two boundaries with orthogonal, sinusoidal, stationary shearing patterns. Numerical resolution 48^3. The brightness is proportional to the electrical current density and is normalized separately for each frame.

pushed closer together. However, as long as the longitudinal component of the magnetic field dominates, its flux density must remain nearly constant. Thus, in a plane perpendicular to the average field orientation, the motion is approximately a stagnation point flow, with the longitudinal field component acting as the "incompressible" medium.

The result is that the distance between any two field lines in the in-

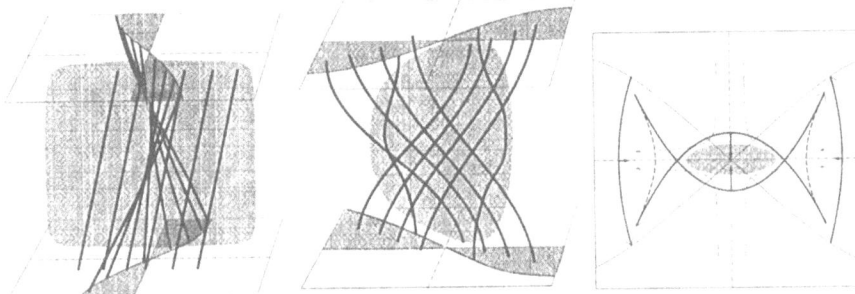

Fig. 6. Schematic illustration of flux interlocking (Galsgaard and Nordlund 1996). The thick lines represent magnetic field lines connecting the two boundaries, the shaded boundary areas show the displacements on the boundaries, and the shaded surfaces between the boundaries indicate the location of the strongest current. The panels illustrate the evolution of the magnetic field from a simple, sheared configuration on the left to a pair of interlocked flux bundles (middle and right)

terlocked flux bundles decreases exponentially with time, if the boundary motions proceed with constant speed. If the boundary motion stops, so does the stagnation point flow (apart from the wave motions that are induced by the deceleration). Thus, it is possible for the system to evolve, at least for a while, through a sequence of quasistatic equilibria, characterized by exponentially decreasing distances between field lines, and hence exponentially increasing current densities in the region between the two flux strands. The total electric current per unit width of the current layer is given by the angle between the magnetic field lines in the two strands, and hence does not vary much (depending on details of the boundary motion pattern the angle may increase with time or may remain more or less constant).

We conclude that boundary motion patterns that result in a topological interlocking of two flux strands cause the formation of a current layer with exponentially growing peak current density in-between the two flux strands. The integrated Joule dissipation of the current layer – being proportional to the current density squared times the thickness – also grows exponentially. The work performed by the boundary motions, on the other hand, only grows approximately linearly with the boundary displacement.

Even if the Joule dissipation is initially very weak, it must soon become comparable to the boundary work, after a time that depends only logarithmically on the resistivity. On the other hand, before that time the Joule dissipation is indeed ignorable, and the situation is quasi-static, with a rate of dissipation that allows the magnetic energy to survive for a very long time. Thus, there is in practice something similar to an off/on behavior of the Joule dissipation, even if in principle it is just a difference in time scales.

3.3   Current sheets – real and numerical

The previous discussion is true for a numerical experiment, as well as for a physical current layer. In a properly designed numerical experiment, the exponential growth of the current density is arrested when the thickness of the current sheet is a few grid zones. Thus the exponential growth leads to numerical discontinuity – a jump that is just marginally resolved.

For a finite conductivity (numerical or real), there is no distinction between an "ideal" discontinuity and the end result of the exponential growth. Therefore, at least from the practical point of view, the discussion about the existence or not of finite time discontinuities (van Ballegooijen 1985, 1986; Parker 1994) is perhaps somewhat esoteric.

3.4   Transition to a dynamic state

The increasing Joule dissipation corresponds to an increasing rate of cancellation of the anti-parallel components of the two flux strands near the line of symmetry. When the level of dissipation reaches and exceeds the level of boundary work, the cancellation begins to have dramatic consequences. The cancellation is then able to remove the anti-parallel components of the magnetic field faster than they are being built up by the boundary motions.

Since no fieldlines are destroyed (cf. the line tying and smooth motions at the boundaries), the cancellation of the anti-parallel components along the current sheet just corresponds to a straightening out of the field lines; i.e., reconnection of the field lines from opposite sides of the current sheet. The straightening-out is made possible when reconnection in the current sheet allows field lines to slip out of the topological constraint of the two interlocked flux strands. Instead of being connected to points that are moving increasingly far apart on the two boundaries, after reconnection in the current sheet field lines become connected to foot points that are more near-by on the two boundaries, and the field lines are free to straighten out (cf. Fig. 7).

With no magnetic forces opposing the straightening out, the dynamics of the reconnected field lines is basically that of a sling shot; a balance between the tension force and the inertia of the fluid "attached to" the field lines. The fluid on the reconnecting field lines is thus accelerated to speeds of the order of the Alfvén speed times the initial angle of bending of the field lines.

Thus, as a consequence of – and an integral part of – the reconnection in the current sheet, a jet sheet forms along the edges of the current sheet (Fig. 8). In projection onto a plane, such as that in Fig. 5, the formation of the jet sheet is indirectly visible also in images of the current density, through the bubble-like features that form as the jets hit the ambient medium. It is obvious from Fig. 5, that once the first current sheet forms, the system enters a very dynamic state; the growing bubbles soon interfere with the current sheets, and within a time interval comparable to that of the formation of the

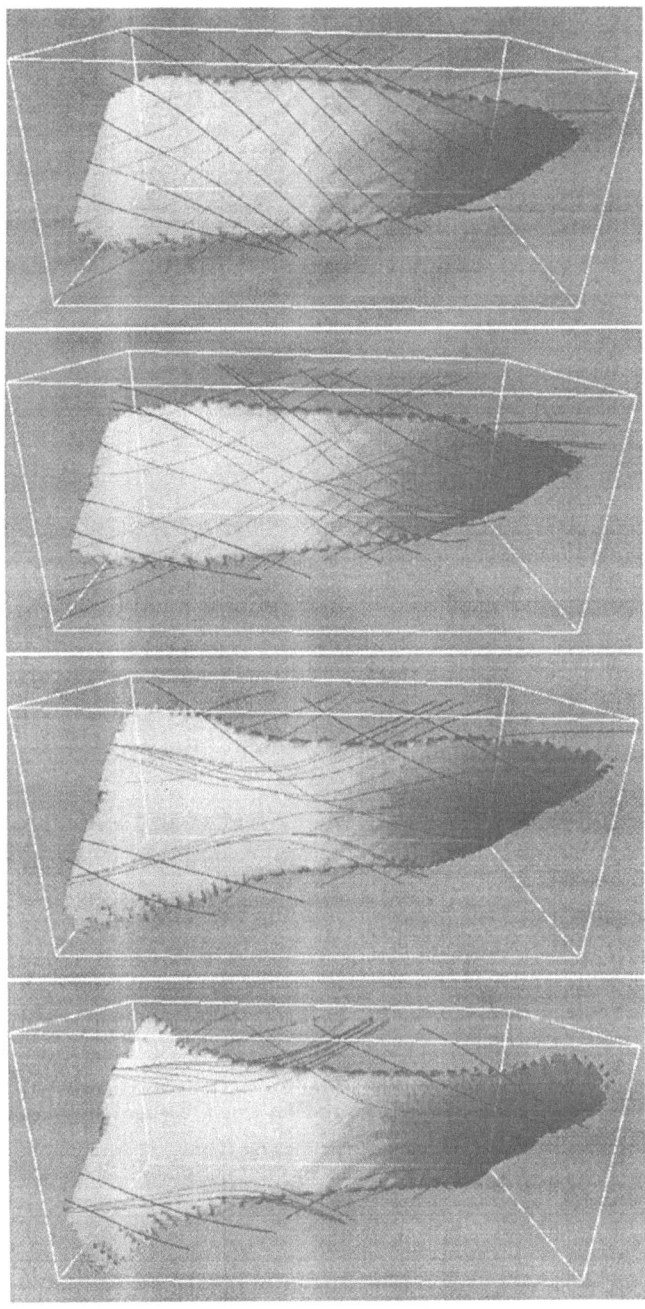

Fig. 7. The evolution of reconnecting magnetic field lines near a current sheet (in color at URL http://www.astro.ku.dk/~aake/papers/spm96).

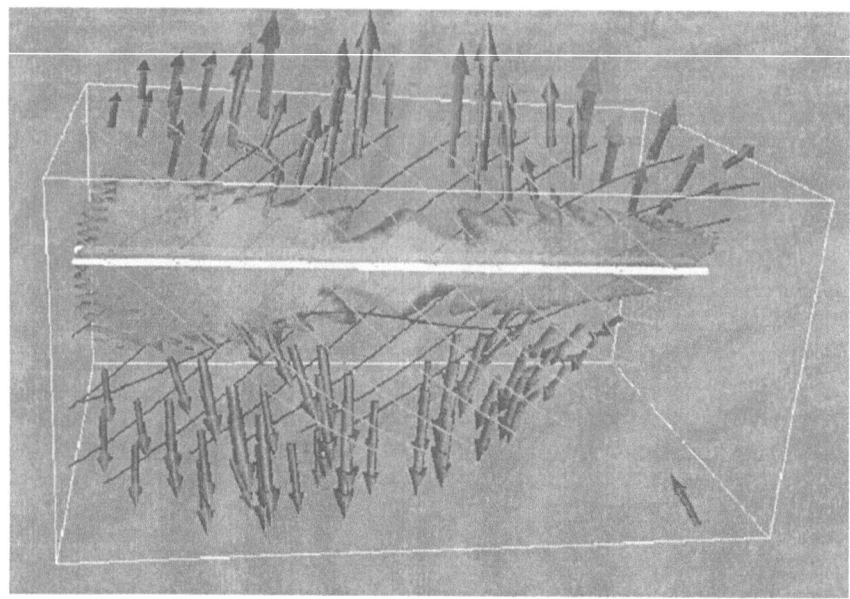

Fig. 8. Reconnecting current sheet with jet sheets extending from the edges (in color at URL http://www.astro.ku.dk/~aake/papers/spm96).

first current sheet, the system has become a hierarchy of interacting current sheets (cf. the last panel of Fig. 5).

Considering for a moment the energy budget of these events, we may identify a number of energy conversion processes. 1) The Joule dissipation associated with the current sheet corresponds to magnetic energy transformed directly into heat. 2) The sling shot force that accelerates the fluid in the jet sheets corresponds to a conversion of magnetic energy into kinetic energy. 3) Some of the kinetic energy is dissipated by viscous forces directly into heat. 4) Part of the kinetic energy is absorbed into magnetic energy as the jet sheet blows a bubble in the ambient medium. 5) The boundaries of the bubbles are sites of enhanced magnetic dissipation, where magnetic energy is again converted into heat. 6) Magneto-sonic waves and Alfvén waves are launched as the bubble is blown, and carry energy away from the source region. 7) The wave energy is also eventually dissipated into heat.

Estimates of the energy conversion rates in the numerical experiments (Galsgaard and Nordlund 1996) show that the order of significance is approximately the same as the numbering of items in the list above; the major part of the energy conversion is direct Joule dissipation, with only a smaller fraction appearing as viscous dissipation (and an even smaller fraction appearing as dissipation of wave energy).

The main dissipation channel thus appears to be the direct Joule dissipation in each current sheet. However, the secondary dissipation that appears because of the work generated by the jets emanating from each sheet is very significant from another point of view; it goes into messing up the surrounding field, and thus contributes to creating a hierarchy of scales.

3.5   The statistically steady state

In any closed system where work is being performed by the boundaries, a balance must eventually be obtained between the boundary work and the dissipation in the interior of the domain. The balance need not be detailed in either space or time; but on the average the volume integral of the dissipation $\int Q dV$ must equal the surface integral of the boundary work $\int W dA$.

When line-tying is efficient, the rate of work per unit boundary area is equal to the Poynting flux

$$W_P = (\mathbf{E} \times \mathbf{B})_\perp = -((\mathbf{u} \times \mathbf{B}) \times \mathbf{B})_\perp = -U B_\mathbf{u} B_\perp \ , \tag{1}$$

where $B_\mathbf{u}$ is the component of $\mathbf{B}$ parallel to the velocity $\mathbf{u}$ and $B_\perp$ is the component of $\mathbf{B}$ perpendicular to the boundary.

Figure 9 displays an example of the volume averaged dissipation and the boundary averaged work as a function of time. (Note that the staircase-like steps visible in the Poynting flux are due to the bouncing of Alfvén-waves between the boundaries. These Alfvén-waves are initiated by the initial acceleration of the boundary motions and are not particularly significant; a smoother start reduces their amplitude without significantly changing the evolution of the current sheets.)

The hierarchy of interacting current sheets corresponds to a spatially very intermittent distribution of Joule dissipation (as well as viscous dissipation). Even after averaging over the whole volume, both the dissipation and the boundary work are quite intermittent in time. Some of the changes in the average boundary work are directly related to the random changes in the boundary motion patters (the times of change are indicated along the time axes in the panels). Others are related to the relaxation of field line tension associated with major reconnection events. Major reconnection events are visible as peaks in the average Joule dissipation, while many smaller ones are not individually distinguishable in the volume average.

3.6   Field line mappings

What factors maintain the hierarchy of current sheets? To understand this, we need to look at how field lines map through the domain. Figure 10 illustrates the mapping from one boundary to the other. In the initial phases of the first boundary shearing motion, the mapping is still simple. It quickly becomes chaotic, however, because of both repeated foldings of the boundary motion patterns, and reconnection of field lines in the interior ("bakers

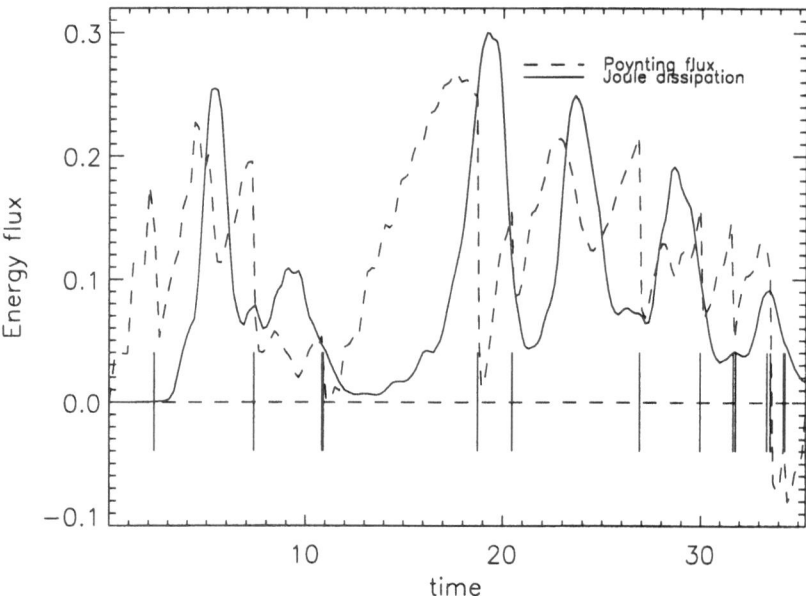

Fig. 9. The boundary averaged Poynting flux and the volume averaged Joule dissipation for one of the experiments of Galsgaard & Nordlund (1996). The vertical lines indicate the times when the boundary shearing patterns were changed. Time is measured in units of Alfvén travel times.

transformations" and current sheet reconnections). In the quasi-stationary asymptotic state, a field line visits the neighborhood of many current sheets, on the way from one boundary to the other. Each current sheet neighborhood consists of twisted and interlocked strands of magnetic flux, and corresponds to a strongly non-linear local mapping of field lines. As a result, neighboring field lines at one boundary may connect to widely different locations on the other boundary.

Now consider what happens to the local current sheets when large scale boundary shearing motions are applied. Flux strands that are already intertwined are tightened, and the general level of Maxwell stress tends to increase, over a range of scales, down to the smallest scales resolved. This provides driving of the current sheets at all scales, similar to the driving of the single current sheet that was discussed above.

In this hierarchical situation, if one focuses attention on a small current sheet somewhere inside the experiment, the "boundary conditions", or "boundary motions" that control its evolution are the patterns of stress from scales slightly larger than that current sheet; as the boundary motion proceeds, a tightening of the "mess" on smaller scales occurs, as long as the flux

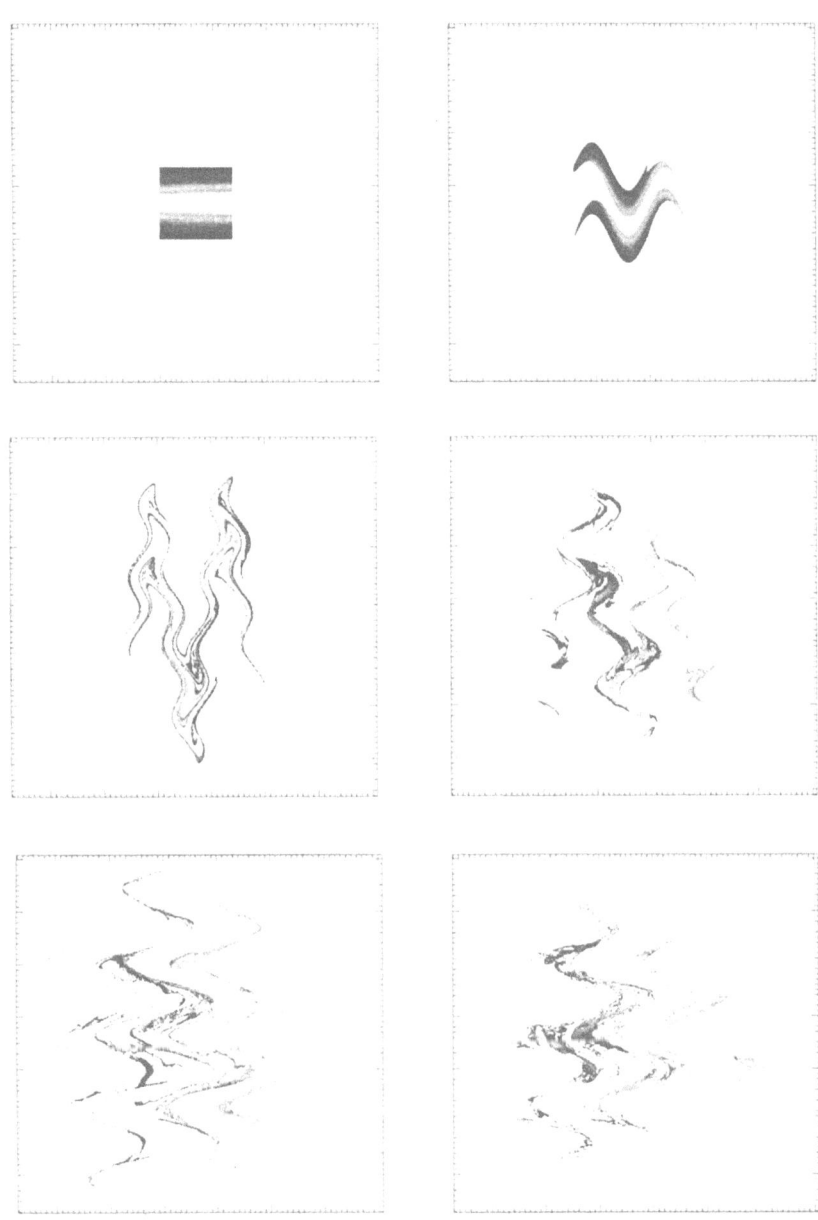

Fig. 10. Magnetic field line mapping patterns. The panels show the evolution of the mappings of a square pattern of field lines from one boundary to the other (the pattern is color keyed in the WWW version of this figure). The linear size of the area shown is five times that of the periodic box (that coincides with the square in the first panel).

bundles on various scales are interlocked. The stress that is taken up by an intermediate scale current sheet also partly goes into the smaller scale current sheets that are attached to it. Hence, if smaller scale current sheets become dynamically active, then the work that it takes to drive them is taken from the free energy of the larger scale configurations.

Conversely, the existence of smaller scale structure prevents the evolution of a larger scale current sheet from proceeding all the way to a single, smooth and thin current sheet before it becomes dynamically active. Thus, when a hierarchy of scales is present, one should not expect even the logarithmic dependence on resistivity that may be deduced for simple, monolithic, current sheets: all we get with more numerical resolution is more detail. If we perform exactly the same shearing motions as the ones that created the initial, solitary current sheet, but on top of the statistically steady state, with its messy mapping, then the work goes into the hierarchy of smaller scales *before* we get to the logarithmic difference.

What local factor determines if and when a current sheet forms? Galsgaard and Nordlund (1996) argue that it is the *the local winding number*; the number of times two neighboring field lines wrap around each other.

3.7   The Local Winding Number

Consider the first, archetype, current sheet formation that was discussed above. If the y and z components of B are scaled down, nothing changes in the relative force balance between the entrance and the exit regions of the stagnation point type motions. Such scalings corresponds to changing the aspect ratio of (numerical or thought) experiments; the angles field lines are making at the boundaries change, but the wrapping patterns are maintained.

Since it is the force balance between the entrance and exit regions that determine when a current sheet forms, we conclude that the time scales for formation of current sheets are invariant with respect to such aspect ratio type scalings, and that it is the wrapping patterns, rather than the actual angles, that are significant.

The wrapping patterns may indeed be characterized by the local winding numbers; each field line at the boundary may be labeled by its local winding number, to create a pattern (or image – cf. Fig. 11). According to the considerations for a single pair of interlocked flux bundles, the current sheet forms when the the flux bundles have been wound around each other about once. For particularly symmetric situations, such as for example a cylindrical flux tube, the winding must be somewhat higher before a kink develops, and flux sheets form (Galsgaard and Nordlund 1997a), but such symmetries only occur in oversimplified and construed setups.

The results from the numerical experiments are consistent with the discussion above; histograms of the winding number (cf. Fig. 11) show that they mostly are of the order of unity or less, and that they only rarely exceed two. The winding numbers reach these values also for cases with small aspect ratios

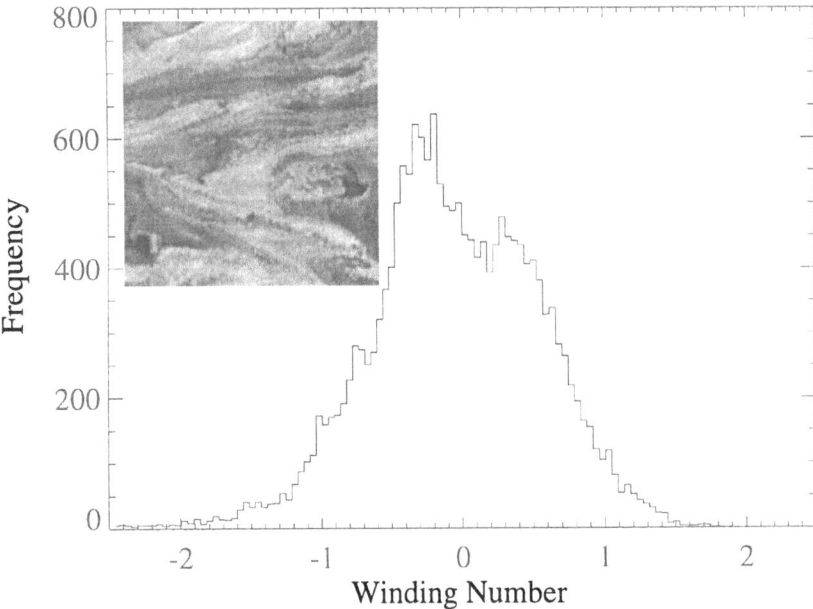

Fig. 11. A histogram of the local winding number, for a snapshot from one of the Galsgaard & Nordlund (1996) experiments. The corresponding image is shown as an inset, with the full range of variation of the local winding number mapped to a gray scale (color version at URL http://www.astro.ku.dk/~aake/papers/spm96). For each point in the image, the local winding number is calculated as the number of windings performed by two neighboring field lines (initial distance = $10^{-3}\Delta x$) as they pass from one boundary to the other, averaged over the initial angle between the two points.

$V_d t_d/L$ ($V_d t_d$ is the "stroke length", or correlation length, of the boundary motions, and L is the length of the flux tube).

So, it is not the angle $\mathrm{atan}(B_{y,z}/B_x)$ itself that decides when the magnetic dissipation Q_J "turns on", but the local winding number. Figure 12 illustrates this result more directly; current sheets form independent of the aspect ratios $V_d t_d/L$ ($V_d t_d$, but the angles that the current sheet structures make relative to the boundary increase with increasing stroke lengths of the boundary motions.

Using an argument by Parker (1988), a remarkable conclusion may be drawn from these results: Parker pointed out, that if there was any dependence of the average dissipation upon the resistivity, then the dependence had to be such that the average dissipation *increases* with decreasing resistivity, because a reduced dissipation level at given angles implies that the angles have to increase in order for the dissipation to match the boundary work. In

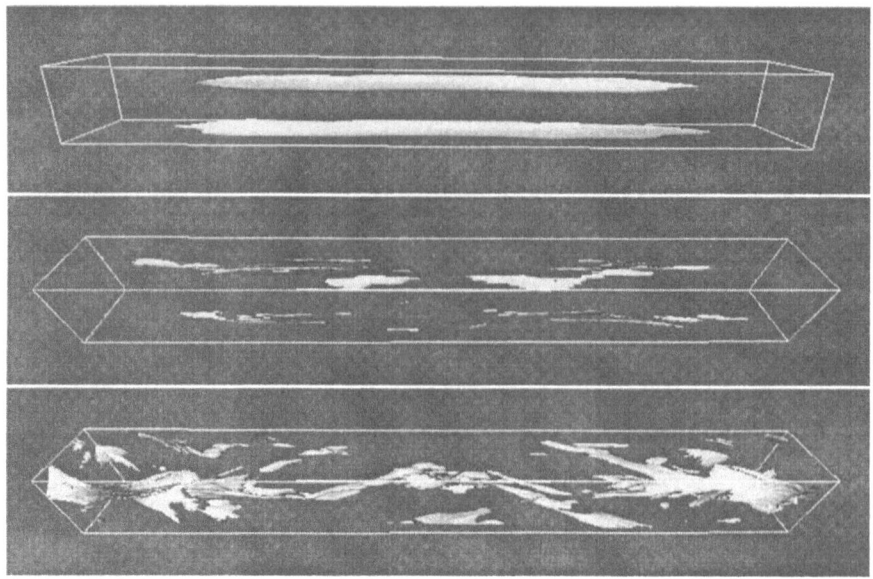

Fig. 12. Isosurfaces of the Joule dissipation for cases with different values of the dimensionless parameter $V_d t_d / L$ (available in color at URL http://www.astro.ku.dk/~aake/papers/spm96). From top to bottom; $V_d t_d / L_x = 0.006$, 0.06 and 0.2, respectively.

our interpretation, the winding numbers would have to increase in order to increase the dissipation to obtain a balance with the boundary work.

But the winding numbers are already of the order of unity, and even the most symmetric and ideal setup becomes MHD unstable at winding numbers of the order of 2–3. This shows that the levels of dissipation obtained in the numerical experiments are indeed quite representative, even when compared to cases with vastly larger magnetic Reynolds numbers.

3.8   A scaling law for magnetic dissipation

In cgs units, the Poynting flux (boundary work) done against a field of perpendicular strength B, inclined an angle Φ, is

$$W_P = V \frac{B^2}{4\pi} \tan \Phi \ , \tag{2}$$

where V is the boundary velocity in the direction opposite to the field inclination.

The conclusion earlier, that it is the local winding number that controls the dissipation, and not Φ by itself, implies that if we perform the same boundary motion, but with a smaller "stroke" $V_d t_d$, all that changes is that the boundary work is scaled down in proportion to the ratio $\tan \Phi = V_d t_d / L$. So, we obtain the overall scaling

$$\langle W_P \rangle \sim \frac{B_0^2}{4\pi} \frac{V_d^2 t_d}{L} \ , \tag{3}$$

where B_0 is a scaling factor for the magnetic field strength. Correcting for the limits when the angle is large ($V_d t_d > L$), or the Alfvén travel time is long ($V_A t_d < L$) gives the more general scaling relation (Galsgaard and Nordlund 1996)

$$\langle W_P \rangle \sim \frac{B_0^2 V_d^2 t_d}{4\pi L} \frac{[1 + (L/t_d V_A)^2]^{\frac{1}{2}}}{[1 + (V_d t_d / L)^2]^{\frac{1}{2}}} \ . \tag{4}$$

Numerical models abide by this scaling over three orders of magnitude in average dissipation (cf. Fig. 14 of Galsgaard and Nordlund 1996).

4   Summary and Concluding Remarks

In summary, the most important conclusions that may be drawn from the experiments of Galsgaard and Nordlund (1996) are that:

– Boundary motions that create interlocked flux bundles lead to exponential growth of the peak current densities in the regions between the flux bundles.
– The exponential growth implies that, for any finite (numerical or real) conductivity, a current sheet forms in a time that scales only logarithmically with the conductivity.
– The formation of the first current sheet leads to a transition to a dynamical regime, where jet sheets extend from the edges of the current sheet, and quickly mess up an initially smooth field into a hierarchy of intertwined flux bundles, with their associated current sheets.
– A statistically steady state ensues, with a hierarchy of current sheets in which the smaller current sheets provide a dissipation path for larger scale stress.
– The average dissipation obeys a scaling law that is independent of the resistivity, and the results may thus immediately be applied to estimate, e.g., solar coronal heating.

Thus, the long-standing problem of "fast reconnection" in a single current sheet, in the limit where the resistivity goes to zero is a "non-problem"; the question is not what happens with the dissipation in a single current sheet as η goes to zero, but what happens to the current sheet itself.

4.1   A unification of coronal heating theories

The results of Galsgaard & Nordlund (1996) may be seen both as a vindi-
cation of the original work by Parker, where he introduced the concept of
"topological dissipation" (Parker 1972) and discussed its application to coronal heating (Parker 1983b), and as a unification of Parkers ideas with the –
apparently contradictory – work by van Ballegooijen (1985, 1986). Indeed,
the scaling law (4) is similar to an equation by Parker (1983b), as well as to
Eq. (61) from van Ballegooijen (1986).

The main difference lies in the interpretations; Parker assumed that dissipation would commence at a certain characteristic angle (that he then went
on to estimate from an estimate of the required heating). van Ballegooijen
interpreted the exponential growth as a "cascade" to smaller wavenumbers,
that would lead to dissipation without the formation of current sheets, and
identified a factor with the same dimension as $V_d^2 t_d$ as a diffusion constant,
that he then estimated from observations.

Galsgaard & Nordlund interpret that factor as part of $V_d \Phi = V_d^2 t_d / L$,
where V_d is a measure of the actual boundary motions, and $\Phi \approx V_d t_d / L$ is an
estimate of the angle, based on the importance of the local winding number;
for a local winding number of the order of unity, the angles that the field lines
make to the boundary are of the order of $V_d t_d / L$ where $V_d t_d$ is the "stroke
length" of the motion, or simply the characteristic scale of the motion.

4.2   Consequences for the coronal heating problem

Given that the scaling law (4) provides a good estimate of the dissipation in a
boundary driven, magnetically dominated plasma, the main uncertainty in estimating the rate of heating of the solar corona lies in estimating the relevant
parameters of the boundary motions. At the surface of the Sun, there are observed motions whose scales range from hundreds of thousands of kilometers
(the scales of active region complexes), over the scales of supergranulation
(tens of thousands of kilometers), to granulation (hundreds to thousands of
kilometers). And, from the huge Reynolds numbers of even the smallest resolved motions, we conclude that there must be an extended inertial range
of turbulent motions on scales smaller than the scales of granules.

From estimates of the order of magnitude of contributions from these
various scales of motions it is not even clear which scales that contribute
most (cf. the concluding discussion in Galsgaard and Nordlund 1996); the
contributions from supergranular and granular scales seem to be of similar
importance.

However, from the abundance of coherent structures in (X-ray) images of
the coronal heating, one may infer that major contributions to the heating
must be associated with large scale, coherent motions. Indeed, it is well known
that the systematic shearing motions that occur in active regions are closely
associated with X-ray activity. Also, the emergence of new magnetic flux in an

already existing active region complex gives rise to obvious activity. Motions on smaller scales must be responsible for additional heating – indeed, the better the spatial and temporal resolution of the X-ray telescopes becomes, the more structure is found.

The actual distribution of the contributions to coronal heating from motions and flux emergence on different scales can only be obtained from a more detailed knowledge of the motion patterns at and near the solar surface. Both improved observations and the steady (exponential) rate of increase of the computing power available for dynamical modeling will contribute to the required knowledge.

Acknowledgments

This work was supported in part by the Danish Research Foundation through its establishment of the Theoretical Astrophysics Center. We thank the Danish Natural Science Research Council and 'Centre National de Calcul Parallel en Sciences de la Terre' in Paris for supporting this work through generous allotments of computing resources.

References

Berger, M. A. (1991): *A&A* **252**, 367–376
Berger, M. A. (1993): *Phys. Rev. Lett.* **70**, 705
Biskamp, D. (1993): *Nonlinear Magnetohydrodynamics*, Cambridge Monograms on Plasma Physics
Brandenburg, A., Nordlund, A., Stein, R. F., and Torkelson, U. (1995): *ApJ* **446**, 741–754
Galsgaard, K. and Nordlund, Å. (1994): in G. H. J. van den Oord (Ed.), *Procs. Utrecht Workshop "Fragmented Energy Release in Sun and Stars"*, Vol. 68, Space Sci. Rev.
Galsgaard, K. and Nordlund, Å. (1996a): *Journal of Geophysical Research* **101(A6)**, 13445–13460
Galsgaard, K. and Nordlund, A. (1997a): *Journal of Geophysical Research* (in press)
Galsgaard, K. and Nordlund, Å. (1997b): *Journal of Geophysical Research* (in press)
Gómez, D. and Fontán, C. (1992): *ApJ* **394**, 662–669
Heyvaerts, J. and Priest, E. R. (1984): *A&A* **137**, 63–78
Heyvaerts, J. and Priest, E. R. (1992): *ApJ* **390**, 297–308
Longcope, D. W. and Sudan, R. N. (1994): *ApJ* **437**, 491–504
Mikić, Z., Schnack, D. D., and van Hoven, G. (1989): *ApJ* **338**, 1148–1157
Nordlund, Å. and Galsgaard, K. (1996): *Journal of Computational Physics* (in preparation)
Parker, E. N. (1972): *ApJ* **174**, 499–510
Parker, E. N. (1975): *ApJ* **201**, 502–508
Parker, E. N. (1979): *Cosmical Magnetic Fields*, Clarendon, Oxford
Parker, E. N. (1982): *Geophys. Astrophys. Fluid Dyn.* **22**, 195–218

Parker, E. N. (1983a): *Geophys. Astrophys. Fluid Dyn.* **23**, 85–102

Parker, E. N. (1983b): *ApJ* **264**, 635–641

Parker, E. N. (1983c): *ApJ* **264**, 642–647

Parker, E. N. (1987): *ApJ* **318**, 876–887

Parker, E. N. (1988): *ApJ* **330**, 474–479

Parker, E. N. (1991): *ApJ* **372**, 719–727

Parker, E. N. (1994): *Spontaneous Current Sheets in Magnetic Fields*, Oxford University Press, Oxford

Priest, E. R. and Forbes, T. G. (1992): *J. Geophys. Res.* **97**, 16757

Ricca, R. L. and Berger, M. A. (1996): *Physics Today* **49(12)**, 28–34

Stein, R. F., Galsgaard, K., and Nordlund, Å. (1994): in J. D. B. et al. (Ed.), *Proc. of the Cornelius Lanczos International Centenary Conference*, Society for Industrial and Applied Mathematics, Philadelphia, 440–442

Strauss, H. R. (1991): *ApJ* **381**, 508–514

Sturrock, P. A. and Uchida, Y. (1981): *ApJ* **246**, 331

van Ballegooijen, A. A. (1985): *ApJ* **298**, 421

van Ballegooijen, A. A. (1986): *ApJ* **311**, 1001–1014

van Ballegooijen, A. A. (1988): *Geophys. Astrophys. Fluid Dyn.* **41**, 181

Energy Release Processes in Active Regions

Arnold O. Benz

Institute of Astronomy, ETH-Zentrum, CH-8092 Zürich, Switzerland

Abstract. A standard model of impulsive energy release has emerged during the recent years: Magnetic energy is dumped into coronal electrons (and possibly ions) accelerating them to some tens of keV. These particles mostly precipitate into the chromosphere, radiate hard X-rays and heat it to millions of degrees. The hot chromospheric material is ejected into the corona and produces the soft X-ray flare. The theory behind the energy release is reconnection, proposed for various geometries.

The standard model is here confronted with observations showing a large variety of energy releases on different time scales from the smallest noise storms and high coronal flares to the large coronal mass ejections. In view of the differences of these phenomena it is unlikely that a simple uniform model can explain them all. Hard X-ray and Hα emissions suggest a total duration of a flare of ten minutes to one hour with individual episodes of contiguous acceleration of one minute. Elementary hard X-ray peaks have 5–10 s duration, corresponding to groups of beams observable as type III radio bursts. The rise time of these beams is of order 0.1 s. The smallest time scale is indicated in narrowband radio spikes of a few 0.01 s duration. They suggest spatial scales between less than 12 and 300 km.

Yohkoh observations have allowed us to measure and locate the thermal energy of the ejected material. It can readily be related to the non-thermal energy input observed in other wavelengths. The geometry of the energy release has been studied in hard X-ray and radio emissions.

New observations of the coherent radiation of *microflares* in the 1–3 GHz region show bursts that are neither visible in synchrotron nor thermal emissions. On the other hand, there seem to be also incoherent radio events with no coherent signatures. Significant differences exist between electron acceleration at high altitudes producing interplanetary electron beams and low-altitude activity in the network of the quiet Sun.

1 Introduction

Solar flares have been an old enigma of astrophysics and remain unexplained today. Nevertheless, observations in the past two decades have revealed fascinating glimpses into some of the processes involved. Since important parts are not yet clear, there is no generally accepted scenario of magnetic energy release in the solar atmosphere. Although a 'standard model' will be described in section 2 of this review, the following two sections hasten to show that this standard is not completely in accordance with the observations.

Several books and major reviews have recently been published on coronal energy releases (e.g. Tandberg-Hanssen & Emslie 1988; Svestka, Jackson & Machado 1992; Kahler 1992; Brown et al. 1994; Benz & Krüger 1995). The emphasis here is on the primary phase of the energy release. It is often called the non-thermal phase, although its main distinction is not the possible deviation of the particle energies from a Maxwellian distribution, but the extremely high mean energy per particle, being twenty or more keV.

2   The Standard Scenario

A 'standard scenario' for flares has emerged in the past decade. In this scenario free magnetic energy builds up in the corona and is impulsively released by reconnection. The energy is first dumped into energetic electrons and possibly ions. They propagate along magnetic fields into the chromosphere where they lose their energy by collisions emitting hard X-ray bremsstrahlung and heating the chromospheric material to millions of degrees. The hot plasma evaporates into the corona and constitutes the thermal flare seen in soft X-rays, in UV and EUV lines and, most recently, in thermal millimetric emissions (Kundu et al. 1990). The succession of the two phases has been conjectured since the systematic delay between centimetric radiation (gyrosynchrotron emission) and soft X-rays (thermal bremsstrahlung) has been discovered by Neupert (1968).

2.1   Basic physical processes

A hot magnetized corona situated on top of a convective layer constantly receives magnetic energy due to the footpoint motion of magnetic loops penetrating into the corona or due to the injection of new magnetic flux into the corona from below. As a result, the magnetic field in the corona deviates increasingly from a potential field, and free magnetic energy builds up. The release of this energy is inevitable as it cannot build up indefinitely. A well-known process, but not the only one, is reconnection (e.g. reviews by Priest 1994; Somov 1994). Still strongly debated are the questions why the energy release occurs impulsively in flares and whether there exists also a quasi-continuous release of the energy.

Since the particle density in the corona is generally small compared to the energy density of the magnetic field, it is quite natural that the energy released per available particle is large. A trivial consideration may illustrate this point:

Let the energy of a magnetic field with strength $B = 100$ gauss be released into a total plasma density of $n = 10^{10} \text{cm}^{-3}$. It results in an average energy per particle of

$$\epsilon = \frac{B^2}{8\pi n} \approx 25 \text{ keV} ,$$

(1)

corresponding to a temperature of $2.8 \cdot 10^8$ K. Although the specific mechanism of particle acceleration is not known, the resulting ubiquitous numbers of electrons (and possibly ions) of some tens of keV is not surprising. The $4 \cdot 10^{29}$ erg, the energy of a medium sized flare, available in the above example assuming a volume of 10^{27} cm^{-3}, first resides in energetic and possibly non-thermal particles.

Continuous progress on our understanding of the acceleration process has been achieved over the past decade (e.g. review by Benz et al. 1994). Most recently, the theoretical work concentrates on fragmented energy release in many small reconnection sites (e.g. Vlahos et al. 1995b). A different line of thought has been followed suggesting stochastic acceleration of protons by Alfvén waves (e.g. review by Miller et al. 1996). Similarly electrons may be accelerated by transit-time damping of cascading fast-mode waves driven by the reconnection turbulence (Miller, LaRosa & Moore 1996).

Narrowband spikes in the short metric and decimetric wavelength ranges are the most fragmented flare emission. They represent an interesting and challenging field of research. The measurements of the bandwidth of individual spikes from 0.3 to 8.5 GHz by Csillaghy & Benz (1993) have revealed that the bandwidths of spikes in a given flares scatters by more than an order of magnitude and differ significantly from event to event at the same center frequency. The bandwidth thus does not seem to be an intrinsic property of the emission process, but appears to be caused by variations in the plasma parameters of the source environment, such as the plasma frequency or the electron gyrofrequency. The source dimension, ℓ, may thus be evaluated from the observed bandwidth $\Delta\nu$ and the relation,

$$\ell \approx H \frac{\Delta\nu}{\nu} \quad , \qquad (2)$$

where ν is the center frequency of the spike and H is the scale length of the relevant parameter to the emission process, either the plasma frequency or the electron gyrofrequency. Both density and magnetic field are expected to change on the order of less than 6000 km, the upper limit of the radius of active region loops. Csillaghy & Benz derive source scales of 15 to 260 km. The small scales are most frequent, and the distribution decreases from small scales to larger scales. Karlicky et al. (1996) find a power-law index of about -1.7 from a Fourier transformation in frequency. Spike sources have been proposed to coincide with the acceleration regions of electrons (Benz 1985).

The initial flare plasma expands along the magnetic field lines at two scales of velocity.

(i) The one defined by the electrons is a considerable fraction of the speed of light, the other one is given by the ions. Electrons at a few tens of keV have mean free path l proportional to the fourth power of the velocity v,

$$l_e = 3.1 \cdot 10^{-20} \frac{v^4}{n_e + Z_i n_i} \quad \text{[cm]} \qquad (3)$$

using the deflection time given in Benz (1993). For solar abundances

$$l_e \approx 1.9 \cdot 10^5 (\frac{\epsilon}{10\mathrm{keV}})^2 (\frac{10^9 \mathrm{cm}^{-3}}{n_e}) \quad [\mathrm{km}] . \tag{4}$$

Protons at the same energy have a very similar mean free path. If accelerated in the corona, most initial flare particles propagate either to the footpoints of the loop, penetrate into the chromosphere and lose their energy. An unknown, possibly considerable fraction is first magnetically reflected, but precipitates after a short time. A small fraction escapes into interplanetary space. The energetic electrons form collisionless beams penetrating the coronal plasma due to a return current that practically inhibits the induction of a magnetic field and balances the electric charge separation.
(ii) The ion population expands adiabatically and defines the second velocity scale. It suffers different electromagnetic effects, such as the fire-hose instability. The electromagnetic instabilities effectively slow down the propagation of the ions to about the Alfvén velocity (e.g. Benz 1993, ch. 7).

2.2   Electron beams

Propagating electron beams may excite characteristic radiation signatures. Longest known are type III radio bursts excited at the local plasma frequency or its harmonic if the velocity distribution is unstable to growing Langmuir waves. Beams may also emit thin target bremsstrahlung emission in hard X-rays (HXR) and possibly gyrosynchrotron radiation in centimetric waves. When the beam hits a region of high density such as the chromosphere, thick target HXR emission is produced. Although these radiations have been observed as separate bursts for several decades, they can not be combined in a straightforward way to reach quantitative results. Often type III bursts do not correlate with HXR events or are not even associated (Simnett & Benz 1986).

Nevertheless, correlated HXR and type III radio bursts have been reported with increasing accuracy of timing, in particular at decimetric frequencies. An occasional coincidence on timescales of minutes has been established very early (Kundu 1961; Kane 1972) and was later improved to 0.1 minute. In the 1980's several studies have demonstrated that groups of meter and decimeter wave type III bursts correlated with a similar number of HXR peaks coinciding within one second (Kane, Pick & Raoult 1980; Kane, Benz & Treumann 1982; Dennis et al. 1984; Savant et al. 1990). However, there are generally no one-to-one correlations and often there are more type III bursts than HXR peaks. An extensive study on the tuning of correlating HXR and type III bursts has recently been published by Aschwanden et al. (1995). They find a delay of the radio peaks at 300 MHz by 0.4 ± 0.6 s.

The reason for this delay may be found in the observation of oppositely drifting type III bursts by Aschwanden, Benz & Schwartz (1993). Such bursts

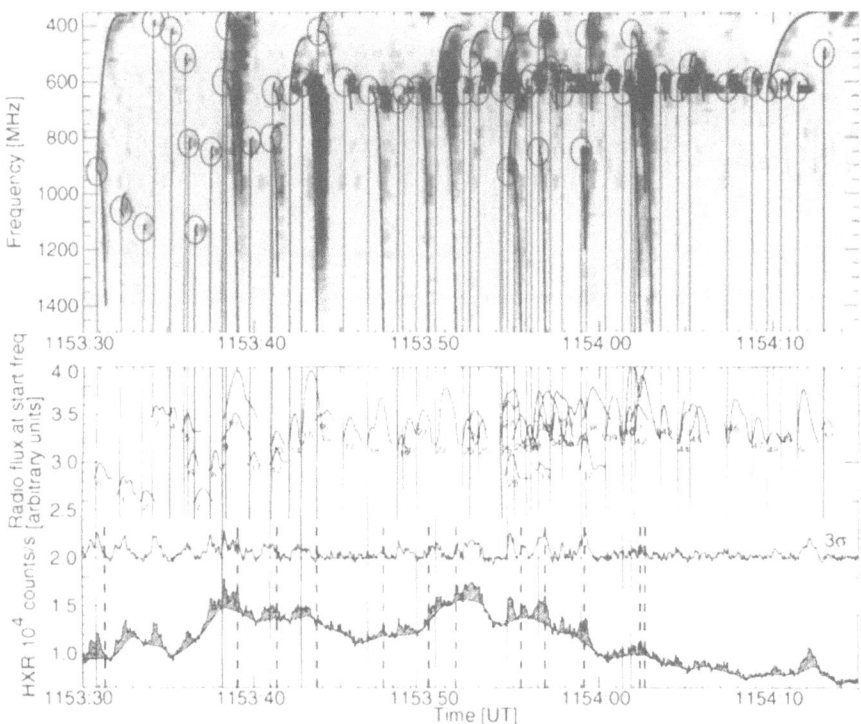

Fig. 1. *Top:* Radio flux density of solar type III emission observed by the Phoenix spectrometer of ETH Zurich and presented as a spectrogram. The start frequency is marked with a circle and connected with a thin line to the HXR time profile observed by BATSE. Drifting radio structures are outlined by thick curves. *Middle:* Radio flux vs. time in the channel of the start frequency. *Bottom:* Hard X-ray counting rate measured by BATSE/DISCSC on the GRO satellite (from Aschwanden, Benz & Schwartz 1993).

start at frequencies between 600 and 800 MHz and drift to both higher and lower frequencies. They are presumably caused by electron beams propagating downward and upward in the corona from a common height of injection. Figure 1 shows that the best correlation with HXR peaks is found at the frequency where the burst is first seen. It will be called 'start frequency' and identified with the injection of the beam into the flux tube of propagation. According to the standard picture, one would expect the HXR to correlate with the radio emission in the highest frequency channel. The most likely interpretations for the observed discrepancy are (i) that the radio emission is delayed by a finite growth time of the unstable waves and their conversion into radio emission and (ii) by the slower propagation of radio waves in a

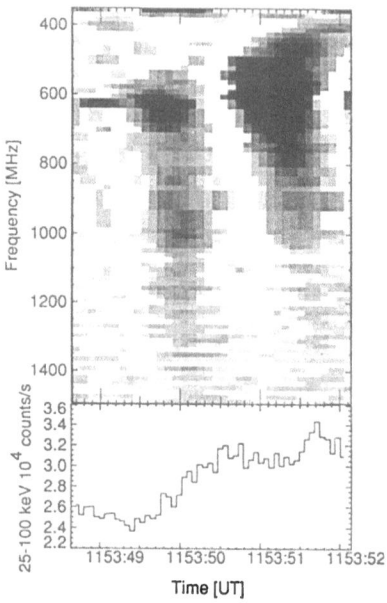

Fig. 2. *Top:* Radio flux density of solar type III emission observed by the Phoenix spectrometer of ETH Zurich. Two reversed slope bursts mark two down-going beams. The second one is accompagnied by a simultaneous up-going beam ($\nu \lesssim$ 550 MHz). *Bottom:* Hard X-ray counting rate measured by BATSE on the GRO satellite (from Benz 1993).

plasma. The drift rates of the radio bursts in Figure 1 suggest a slow speed of the radio emitting beam, corresponding to energies of $\lesssim$ 5 keV if electrons. It is well below the $\gtrsim$ 25 keV necessary to emit the HXR photons presented in Fig.1, bottom. The particles thus do not seem to be identical. It may be noted here that the delays are quite variable in a flare and that undelayed radio emission has also been observed (Fig.2). It does not support an interpretation of the radio emission by proton beams.

The observed injection frequency indicates a density of $6.2(\pm1.6)10^9$ cm^{-3} (assuming fundamental plasma emission). This density clearly refers to an acceleration region outside of the flare loops, from where typical values of order 10^{11} cm^{-3} are reported (e.g. recently Tsuneta 1996; Silva et al. 1996).

2.3   Time-of-flight effect

If the energetic electrons emit at a different location from the acceleration site, the radiation of the faster particles should be observed first. This has indeed been discovered in short HXR peaks by Aschwanden et al. (1995) from

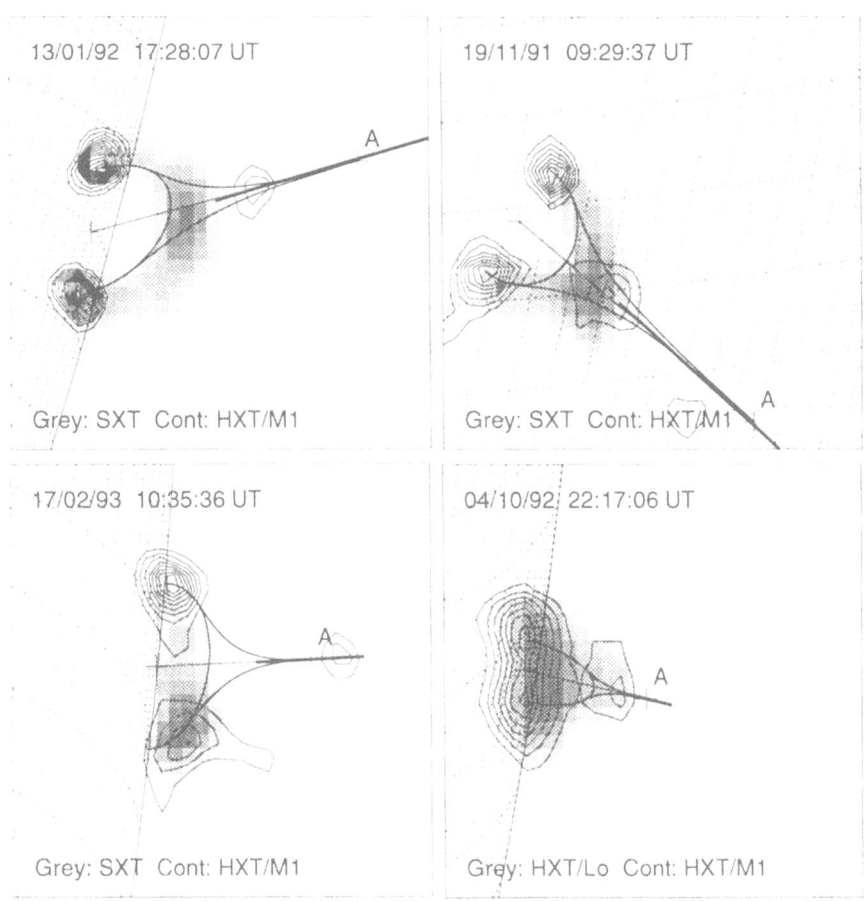

Fig. 3. Overlays of hard X-ray (contours), soft X-rays (grey), and time-of-flight distance marked by a letter *A* and error bars. The four flares were observed with Yohkoh (HXT and SXT) and BATSE (time-of-flight). An approximate circular loop is fitted into the SXR data and extended into a cusp (from Aschwanden et al. 1996).

SMM/HXRBS data. The delays were obtained from the cross-correlation of different energy bands of HXR emission. A formal distance between acceleration site and target can be derived, assuming that electrons at different energies have been accelerated at the same time. In a recent analysis of 8 large flares observed simultaneously with GRO/BATSE and Yohkoh, Aschwanden et al.(1996) find good fits of a single distance to all energy bands, suggesting that the assumption is justified. Figure 3 suggests that the derived distance correlates with the size of the associated soft X-ray loop. This has been confirmed using more data and a rigorous analysis. Thus the derived distance seems to be a geometrical parameter. Its value was found to exceed the half-length of the soft X-ray loop in all cases. Furthermore, the projected height of the acceleration site coincides with the 'above-the-loop' source observed in HXR within the error bars (Fig.3).

3   Deviations from the Standard Scenario: Flares at High Altitudes

Considering the complexity of coronal magnetic fields as well as the non-linearity of Maxwell's equations and the fluid equations, it is not surprising that observations are not as simple as the standard scenario. There are certainly many deviations from the standard, and the selection presented here is not meant to be complete.

Best known and widely discussed are coronal mass ejections (CME). Without going into mythology and ideology, it may be noted here that CMEs are not only the most energetic phenomena known to occur in the solar system claimed to reach 10^{34} ergs (Kane et al. 1995), they also are known to convert about half their primary energy in outward motion of coronal plasma (e.g. review by Kahler 1992). This is clearly not part of the standard flare scenario, where most of the initial energy is thermalized in the chromosphere. Cliver et al.(1986), among many other authors, find the CME associated energy release at considerably higher altitude than ordinary (impulsive) flares and interpret them as taking place in a streamer-like magnetic configuration. The high altitude may not be the only reason for the large energy and outward motion, however. Qualitatively different processes may be involved as proposed e.g. by Gosling (1993).

3.1   Narrowband metric spikes

High-altitude energy release at much smaller scales has recently been studied by Krucker et al. (1995). Figure 4 shows the imaging information of a metric spike event associated with type III bursts of electron beams propagating upward in the corona. The spike emission process must be coherent to explain the high brightness temperature of the emission as expected from the interaction of nonthermal electrons with collisionless waves.

The spike source in Fig. 4 is located at an altitude of about $4.5 \cdot 10^5$ km. With a delay of 42 seconds, a thermal source appears at 1446 MHz on the same potential field line, extrapolated from the longitudinal photospheric magnetic field. The location and the timing of this thermal emission suggests that energy has propagated from a high altitude, consistent with an energy release close to the spike source. No Hα flare has been reported in the *Solar and Geophysical Data*, suggesting that the energy release was small, or not much of it has reached the chromosphere. A statistical analysis of metric spikes finds an Hα-flare association rate of 30%, confirming the weakness of

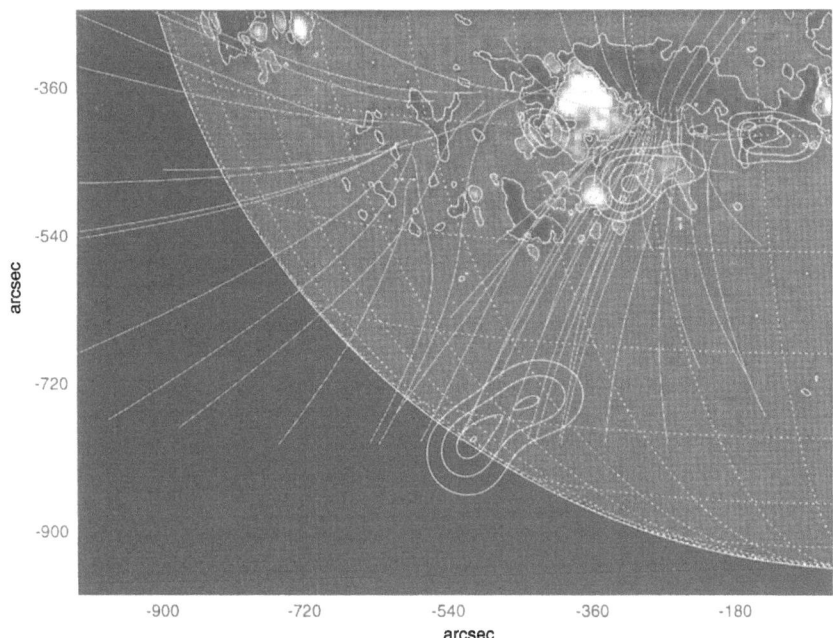

Fig. 4. Overlays of radio contours (VLA), magnetogram (grey, KPNO), and high reaching potential field extrapolation (lines) for metric spike event on 07/08/89 17:30 UT. The spike source at 333 MHz is below the center of the image near the limb. The 1446 MHz sources are near the active region (from Krucker et al. 1995a).

chromospheric signatures of such events (Benz et al. 1996). These authors predict a lower limit of 2000 metric spike/type III events per year at high solar activity.

Krucker et al. (1996a) have also observed a similar event together with Yohkoh/SXT and have identified a weak soft X-ray source at the footpoint of a potential field line through the spike source with a similar delay. At the projected position of the radio spikes, they have found a stationary jet-like structure. The new observation is consistent with a high altitude of the spike source. In addition they found the associated metric type III burst to propagate to 40 MHz and below, strongly suggesting that the electron beam has escaped into interplanetary space. The outward flowing energy may be half of the inward propagating energy. The metric spike/type III events thus have many features in common with 'high coronal flares' proposed for the acceleration of solar energetic electron events observed in interplanetary space.

3.2 Soft X-ray jets

The extremely weak stationary jet-like structures observed in association with spikes are different from the regular SXR jets observed by Shibata et al.(1992, 1996) and Strong et al. (1992). Such jets have been reported to exist for up to one hour. Their energy release seems to be at low altitude and material is observed to move upward. For the type III bursts accompanied with SXR jets (Aurass et al. 1994; Kundu et al. 1994, 1995a) no associated metric spikes have been reported. The type III bursts were found in the direction of the jet extrapolated to higher altitude, and the SXR emission often starts long before the radio bursts. Only recently Raulin et al. (1996) have found type III bursts at the beginning of a jet, more consistent with the standard flare scenario (cf. Fig.5).

3.3 Radio noise storms

Noise storms constitute a different coronal meterwave radio emission. They are usually located at high altitudes with typical heights of $2\text{-}7\cdot10^5$km (El-garøy 1977). The emission is generally assumed to be caused by interactions between non-thermal electrons and waves, but it is not related to chromospheric signatures of flares, such as $H\alpha$ brightenings, nor hard X-rays. Thus the standard scenario - energy release in the corona and thermalization in the chromosphere - needs to be tested.

It has been known for some time, and was confirmed by Raulin et al. (1994), that the start of a noise storm is often accompanied by soft X-ray emission. Combining VLA radio maps and Yohkoh soft X-ray information, Krucker et al. (1995b) have recently found the continuum of an ongoing noise storm source near the top of a bright SXR loop. This source is displayed in Fig.6. It is quasi stationary in time with a few occasional bursts. Although this observation is suggestive of an association of the noise storm with the heating of the loop, it remains unclear why the more bursty sources of the noise strom were found in a region of the corona without SXR loops and why there were SXR loops in the same active region apparently without noise storm emission. The noise storm radio emission seems to point to a particular location of special loops, but we do not yet understand its message.

In conclusion, energy release at high altitudes may include all the ingredients of the standard scenario, but does not seem to be sufficiently described by its elements alone.

4 Microflares

Small flares, also called microflares or nanoflares, are of great interest as they have been speculated to provide the major energy input into the corona. Many fluctuations are indeed observed in chromospheric and transition region

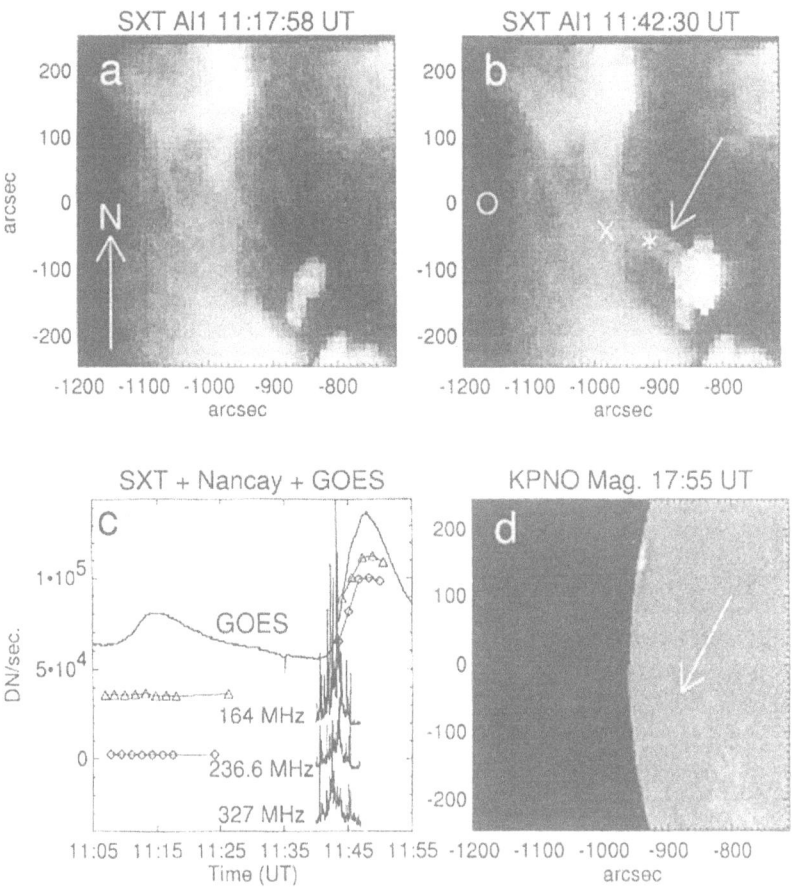

Fig. 5. *a:* Yohkoh/SXT image in soft X-rays before the jet. *b:* Jet marked by arrow and centroid positions of radio type III bursts observed with the Nancay radioheliograph at 327 MHz (asterisk), 236 MHz (cross), and 164 MHz (circle). *c:* Total flux in SXR (GOES) and in metric radio emission (MHz) of type III bursts (from Raulin et al. 1996).

lines, soft X-ray continuum and high-frequency radio emission. There is no question that this variability expresses changes in the solar atmosphere, but it is not at all clear whether they are signatures of flare-like processes since there are other forms of energy input. The standard flare scenario provides an observational definition of what to consider a flare: A microflare may be considered an event that is driven by magnetic energy release, possibly initiated by reconnection. The energy first is dumped into a few particles reaching high energies. These particles are transported to denser regions and heat a remote plasma.

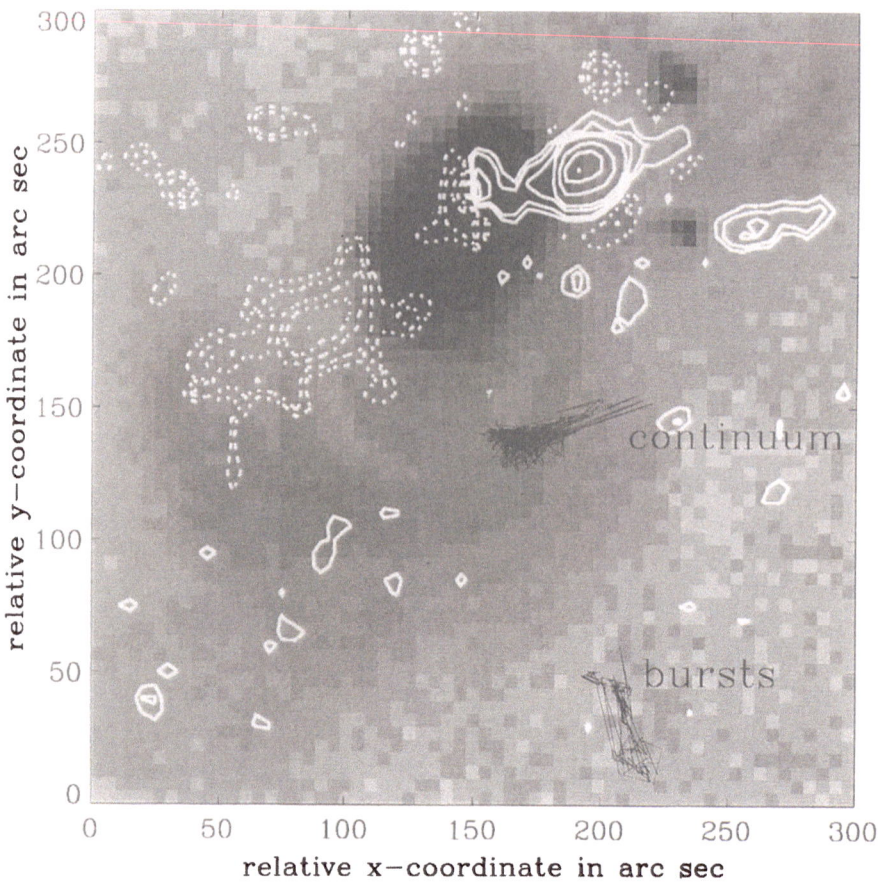

Fig. 6. Yohkoh/SXT image (grey), KPNO magnetogram (contours), and centroid positions of radio noise strom observed by the VLA at 333 MHz. The storm positions at different times are connected by lines. The two radio sources are labeled 'continuum' (steady, few bursts) and 'bursts' (short sequence of isolated bursts). (from Krucker et al. 1995b).

Saturation of CCD arrays has made it difficult in the past to detect the emission of the quiet regions in soft X-rays except for coronal bright points. They consist of several parallel loops, typically 10^4 km long with a mean lifetime of eight hours. The footpoints have been found in photospheric magnetograms to form small bipolar regions. They are part of the magnetic network of the chromosphere. The X-ray bright points and their associated He I absorption in the chromosphere occur primarily above newly emerged bipolar regions and even more frequently in disappearing elements of the network (e.g. Harvey 1985).

The first direct comparison between X-ray bright points and *centimeter radio waves* were made only recently by Nitta et al. (1992). Enhanced radio emission during optically observed coronal brightenings has previously been observed at 20 cm wavelength (Habbal et al. 1986, Habbal & Harvey 1988), at 6 cm (Fu et al. 1987) and more recently at 2 cm (Kundu et al. 1994). Rapid variations observed in the radio emission suggest that individual loops brighten and fade on time scales of minutes. The total duration of the radio events is similar to the X-rays. In the quiet Sun, centimeter radio emission originates mostly in the chromosphere. If thermal, the radio counterparts thus constitute a chromospheric signature of coronal bright points. The detection of associated type III bursts (Kundu et al. 1995b) indicates the presence of non-thermal electrons and demonstrates the existence of a high-energy phase. A reconnection model of converging magnetic flux, following the standard scenario, has been developed by Priest, Parnell & Martin (1994).

It is well-known that the quiet Sun is variable outside of coronal bright points. UV line observations show continuous variability on a timescale of minutes (Brueckner & Bartoe 1981; Porter et al. 1987). Gary et al. (1990) examined the time variability of the network at 3.6 cm radio waves and found that the *chromospheric* radio brightness distribution underwent slow morphological changes due to the slow evolution of the magnetic field structure. They noted that the brightest locations at 3.6 cm corresponded to areas containing approaching magnetic elements of opposite magnetic polarity, which eventually cancel. In addition to slow morphological changes, the 3.6 cm brightness distribution displayed several discrete temporal changes. These did not appear to correlate with changes in the magnetic field morphology or with small surge-like events seen in Hα filtergrams. The authors conclude that the variations in 3.6 cm brightness were not associated with structural changes of the photosphere and were more likely associated with changes in coronal conditions.

Long SXR exposures have recently become possible with Yohkoh. Benz et al. (1996) have found the *coronal* SXR emission of quiet regions to be structured and correlated with the magnetic network as observed in magnetograms from the photosphere. The structure has spatial scales down to the limit of resolution of 4". The SXR network elements are, however, variable and show brightenings of typically 10 minute durations (Fig.7). The brightenings are two orders of magnitude both fainter and shorter than coronal bright points. The association with the network suggests that the location of the X-ray bright plasma is in the low corona. The energy of X-ray emitting plasma is of the order of 10^{25}. Their rate of more than 10^3 per hour thus constitutes a considerable energy input into the corona.

Most interesting is the close correlation of radio emission at 2 cm with the SXR light curve in Fig. 7. Note that the radio brightenings are three orders

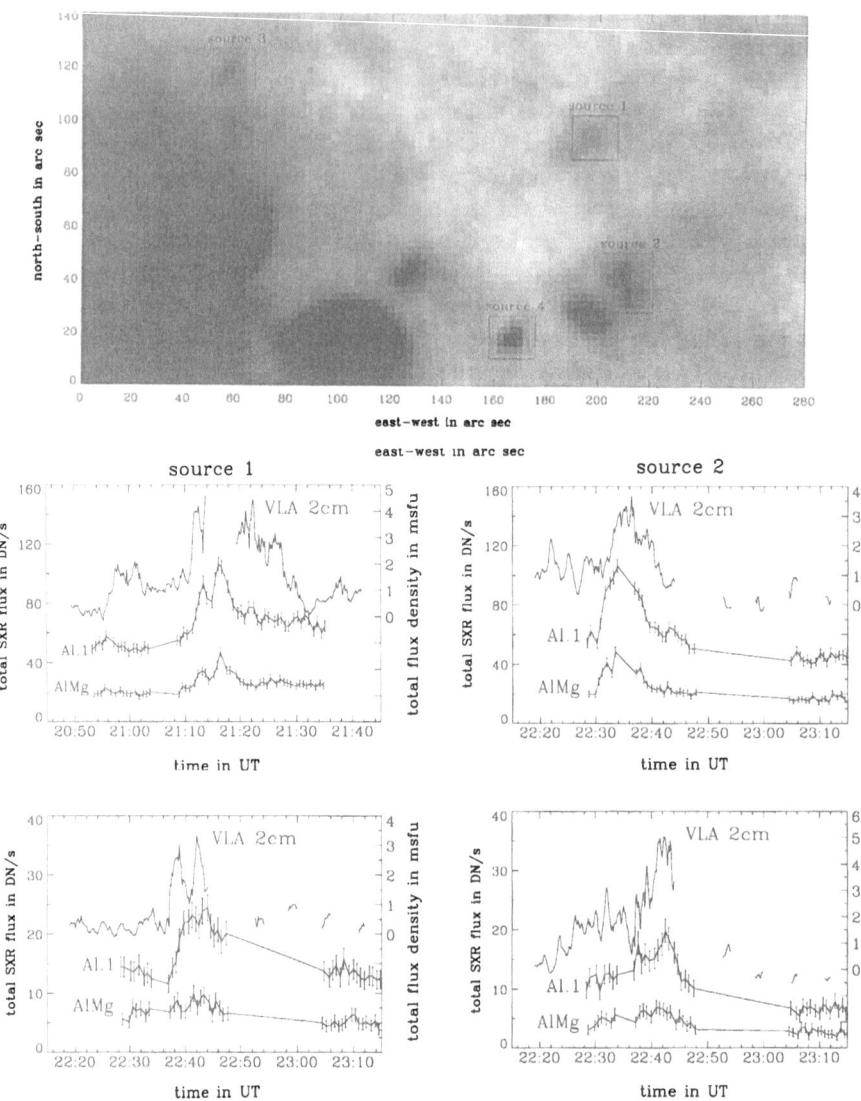

Fig. 7. *Top:* Soft X-ray image observed by Yohkoh/SXT of a quiet region. In the left half of the image, some high-reaching loops from a nearby active region produce a foreground. The numbered squares indicate weak sources that have been analyzed in time. *Bottom:* Time profiles of the sources selected in the top image. For each source the variation is given at 2 cm wavelength as observed by the VLA and in the filters, Al.1 and AlMg, used by Yohkoh. The former filter has a broader range and also includes plasma of lower temperature than the one observed with the second filter (from Krucker et al. 1996).

of magnitude fainter than the smallest flares reported by survey instruments. The timing of the radio peak in relation to the SXR emission is not clear. In 3 out of the four cases in Fig. 7 it occurs before the SXR maximum. This is similar to regular flares, where the radio emission is non-thermal gyrosynchrotron emission and the thermal SXR are delayed according to the Neupert effect. However, the non-thermal nature of the radio emission of the network brightenings still needs to be proven. Even in the case, the radio emission will turn out to be thermal, the network brightenings may be flare-like according to the standard scenario.

5 Conclusions

Recent advances in coronal physics clearly show that magnetic energy is released at all altitudes in the solar corona. In the widely, but not universally accepted view the magnetic energy is build up in the corona by a slow evolution. The release of the energy is triggered by reconnection and has been modeled by MHD in the past decades. A 'standard scenario' has become widely accepted in which energy is relased in the corona and the hot thermal, SXR emitting flare plasma is supplied by the secondary heating and evaporation of the chromosphere.

The kinetic part of the flare physics is less known. It is related to particle acceleration and fragmentation of the energy release. As the energetic particles escape from the acceleration region, they unify to beams (as e.g. simulated by Vlahos & Raoult (1995a), which again unify by loosing their energy in a few targets in the chromosphere. Chromospheric conduction and evaporation are again processes that may follow MHD descriptions.

There are obvious deviations from the standard scenario. Apparently different types of energy release seem to operate. In high-altitude energy release the chromosphere plays a less important role. At low altitudes and in quiet regions, a non-thermal signature of network brightenings has not yet been found.

Much of the future in flare physics should be devoted to establish more elements of the magnetic energy release processes. They may have different importance at different altitudes and in possibly different magnetic geometries. Outstanding problems include (i) the geometry of reconnection, in particular the question whether it is helmet-like or more the type of converging magnetic loops. (ii) On the question of particle acceleration, the rivalry between DC electric field acceleration and stochastic acceleration persists and does not seems to come to definite answers soon. (iii) In both cases the kinetic situation is not yet clear in view of collisional waves in the plasma. Why are they not visible in radio emission or, if they are in spikes and type I bursts, why are they not always observable? (iv) Finally the role of protons as likely carriers of a considerable fraction of the flare energy remains unclear.

References

Aschwanden M.J., Benz A.O. Schwarz R.A.,1993, ApJ 417, 790

Aschwanden M.J., Montello M.L., Dennis B.R., Benz A.O.,1995, ApJ 440, 394

Aschwanden M.J., Schwarz R.A., Kosugi T., Hudson H.S., Wills M.J., 1996, ApJ, submitted

Aurass H., Klein K.L., Martens P., 1994, Solar Phys. 155, 203

Benz A.O., 1985, Solar Phys. 96, 357

Benz A.O., 1993, *Plasma Astrophysics*, Kluwer, Dordrecht

Benz A.O., Kosugi T. and 10 coauthors, 1994, Solar Phys. 253, 33

Benz A.O., Krüger A. (eds.), 1995, *Coronal Magnetic Energy Releases*, Lecture Notes in Physics 444

Benz A.O., Csillaghy A., Aschwanden M.J., 1996, A&A 309, 291

Brown J.C. et al., 1994, Solar Phys. 153

Brueckner G. E., Bartoe J. D. F., 1983, ApJ 272, 329

Cliver E.W., Dennis B.R., Kiplinger A.L., Kane S.R., Neidig D.F., Sheeley N.R.Jr., Koomen M.J., 1986, ApJ 305, 920

Csillaghy A., Benz A.O., 1993, A&A 274, 487

Dennis B.R., Benz A.O., Ranieri M., Simnett G.M.,1984, Solar Phys. 90, 383

Elgarøy Ø., 1977, it Solar Noise Storms, Pergamon Press, Oxford

Gary D. E., Zirin H., Wang H., 1990, ApJ 355, 321

Gopalswamy, N. et al., 1995, ApJ 455, 715

Gosling J. T., 1993, JGR 98, 949

Kahler S.W., 1992, ARA&A 30, 113

Kane S.R., 1972, Solar Phys. 27, 174

Kane S.R., Pick M., Raoult A., 1980, ApJ 241, L113

Kane S.R., Benz A.O., Treumann R.A., 1982, ApJ 263, 423

Kane S.R., Hurley K., McTiernan J.M., Sommer M., Boer M., Niel M., 1995, ApJ 446, L47

Karlicky M., Sobotka M., Jiricka K., 1996, Solar Phys. in press

Krucker S., Benz A.O., Aschwanden M.J., Bastian T.S., 1995a, A&A 302, 551

Krucker S., Benz A.O., Aschwanden M.J., Bastian T.S., 1995b, Solar Phys. 160, 151

Krucker S., Benz A.O., Aschwanden M.J., 1996a, A&A in press

Krucker S., Benz A.O., Acton L.W., 1996b, ApJ submitted

Kundu M.R., 1961, JGR 66, 43308

Kundu M.R., White S.M., Gopalswamy N., Bieging J.H., Hurford G.J., 1990, ApJ 358, L69

Kundu M.R. et al., 1994, it Proc. Kofu Symposium, NRO Rep. 360, 343

Kundu M.R. et al., 1995a, ApJ 447, L135

Kundu M.R., Raulin J. P. Pick M., Strong K. T., 1995b, ApJ 444, 922

Miller J. A., LaRosa T.N., & Moore R.L., 1996, ApJ 461, 445

Miller J. A. et al., 1996, JGR, in press

Neupert W.M., 1968, ApJ 153, L59

Nitta N., Bastian T.S., Aschwanden M.J., Harvey K.L., Strong K.T., 1992, PASP 44, L167

Porter J. G., Moore R. L., Reichmann E. J., Engvold O., Harvey K. L., 1987, ApJ 323, 380

Priest E.R., 1994, in: Plasma Astrophysics, Saas-Fee Advanced Course 24, Springer-Verlag

Priest E.R., Parnell C.E., Martin S.F., 1994, ApJ 427, 459

Raulin J. P., Klein K. L., 1994, A&A 281, 536

Raulin J. P., Kundu M. R., Hudson H. S., Nitta N., Raoult A., 1996, A&A 306, 299

Savant H.S., Lattari C.J., Benz A.O., Dennis B.R., 1990, Sol. Phys. 130, 57

Simnett G.S., Benz, A.O., 1986, A&A 165, 227

Shibata K. et al., 1992, Publ.Soc.Japan 44, L173

Silva A.V.R. et al., 1996, ApJ 458, L49

Somov B.V., 1994, *Cosmic Electrodynamics*, Kluwer, Dordrecht

Strong K. et al., 1992, Publ.Soc.Japan 44, L161

Svestka Z., Jackson B.V., Machado M.E. (eds.), 1992, *Eruptive Solar Flares*, Lecture Notes in Physics 399

Tandberg-Hanssen E., Emslie A.G., 1988, *The Physics of Solar Flares*, Cambridge University Press

Tsuneta S., 1996, ApJ 456, 840

Vlahos L., Raoult A., 1995a, A&A 296, 844

Vlahos L., Georgoulis M., Kluiving R., Paschos P., 1995b, A&A 299, 897

Production of Flare Accelerated Particles at the Sun

Gérard Trottet and Nicole Vilmer

Observatoire de Paris, Section d'Astrophysique de Meudon, LPSH-DASOP (URA 2080), F-92195 Meudon Cedex, France

Abstract. Energetic particles take a fundamental part in the physics of solar flares because they carry a large fraction of the energy released. The most direct and quantitative information on the physical processes involved in the production of non-thermal electrons and ions has been taken from the wealth of hard X-ray (HXR) and gamma-ray (GR)observations performed over the last fifteen years. This review intends to summarize statistical, temporal and spectral characteristics of HXR/GR flares which provide strong constraints on the acceleration/transport process(es) at work during solar flares. The main highlights are: (i) there are no distinct classes of flares; (ii) acceleration time-scales are short (1s or less) but continuous acceleration may occur on long time-scales (tens of minutes to hours); (iii) the energy in ions is comparable to the energy in electrons though there is some variability of electron versus ion content from flare to flare and within one given flare and (iv) there is a preferential acceleration of heavy ions. At present, the different models of particle acceleration have not reached the point where they can account for the ensemble of observational constraints. However, the "avalanche" theory may provide a potential way to progress towards a global understanding of flare energy release and acceleration processes.

1   Introduction

The energy carried by flare-accelerated electrons ranges from $\sim 10^{26}$ ergs for microflares (Lin et al. 1984) to $\sim 10^{34}$ ergs for the largest flares detected so far (Kane et al. 1995). Simnett (1986) suggested that the bulk of the energy may reside in non-thermal ions Such a suggestion is supported by recent estimates, from gamma-ray line spectroscopy, which indicate that > 1 MeV ions may carry a comparable or even higher amount of energy than > 20 keV electrons (Ramaty et al. 1995). Such an energy in accelerated particles represents a large fraction of the total amount of energy released during a solar flare. Thus we cannot understand flares without understanding acceleration i.e. (i) the energy release process(es) which operates on a macroscopic scale such as that of an active region magnetic field and (ii) the acceleration process(es) which most likely operates on a microscopic scale such as that of a reconnection region. The evolution of our understanding of solar flare particle acceleration has been reviewed by Chupp (1996). Significant progress has been made over the last decade in investigating these complex problems because: (i) new theoretical approaches have been proposed and (ii) joint observations of particle related emissions have been extensively performed over a large wavelength range going from the gamma-ray (GR) to the radio domains.

In this review we focus on the observational aspects and we restrict our presentation to the hard X-ray (HXR) and GR measurements because they constitute the most quantitative diagnostics of particle numbers, spectra and chemical composition. A complementary overview of HXR and radio diagnostics which provides information on fast time variations and of the morphology of magnetic structures into which particles are accelerated and/or propagate can be found in Vilmer and Trottet (1996). In the next section we discuss the statistical properties of HXR events and compare them with the predictions of macroscopic models of energy release. In section 3 we investigate how ion acceleration relates to electron acceleration. Section 4 is devoted to GR line (GRL) spectroscopy which allows us to determine the chemical composition of the ambient medium in the GRL production region and of the accelerated ions as well as to estimate the number of accelerated ions. A brief overview of diagnostics of high energy ions and neutrons is given in section 5. The final section summarizes the observational constraints that have to be met by any realistic acceleration/energy release models and briefly outlines to what extent the predictions of existing theories are consistent with these constraints.

2   The Frequency Distribution of HXR Flare Parameters

Frequency distributions have been extensively reported for various kinds of energy release events observed in the soft X-ray (SXR) (Hudson et al. 1969; Drake 1971; Lee et al. 1993; Shimizu 1995), in the HXR (Datlowe et al. 1974; Lin et al. 1984; Dennis 1985; Crosby et al. 1993; Lu et al. 1993; Bromund et al. 1995) and in the radio (Akabane 1956; Fitzenreiter et al. 1976; Kosugi 1985) domains. All these distributions can be represented by a power law of index $\alpha > -2.5$.

This paper deals with non-thermal particles and the following presentation has been restricted to results obtained for HXR events. Indeed, it

Table 1. Slopes of the frequency distributions of HXR flare parameters

	Crosby et al.	Bromund et al.
D (s)	-2.17 ± 0.05	-2.40 ± 0.04
P (ph cm^{-2} s^{-1})	-1.59 ± 0.01	-1.86 ± 0.01
F (ergs s^{-1})	-1.67 ± 0.04	-1.92 ± 0.02
W (ergs)	-1.53 ± 0.02	-1.67 ± 0.02

is widely accepted that the hard X-ray emission above $\sim$ 20 keV is due
to bremsstrahlung radiation of accelerated electrons. The most recent studies have been performed on two independent sets of more than 6000 events
recorded by the Hard X-Ray Burst Spectrometer (HXRBS) on SMM from
1980 to 1989 (Crosby et al. 1993) and by the ISEE3-ICE X-ray spectrometer
from 1978 to 1986 (Bromund et al. 1995). The studied parameters are the
event duration (D), the peak photon flux (P), the peak energy flux in accelerated electrons (F) and the total energy in accelerated electrons (W). F and
W have been calculated for a single power law photon spectrum and for a
thick target interaction model. Table 1 displays the power law slopes derived
in both studies. It shows that the results of Bromund et al. lead to significantly steeper distributions. This may be due to differences in instrument
sensitivity or/and to different methods of estimating F and W.

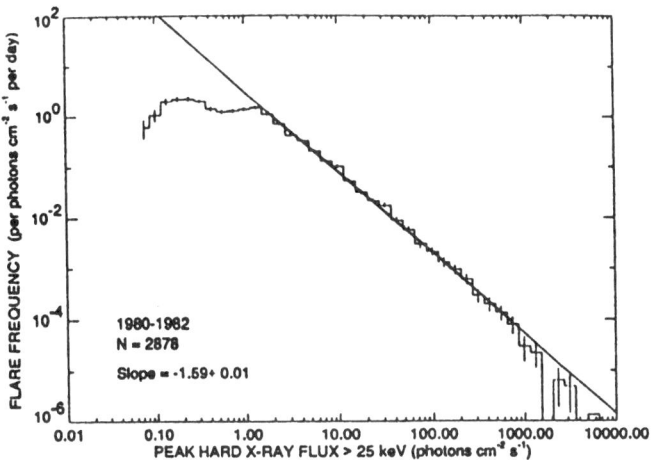

Fig. 1. The frequency distribution of the peak hard X-ray flux above 25 keV
observed by HXRBS/SMM. The solid curve is the power law which best fits
the binned data (from Crosby et al. 1993).

As an example Figure 1 displays the frequency distribution of P(> 25
keV) (from Crosby et al. 1993). This distribution is well represented by a
power law (solid line in Figure 1), with a slope of $\sim$ -1.6, for P ranging from
$\sim$ 1 to 1000 ph cm^{-2} s^{-1}. Furthermore Crosby et al. pointed out that the
distribution obtained from the microflares observed by Lin et al. (1984) agrees
closely with the extrapolated distribution of all flares observed by HXRBS
near the solar maximum. Thus the same power law can be extended down to
smaller values of P (at least one order of magnitude) so that the deviation
from the power law for P < 1 ph cm^{-2} s^{-1} observed with HXRBS only reflects

the effect of the instrumental sensitivity threshold. Looking at the frequency distributions in durations and fluxes, there also does seem to be a paucity of the strongest or longest flares. It is possible that observations miss more large flares in proportion to smaller ones (data gaps, saturation effects), however it cannot be excluded that this roll-off may be real (Bromund et al. 1995).

Most of the flare parameters are not independent but are correlated with each other to some extent. In particular, W and F and W and F×D are highly correlated (correlation coefficient > 0.8). On the other hand, the correlation between F and D is weak. Thus: (i) the size of the peak more directly determines the size of an event than does its duration; (ii) there is no predominant rate of rise or decay of the flare energy (Bromund et al. 1995).

It has to be emphasized that none of the HXR flare parameters shows a binomial distribution. Thus **there are no physically distinct classes of flares** such as impulsive and gradual as was often considered in the past. Moreover there is little long term variation of the frequency distributions though the flare rate may vary by more than one order of magnitude from solar maximum to solar minimum (Crosby et al. 1993, Bromund et al. 1995). Significantly steeper frequency distributions are however observed during short periods of minimal activity corresponding to the minimum phases of the 154 day cycle (Bromund et al. 1995).

Provided that the total energy in non-thermal electrons is correlated to the total energy released during a flare, the above results can be confronted with the predictions of flare models. Two classes of models have been investigated so far, i.e "stochastic relaxation" models and the "avalanche model" (see Bastian and Vlahos 1996 for an extended discusion).

One of the earliest attempts to account for the flare size distribution, in the framework of "stochastic relaxation" models, was that of Rosner and Vaiana (1978). They considered a system in which the energy is stored and which has a constant probability per unit time of releasing that stored energy. Each of the flaring events is assumed to completely release the energy in the system down to its ground level. It is further assumed that the rate at which energy is stored is proportional to the total energy E_T in the system. When the energy released is large compared to the unperturbed energy state this leads to a power law frequency distribution as a function of energy with a slope $-(1+r_f/r_b)$ where r_f is the mean flare rate and r_b is defined by $dE_T/dt=r_b E_T$. Though this result is consistent with the observations, Lu (1995) has criticized Rosner and Vaiana's model on a number of grounds. In particular because the slopes of the size distribution of flare parameters are quite insensitive to the variation of the flare occurrence rate, e.g. when the level of activity rises and falls with the cycle, the value of r_f/r_b should remain constant and the driving and triggering mechanisms must be coupled. The energy storage rate should be exponential over more than 6 orders of magnitude in order to produce the power law distribution of flare energies over the observed range ($\sim 10^{26}$ to $> 10^{33}$ ergs). More recently Litvinenko (1994) has suggested a variant of the Rosner and Vaiana's model. His model of energy storage and release considers one or more physically independent reconnecting current sheets which experience recurrent instabilities. Rather than assuming that the energy storage rate is proportional to E_T, he shows, under a number of assumptions, that it is proportional to $E_T{}^{7/4}$. The frequency distribution of flare energies which arises from this model is a power law with index -7/4. Though this is consistent with the observations (see Table 1), this model needs to be further investigated in order to establish that it constitutes a physically realistic model of flares.

In contrast to the "stochastic relaxation" models, cellular automaton models, in which the fragmentation of the energy release plays a fundamental role, have been investigated. Among these, the "avalanche" model, which was first proposed by Lu and Hamilton (1991) and which was subsequently developed by Lu et al.(1993), Vlahos et al. (1995) and Georgoulis and Vlahos (1996), has been extensively explored. The power law size distribution of flare energies is related to the idea that flares can be understood as avalanches of many small reconnection events with the magnetic field in state of "self-organized criticality". The critical state is insensitive to the initial conditions. The system has a distribution of minimally stable regions of all sizes and small perturbations give rise to avalanches of all sizes from the smallest possible avalanche to the size of the system. The power law distribution of avalanche sizes results from the lack of a characteristic length scale between the size of a single unstable site and the overall size of the system.

Lu et al. (1993) have compared the predictions of the "avalanche" model with the observations. Figure 2 (top) displays a scatter plot showing the relationship between W and F derived from the ISEE 3/ICE data. The solid line is from the numerical model. The slope of the predicted curve is fixed by the model, but the normalization depends on the values of physical variables in each individual reconnecting cell (magnetic field B, length scale dL, electron density N_e and reconnection rate ξ) through two adjustable parameters: the energy released, dW, in an elementary reconnection event and the duration, dD, of such an event. Lu et al. obtained a reasonable fit to the data by assuming that all reconnection events are identical with dW $= 2.8 \ 10^{25}$ ergs and dD $= 0.26$ s. The same values of these parameters also give good fits to the observed relationships between W and D and F and D. There is a range of values of B, dL, N_e and ξ which can lead to these values of dW and dD. Taking B $= 400$ G, $N_e = 10^9$ cm^{-3} and $\xi \approx 10$, values which are consistent with those expected for solar flares, Lu et al. found dL to be ≈ 400 km. The predicted size distribution of flare energies depends on a further parameter: the overall size of the system i.e. the characteristic length scale of an active region magnetic field L_A. Figure 2 (bottom) shows that the observed and predicted size distributions of W, are in good agreement. The three theoretical curves were computed for $L_A = 3$, 5 and 7 10^9 cm i.e. for reasonable values of an active region size. The frequency distributions of F and D are

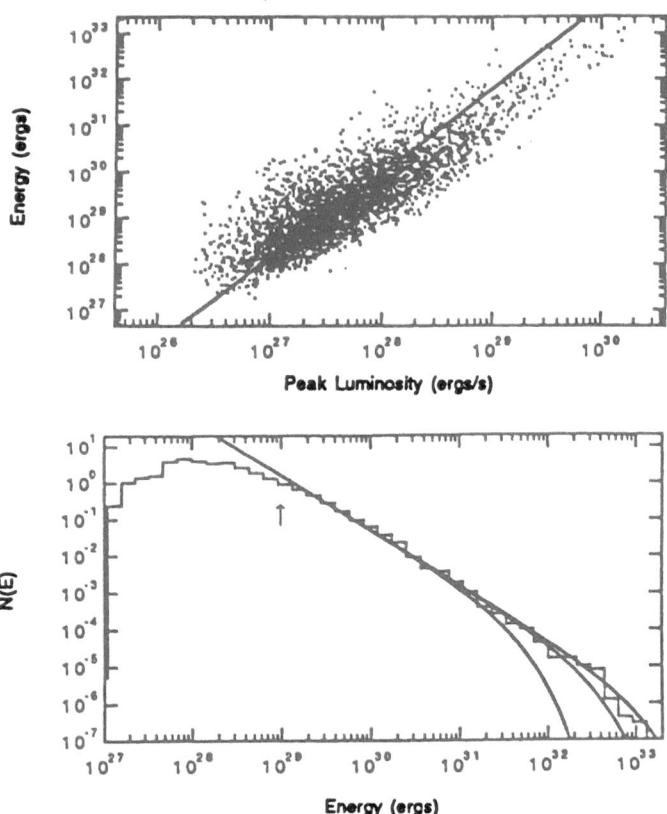

Fig. 2. Top: Scatter plot of the energy vs peak luminosity observed by ISEE-3/ICE. The solid line is the theoretical prediction of the avalanche model with $dW = 2.8\ 10^{25}$ ergs and $dD = 0.26$ s. Bottom: Comparison between the frequency distribution of flare energies observed by ISEE-3/ICE (histograms) with the theoretically predicted distributions for $L_A = 3, 5$ and $7\ 10^9$ cm (see text) (from Lu et al. 1993).

also well reproduced by the model with the same parameters. In addition to this consistency between theoretical predictions and observations, there are a number of other of observational features which provide strong support to the "avalanche model" (see Vilmer and Trottet 1996). In particular:

1. Statistical studies of sub-second HXR pulses indicate: (i) that energization of electrons up to energies around 100 keV occurs on timescales of 1s or less but that there is no characteristic acceleration timescale either within a flare or from one flare to the other (Vilmer et al. 1996); (ii) the frequency distribution of the peak count rate of HXR sub-second pulses obtained by Aschwanden et al. (1995) is similar to that of flares: this is

just what is expected from a scale invariant model like the "avalanche" model.

2. Imaging and spectral radio observations strongly support the idea of fragmented energy release in both space and time (e.g. Trottet 1994; Benz 1985)

3. The existence of a roll-off of the frequency distribution at the largest flare energies, which is a consequence of the finite size of the system and which is marginally statistically significant in the SMM and ISEE3-ICE data (see Figure 2 bottom), is consistent with the results of a recent study by Kucera et al. (1997) which has explored the nature of that distribution as a function of active region sizes. However, Kane et al. (1995) have reported observations of some giant flares, releasing up to $\sim 10^{34}$ ergs, for which the resources of a single active region are inadequate. For these events, the energy release process may thus operate on a spatial scale which is larger than that of a single active region. This is consistent with earlier suggestions, made from joint radio and HXR observations that strong particle acceleration involves the energization of large-scale (10^5 - 10^6 km) magnetic structures (e.g. Trottet 1986).

3 Ion versus Electron Acceleration

The problem we have to face when studying ions is that they are almost undetectable below energies of a few MeV where most of the energy resides. Possible diagnostics of low energy ion beams, such as the enhancement in the red wing of the Ly_α line (e.g. Canfield and Chang 1985) or the impact polarization of chromospheric lines (e.g. Vogt and Hénoux 1996 and references therein), either did not provide convincing detection of the low energy ions or lacked the many relevant observations necessary to provide unambiguous diagnostics. The most direct and unambiguous signatures of accelerated ions are: (i) the gamma-ray line emission (GRL), mostly observed in the 1-8 MeV photon energy range, which is produced by $\sim$ 1-100 MeV/nucl. interacting ions and (ii) the pion related GR emission, observed at photon energies above a few tens of MeV, which is produced by $\gtrsim$ 300 MeV/nucl. ions. However, within the sensitivity of present GR burst detectors, these high energy emissions could only be detected during the strongest flares. Only $\sim$ 50 events with significant GRL emission have been observed since 1980 by spectrometers on board the SMM, HINOTORI, GRANAT, CGRO and YOHKOH spacecrafts. Such a small number of events precludes us from performing statistical studies similar to those discussed in the previous section. Nevertheless they bring unique pieces of information which allow us to investigate how ion acceleration is related to electron acceleration.

Since the early observations of flares by the gamma-ray burst (GRS) spectrometer on SMM it has become clear that the time evolution of the prompt GRL emission in the 4-7 MeV band and of the > 10 MeV emission is similar

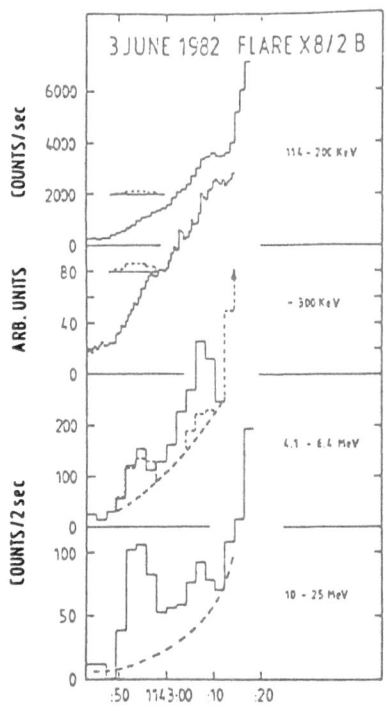

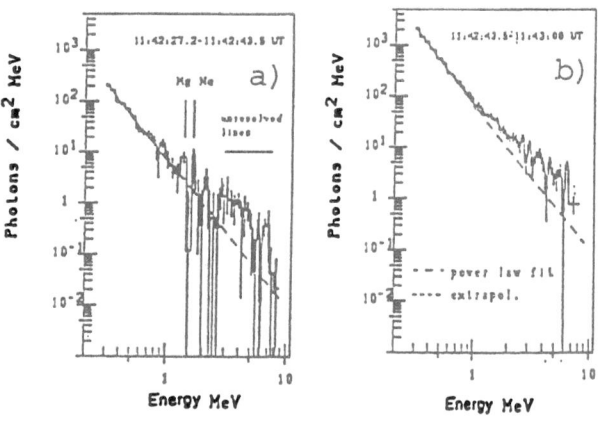

Fig. 3. Top: Temporal behavior of the 3 June 1982 event before the main intense flare maximum (from Rieger (1994)). Bottom: The gamma-ray spectra during two intervals of times at the beginning of the 3 June 1982 flare (from Trottet et al. 1994).

to that of the HXR emission to within the instrument time resolution of 1-16 sec (Forrest and Chupp, 1983). This was subsequently confirmed by observations made by other instruments such as e.g. PHEBUS on GRANAT (Talon et al. 1993). Furthermore, Figure 3 illustrates that when the statistics are sufficient, canonical GRL spectra can be detected before the fast rise of the HXR emission (Chupp et al. 1993; Trottet et al. 1994). These findings clearly indicate that electrons and ions are accelerated simultaneously since the very beginning of a flare. It should be emphasized that these observations rule out the two-phase acceleration model first proposed by Wild et al. (1963).

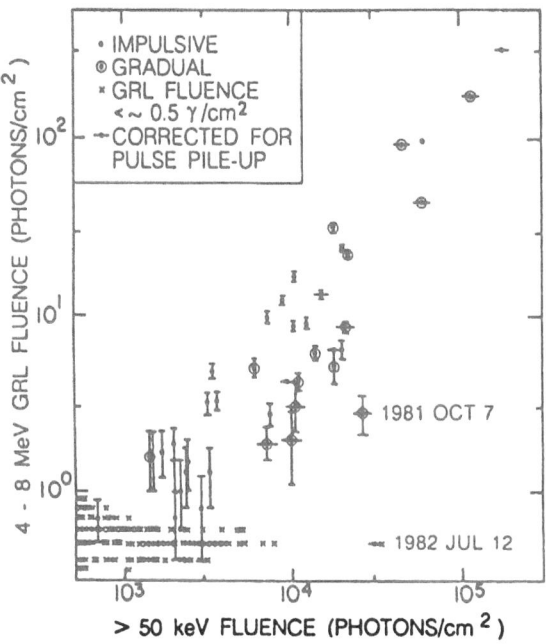

Fig. 4. Plot of GRS/SMM 4-8 MeV GRL fluence vs HXRBS/SMM > 50 keV fluence (from Cliver et al. 1994).

Forrest (1983) first showed that the > 300 keV HXR fluence is correlated with the 4-7 MeV GRL fluence. This was subsequently confirmed by Vestrand (1988). A similar correlation was also obtained between the 40-140 keV fluence and the 2.2 MeV neutron capture line fluence (Vestrand 1991). Figure 4 (from Cliver et al. (1994)) displays a plot of GRS 4-8 MeV GRL fluence versus HXRBS > 50 keV fluence. The figure shows that both quantities are reasonably well correlated. In particular: (i) there is no "population" of large > 50 keV fluence events that lack detectable GRL emission; (ii) the correlation holds down to the nominal detection threshold (~ 0.5 ph cm^{-2}) of GRS for GRL emission. This supports the idea that > 50 keV electrons and $\gtrsim 1$ MeV ions that interact in the solar atmosphere to produce > 50 keV X-rays and gamma-ray lines arise from a common accelerator. Moreover, within the GRS sensitivity, GRL flares do not appear to constitute a special class of flare. Consequently all flares may possibly accelerate > 1 MeV ions, as was first suggested by Forrest (1983). Electron-dominated events (Rieger and Marschhäuser (1990)), in which line emission is apparently missing or muted, does not contradict the above statement. Indeed, Cliver et al. (1994) noticed that, even if the GRL emission observed in these events (upper limit of 2.5 ph cm^{-2}) is reduced to the GRS instrumental background of ~ 0.5 ph cm^{-2}, the altered data points remain within the scatter of points lying in the lower left corner of Figure 4.

If the three points discussed above (i.e. similar HXR and GRL time evolution; detection of GRL emission at the very beginning of flares; correlation between the HXR and GRL intensities) are taken together they strongly suggest that ion and electron acceleration should proceed through the same macroscopic scenario of flare energy release although both species of particles may be accelerated by different processes.

4   Ion Chemical Composition and Energy Content

GRL spectroscopy constitutes our unique source of information on the chemical composition, energy spectrum and energy content of 1-100 MeV/nucl. ions. The theory of GRL production was well established (Lingenfelter and Ramaty 1967) before the first observations of significant GRL emission from solar flares (Chupp et al. 1973). A canonical GRL line spectrum comprises:

1. Prompt narrow lines produced by nuclear interactions of accelerated protons and α particles in the ambient solar atmosphere. The strongest lines at 4.44, 6.13, 1.63, 1.37, 1.78 and 0.85 MeV result respectively from deexcitations in ^{12}C, ^{16}O, ^{20}Ne, ^{24}Mg, ^{28}Si and ^{56}Fe. The relative intensities of these lines depend on the ambient gas composition and on the spectra of incident protons and α particles (α/p).

2. Prompt broad lines produced by the inverse reactions, i.e. by the interactions of accelerated C, O, Ne, Mg, Si and Fe on ambient H and He. The relative intensities of these lines depend on the chemical composition and the energy spectra of the accelerated ions and on the H/He ratio in the ambient medium.

3. Delayed lines at 511 keV and 2.22 MeV from electron-positron annihilation and from radiative capture of thermalized neutrons by H atoms respectively. The 2.22 MeV line is a signature of ~ 1-100 MeV neutrons which are produced by the incident interacting particles and which stream downward towards the photosphere where they are thermalized. The 2.22 MeV to the 4.44 MeV line fluence ratio ($\Phi_{2.2}/\Phi_{4.4}$) is sensitive to the

hardness of the incident particle spectrum but, due to its photospheric origin, it is strongly attenuated or not detected for flares occurring close to the solar limb (e.g. Ramaty and Murphy 1987).

Theoretical computations of GRL spectra have been carried out for both thin and thick target interaction models (Ramaty 1986) by using numerical codes developed by Ramaty et al. (1979). In principle estimates of the incident particle spectrum, the chemical composition of ambient gas in the interaction region and the accelerated particles can be obtained by fitting a theoretical spectrum to an observed one and by measuring $\Phi_{2.2}/\Phi_{4.4}$ when significant 2.22 MeV line emission is detected.. A thick target interaction model is generally assumed because nuclear line emission is thought to be produced in rather compact and dense ($> 10^{12}$ cm^{-3}) sources located between the corona and the photosphere (e.g. Chupp 1984; Ramaty and Murphy 1987).

4.1   Chemical Compositions of the Ambient Gas and of the Accelerated Ions

Detailed studies of GRL spectra were first performed on the 27 April 1981 flare detected by the GRS/SMM instrument (Murphy et al. 1990; Murphy et al. 1991). This event was located close to the west limb, so that the 2.22 MeV line could not be used to estimate the hardness of the incident particle spectrum. This spectrum was represented by a Bessel function, and its hardness, assumed the same for all particle species, was left as a free parameter. The main results of these studies were:

1. The ambient composition of the GRL-producing region is consistent with coronal abundances: the abundances of elements such as Mg, Si and Fe, which have low (< 10 eV) first ionization potential (FIP) relative to the abundances of C, N and O which have FIPs > 11 eV are enhanced with respect to the photospheric composition. However, the abundance ratio Ne/O ≈ 0.38 was estimated to be more than two times larger than the coronal value, 0.15, derived from impulsive solar energetic particle (SEP) events, although Ne has a FIP of 21.6 eV. The Ne/O abundance ratio is known to vary from flare-to-flare or from one active region to another but its value seldom exceeds 0.2 (Saba and Strong 1993; McKenzie and Feldman 1992).

2. The abundances of accelerated ions with low FIPs and of ^{20}Ne relative to the abundance of O are enhanced with respect to the coronal composition. Such an enrichment in heavy ions is also observed in SEP impulsive events (Reames et al. 1994).

This kind of study has been recently performed on a larger sample of events. Share and Murphy (1995) have measured the relative intensities of narrow lines detected in 19 GRL flares observed by the GRS/SMM instrument. They assumed that the broad line component keeps the same shape as that derived for the 27 April 1981 event. Flare-to-flare variations in line fluxes revealed that line fluxes from elements with similar FIPs correlate well with one another and that Ne behaves as a high FIP element. In contrast the low-FIP to high-FIP line ratio, R_{FIP} was found to vary by a factor of ~ 4 from flare to flare. This range is similar to the enhancement in low-FIP relative to high FIP element abundances found in the corona as compared to the photosphere and with the varying abundance ratios inferred in active regions from EUV and X-ray observations (e.g. Meyer 1993; Widing and Feldman 1995). Moreover Share et al. (1996), using measurements (7 of the 19 flares) of the positronium continuum and of the electron-positron annihilation line at 511 keV, have provided some evidence that GRL emission from flares with the lowest values of R_{FIP} tend to be produced deep in the chromosphere, where the abundances should be close to photospheric. They concluded that protons and α particles accelerated in different flares may interact at significantly different depths. Share and Murphy (1995) have also shown that the Ne/O line ratio depends significantly on the low energy spectrum of accelerated protons and α particles. This effect is expected from the excitation cross sections of Ne when the α/p ratio is small ($\lesssim 0.1$). Consequently: (i) the Ne/O or Ne/C abundance ratios can be used to estimate the hardness of the incident particle spectrum; (ii) the use of a power law, which produces a steeper particle spectrum at low energies than does a Bessel function, will lower the enhanced Ne abundance derived by Murphy et al. (1991).

The narrow line intensity measurements provided by Share and Murphy (1995) have been further analyzed by Ramaty et al. (1995) and Ramaty et al. (1996) who reached somewhat different conclusions. They assumed that the incident particle spectrum is an unbroken power law extending down to 1 MeV/nucl. and that the chemical composition of accelerated ions is the average composition measured in impulsive SEP events i.e. Ne/O, Mg/O, Si/O and S/O have been enhanced by a factor $f_{Ne-S} = 3$ and Fe/O by a factor $f_{Fe} = 10$ relative to the coronal composition. They used the nuclear code of Ramaty et al. (1979) to calculate the line fluences and compared these theoretical values with the observed ones to derive incident particle spectra and the ambient composition of the GRL-producing region. Their results indicate that the power law index of the incident particle spectrum varies from flare-to-flare, ranging from $\sim$ -3 to $\sim$ -4.5, and that R_{FIP} is always enhanced relative to its photospheric value, but its variability from flare-to-flare is limited to a narrower range than that inferred from EUV and X-ray observations.

Despite the differences in some of their conclusions, the works of Share et al. (1995, 1996) and of Ramaty et al. (1995, 1996) converge on two issues: (i) the incident particle spectrum should be a power law extending down to 1 MeV/nucl. in order to avoid too high values of Ne/O; (ii) there is some variability of R_{FIP}, i.e. of the composition of the GRL-producing region, from flare-to-flare. A preliminary study of the evolution with time of target abundances (Murphy et al. 1996) provided some indications that the ambient composition may even vary during one given flare. If this is confirmed, this is consistent with the results of multi-wavelength studies of some GRL flares which indicates that particle related radiations are successively emitted from different sites (Chupp et al. 1993; Trottet et al. 1994).

4.2   GRL Emission During the 1 June 1991 Giant Flare

Despite the fact that the 1 June 1991 event was located behind the limb, strong HXR and GRL emission has been observed by the PHEBUS instrument on GRANAT (Barat et al. 1994). This flare, which was in full view for the instrument on ULYSSES, is one of the giant flares reported by Kane et al. (1995). It was estimated that the HXR and GRL emission was produced in the corona, at an altitude of at least 3000 km above the photosphere. This

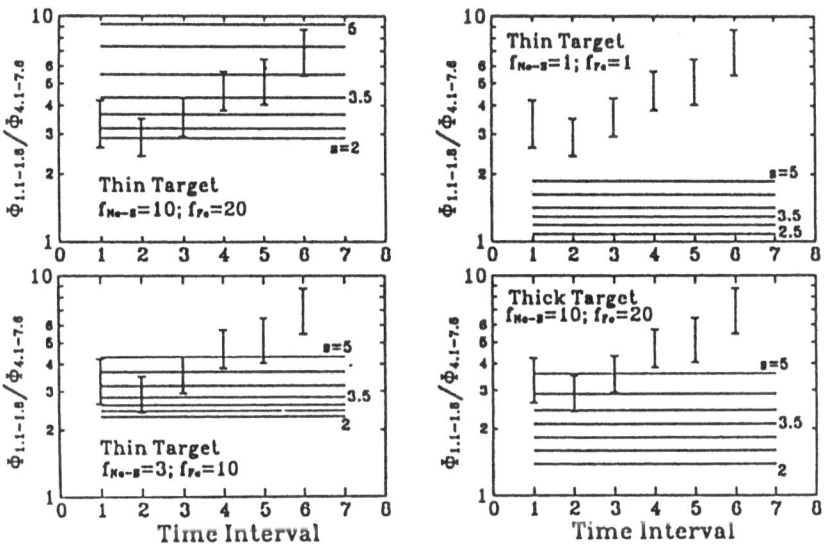

Fig. 5. Comparison between observed (vertical error bars) and computed (horizontal lines) values of R during the 1 June 1991 behind the limb GRL flare observed by PHEBUS/GRANAT (see text) (from Trottet et al. 1996).

is the first instance that it is possible to associate observed GRL emission with accelerated particle interactions taking place in the corona. Trottet et al. (1996) have performed an analysis of the temporal evolution in the flare of the $\Phi_{1.1-1.8}$ and $\Phi_{4.1-7.6}$ GRL fluences measured respectively in the 1.1-1.8 MeV and 4.1-7.6 MeV energy bands. $\Phi_{1.1-1.8}$ includes narrow and broad lines from low-FIP elements and Ne (i.e. from heavy elements) and $\Phi_{4.1-7.6}$

includes narrow and broad lines from low-FIP elements (i.e. from lighter elements). Figure 5 shows the comparison of the measured fluence ratio, R $= \Phi_{1.1-1.8}/\Phi_{4.1-7.6}$ (vertical errors bars), with theoretical values of R (horizontal lines) computed with the numerical codes of Ramaty et al. (1979). The computations were made for thick and thin target interactions and for a range of values of the index -s of the incident particle spectrum, which was assumed to be a power law. The target medium was taken to be of coronal composition and two sets of heavy element enhancement factors (see above) were considered for the accelerated particle composition: $f_{Ne-S} = 3$ and f_{Fe} $= 10$ and $f_{Ne-S} = 10$ and $f_{Fe} = 20$ which represent respectively average and upper limit values of the enhancement factors measured for impulsive SEP events (Reames et al. 1994). The examination of Figure 5 reveals that: (i) a thick target model cannot account for the largest values of R, unless extremely steep spectra are considered (s < -5), even if the upper limits f_{Ne-S} $= 10$ and $f_{Fe} = 20$ are considered (Figure 5 lower right); (ii) a thin target model can account for the observations when the upper limits of the enhancement factors are considered (Figure 5 upper left); (iii) a coronal composition for the accelerated particles ($f_{Ne-S} = 1$, $f_{Fe} = 1$) is definitively inconsistent with the data.

This analysis has brought, for the first time, evidence for the existence of a substantial thin target GRL emitting source located in the corona. This is consistent with the predictions of model calculations of Hulot et al. (1992). Moreover, it suggests that the GRL emitting region, for this flare, is most probably extended. HXR observations of this flare also suggest that a large coronal volume is involved (Kane et al. 1995). It also constitutes the unique confirmation that the accelerated particles which produce gamma-ray lines exhibit heavy element enhancements. Finally the comparison of Figure 5 bottom left and upper left panels indicates that the increase of R with time reflects either a steepening of the particle spectrum with time or an enrichment in heavy ions in the course of the flare, or a combination of both effects. As transport effects do not play any role in the thin target source, this possible increase of heavy element enrichment with time would reflect an intrinsic property of the acceleration process itself.

4.3   Energy Content in Ions

Figure 6 shows the energy content W_{ion} in ions that has been estimated by Ramaty et al. (1995) for the 19 SMM GRL flares and for both thin and thick target interactions. W_{ion} ranges from about 10^{30} to well over 10^{32} ergs. W_{ion} is comparable or even exceeds the energy contained in nonrelativistic $\gtrsim 20$ keV electrons that produce the HXR emission (see section 2). This supports the suggestion of Simnett (1986) that ions may be energetically dominant in solar flares. Most of the energy resides in protons and α particles. These estimates depend however on the composition of the medium and of the heavier

accelerated nuclei. Estimates shown in Figure 6 which were made for coronal ambient composition and impulsive SEP composition for the accelerated
heavy nuclei (see 4.1) have to be taken as lower limits of W_{ion}. Indeed if photospheric composition is considered for both the target and the accelerated
particles, ~ 5 to 10 times more protons would be needed to produce the same
GRL yield (Ramaty et al. 1993). The reason is that the enrichment in heavy
elements increases the contribution of broad lines to the total GRL yield.

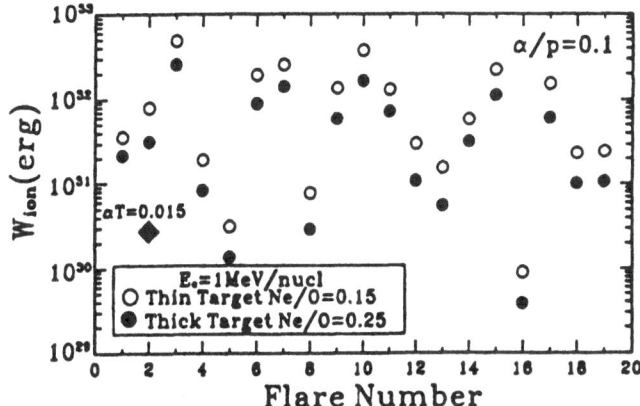

Fig. 6. Energy content in ions required to produce the GRL emission from 19
GRS/SMM flares; the assumed spectra are power laws down to 1 MeV/nucl.
with a flat extension to lower energies (from Ramaty et al. 1995).

Though the energy contained in flare accelerated electrons and ions appear to be globally comparable, variations of the ratio e/p of the number of
> 500 keV electrons to the number of > 30 MeV protons, from flare-to-flare
or even within a given flare have been reported in the literature (e.g. Chupp
et al. 1993; Marschhäuser et al. 1994; Ramaty et al. 1993). Such variations,
which do not exceed a factor of ~ 10, do not rule out the fact that protons
are energetically important, but rather indicate that the relative efficiency
of electron vs proton acceleration varies not only from flare-to-flare but also
within a given flare. It has been suggested that this could be understood
in terms of acceleration in magnetic structures with different characteristics
(Chupp et al. 1993, Trottet et al. 1994). Similarly the apparent lack of significant GRL emission in electron-dominated events does not necessarily mean
that ions are unimportant in these flares: (i) these events are small GRL
events that do not seem to constitute a special class of events (see section 3);
(ii) for one such flare detected by PHEBUS on 11 June 1990 (Trottet et al.
1993; Vilmer and Trottet 1996), we have estimated an upper limit of W_{ion}

$\sim 10^{29}$ ergs above 1 MeV which is in the range of the total energy in > 20 keV electrons ($\sim 4\ 10^{29}$ ergs).

5 High Energy Particles

Any realistic flare acceleration model has to account for the production of high energy particles, relativistic electrons and up to GeV/nucl. ions, even though these particles are not energetically important. The most direct, rather unique, sources of information on high energy interacting ions come from measurements of the high energy (> 10 MeV) pion-related GR emission they produce and from direct detection of high energy escaping ions or neutrons either in space or by ground-level neutron monitors on the Earth. These emissions have been only detected during a few large flares. The observational and theoretical aspects have been extensively discussed in the literature (see e.g. Chupp 1984, 1996; Mandzhavidze and Ramaty 1993 and references therein). The following presentation will be focussed on timescales involved in the production of high energy particles.

Figure 7 (top) shows the time evolution of the 10-24 MeV and 60-95 MeV count rates observed by the PHEBUS instrument on GRANAT during the 24 May 1990 event (Debrunner et al. 1996). The event comprises two impulsive peaks of $\sim$ 20-30 s duration. In the highest energy band these two impulsive peaks are followed by a long duration tail. A study of the PHEBUS response function indicates that most of the tail emission is due to high energy GR photons and that $\lesssim 30\%$ of the count rate recorded above 75 MeV is due to the contribution of > 500 MeV neutrons after 20:53 UT. The bump at $\sim$ 20:50 UT, which corresponds to the onset of neutrons in the neutron monitor data, is however found to be the signature of the arrival of these high energy neutrons at GRANAT. Figure 7 (bottom) displays count spectra above 10 MeV measured during the first (left) and second (right) impulsive peaks. The solid line represents the count spectrum expected from a power law photon spectrum, which was fitted to the data up to 40 MeV and extrapolated to 100 MeV. During the first peak the observed spectrum is steeper than the power law above $\sim$ 50 MeV so that the GR emission is essentially due to bremsstrahlung radiation from relativistic electrons. In contrast during the second peak there is a clear excess of photons $\gtrsim$ 50 MeV, when compared with the extrapolated power law spectrum. This is characteristic of neutral pion decay radiation produced by $\gtrsim$ 300 MeV/nucl. ions. GR emission during the tail was also found to be due to neutral pion decay radiation. Debrunner et al. (1996) showed that the ions at the origin of pion decay radiation also produced the neutrons observed at Earth up to an energy of $\sim$ 2 GeV. Thus GeV ions have been accelerated on a timescale comparable to the time interval between the two impulsive peaks, i.e. in less than 20 s. In fact a detailed examination of spectra accumulated every 4 s reveals that the acceleration timescale of these GeV ions is less than 10 s.

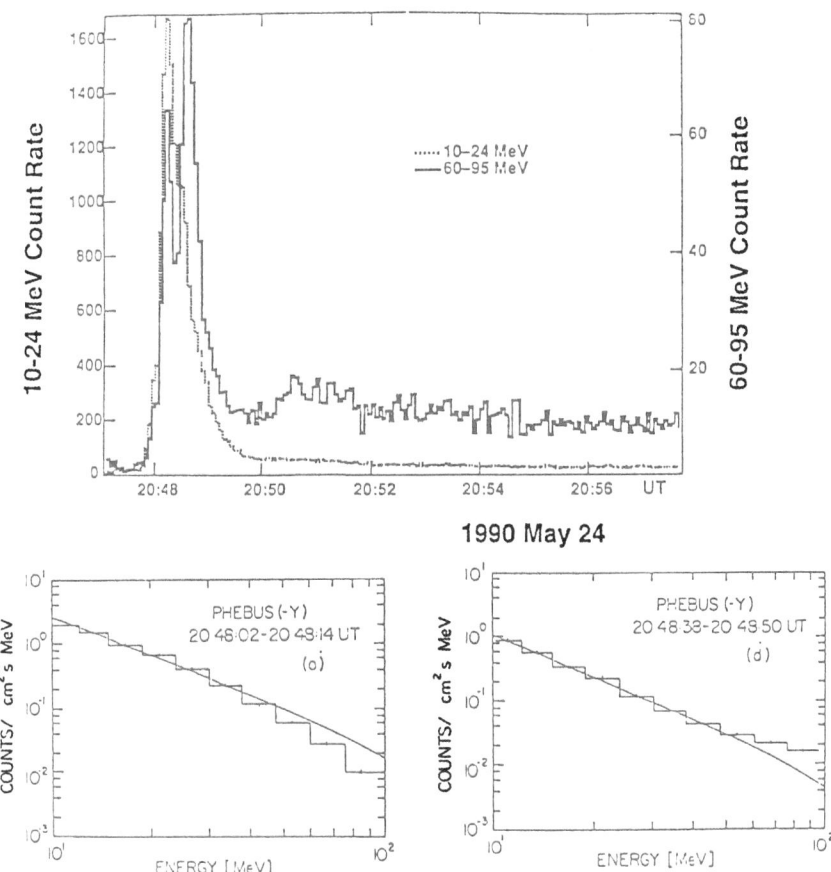

Fig. 7. Top: The count rates (cps) in the 10-24 MeV and 60-95 MeV gamma-rays measured by PHEBUS/GRANAT during the 24 May 1990 flare. Bottom: The count rate spectra from 10 to 100 MeV (histrograms) and a power law photon spectrum convolved with the PHEBUS instrumental response, giving the best fit to the 10-48 MeV count rate spectrum and extrapolated to 100 MeV (solid lines), during the first peak (left) and during the second peak (right) of the 60-95 MeV count rate (from Debrunner et al. 1996).

In contrast to this fast acceleration of high energy ions the existence of the long duration tail from neutral pion decay photons indicates that these ions must also interact for a long time in the solar atmosphere. There are other observations of long lasting pion-related GR emission. The most striking example is the 11 June 1991 flare for which 50 MeV to 2 GeV pion decay GR were observed with the EGRET/CGRO spark chamber telescope for eight hours after the main flare phase (Kanbach et al. 1993). Pion decay GR up to GeV energies were also observed for about two hours for the 15 June 1991 flare with GAMMA-1 (Akimov et al. 1991). Two alternative interpretations of these long lasting high energy GR emissions have been proposed in the literature: (i) the emission is produced by particles accelerated during a few minutes in the main phase of the flare and then trapped in magnetic loops from which they precipitate into the lower atmosphere, (ii) the particles are continually accelerated. This problem has been recently reviewed in Hudson and Ryan (1995). It can be concluded that:

1. Though the particle storage can in principle explain the time evolution of neutral pion decay photons during the 11 June 1991 flare (Mandzhavidze and Ramaty 1992), a more detailed analysis, including GRL and cm-λ radio observations, indicates that it must be ruled out (Mandzhavidze et al. 1996). This work concludes that a model of episodal acceleration and consequent trapping could be consistent with the data, but this model has not yet been developed.
2. Using correlated time histories in the Hα, soft X-ray, m-λ and cm-λ radio emissions with the GRL histories to determine the behaviour of the particles producing these high emissions, Akimov et al. 1996 conclude that continual acceleration should be at work during the 15 June 1991 flare.

In conclusion, it seems that particle storage alone cannot explain the long-lasting high energy GR emission and that acceleration of high energy ions should be at work, well after the main phase of flares. Whether this acceleration is continual or not in the late phase remains an open question.

6 Summary and Concluding Comments

We have summarized observations of HXR and GR signatures of flare accelerated particles which interact with the solar atmosphere. These observations provide constraints which must be met by any proposed flare energy storage, release and acceleration model. We have shown that:

1. There is no evidence for physically distinct classes of flares.
2. Energetic electrons and ions must not be considered separately.
3. Electrons and ions have to be accelerated, up to very high energies, on fast timescales, from less than one second to a few seconds, but acceleration may occasionally also go on for tens of minutes to hours.
4. The energy contained in > 1 MeV/nucl. ions is on the average comparable to the energy contained in > 20 keV electrons, so that ions may play a dominent role in flares. However, there is a strong variability, from flare-to-flare and even during one given flare. An extreme case of such

a variability is provided by "electron dominated" transient bursts which can occur any time during a flare.

5. Finally we have emphasized that there is a preferential acceleration of heavy ions.

A model which meets all the above constraints has not been developed yet. Three different acceleration mechanisms are usually considered in the literature: stochastic, direct electric field and shock acceleration (e.g. Heyvaerts 1981). It is clear from points (3) and (4) that all the three basic acceleration mechanisms should be considered simultaneously in a realistic flare model. Point 3 suggests that the relative efficiencies of these processes vary from flare-to-flare, probably in relation to the characteristics of the magnetic structures into which acceleration takes place. Point 4 is consistent with stochastic acceleration by resonant wave-particle interactions (e.g. Miller and Vinas 1993, Miller and Reames 1996). It has been emphasized that the predictions of the modeling of an active region as a complex system in a self-organized critical state ("avalanche" model) are consistent with the measured size distribution of flares and with the statistical properties of fast pulses frequently observed during HXR events. Moreover Anastasiadis and Vlahos (1994) have shown that, in such complex systems, randomly-distributed energy release sites can set up rapidly and accelerate particles up to high energies on adequate timescales. The "avalanche" model thus appears as an attractive and promising approach for the understanding of flare particle acceleration, as well as energy storage and release, in that it provides a way to relate the physics of acceleration to that of individual small-scale explosive events. It also accounts for the dynamics of the active region magnetic fields. Nevertheless the "avalanche" theory and its relation to the true solar magnetic field is still in its infancy. In this respect Bastian and Vlahos (1996) have pointed out that combining three-dimensional MHD simulations with cellular automaton model will be an important research tool for the future.

References

Akabane K. (1956): PASJ **8**, nos 3-4

Akimov V.V. et al. (1991): 22nd ICRC **3**, 73

Akimov V.V. et al. (1996): Solar Physics **166**, 107

Anastasiadis A., Vlahos L. (1994): ApJ **428**, 819

Aschwanden M.J., Schwartz R.A., Alt D.M. (1995): ApJ **447**, 923

Barat C., Trottet G., Vilmer N. et al. (1994): ApJ **425**, L109

Bastian T.S., Vlahos L. (1996): in Coronal Physics from Radio and Space Observations, ed. by G. Trottet, Lecture Notes in Physics, submitted

Benz A.O. (1985): Solar Phys. **96**, 357

Bromund K.R., McTiernan J.M., Kane S.R. (1995): ApJ **455**, 733

Canfield R.C., Chang C.C.. (1985): ApJ **295**, 275

Chupp E.L., Forrest D.J., Higbie P.R. et al. (1973): Nature **241**, 333

Chupp E.L. (1984): ARA&A **22**, 359

Chupp E.L., Trottet G., Marschhäuser H., et al. (1993): A&A **275**, 602

Chupp E.L. (1996): in High Energy Solar Physics, ed. by R. Ramaty, N. Mandzhavidze, X.-M. Hua, AIP Conf. Proc. **374**, 3

Cliver E.W, Crosby N.B., Dennis B.R. (1994): ApJ **426**, 767

Crosby N., Aschwanden M.J., Dennis B.R. (1993): Solar Phys. **143**, 275

Datlowe D.W., Elcan M.J., Hudson H.S. (1974): Solar Phys. **39**, 155

Debrunner H., Lockwood J.A., Barat C. et al. (1996): ApJ submitted

Dennis B.R. (1985): Solar Phys. **100**, 465

Drake J.F. (1971): Solar Phys. **16**, 152

Fitzenreiter R.J., Fainberg J., Bundy R.B. (1976): Solar Phys. **46**, 465

Forrest D.J. (1983): in Positron-Electron Pairs in Astrophysics, ed. by M.L. Burns, A.K. Hardings, R. Ramaty, AIP Conf. Proc. **101**, 3

Forrest D.J., Chupp E.L., (1983): Nature **305**, 291

Georgoulis M., Vlahos L. (1996) ApJ in press

Heyvaerts J. (1981): in Solar Flare Magnetohydrodynamics, ed. by E.R. Priest, (Gordon and Breach:New York), p.429

Hudson H.S., Peterson L.E., Schwartz D.A. (1969): ApJ **157**, 389

Hudson H.S., Ryan J.M. (1995): ARA&A **33**, 239

Hulot H., Vilmer N., Chupp E.L. et al. (1992): A&A **256**, 273

Kanbach G.O., Bertsch L., Fichtel C.E. et al. (1993): A&AS **97**, 349

Kane S.R., Hurley K., McTiernan J.M. et al. (1995): ApJ **446**, L47

Kosugi T. (1985): PASJ **37**, 575

Kucera T.A., Dennis B.R., Schwartz R.A., Shaw D. (1997): ApJ, in press

Lee T.T., Petrosian V., McTiernan J.M. (1993): ApJ **412**, 401

Lin R.P., Schwartz R.A., Kane S.R., Pelling R.M., Hurley K. (1984): ApJ **283**, 421

Lingenfelter R., Ramaty R. (1967): in High-Energy Nuclear Reactions in Astrophysics, ed. by B.S.P. Shen, (Benjamin:New York), p.99

Litvinenko Yu.B. (1994): Solar Phys. **151**, 195

Lu E. (1995): ApJ **447**, L416

Lu E.T., Hamilton R.J. (1991): ApJ **380**, L89

Lu E.T., Hamilton R.J., McTiernan J.M., Bromund K.R. (1993): ApJ **412**, 841

Mandzhavidze N., Ramaty R. (1992): ApJ **396**, L111

Mandzhavidze N., Ramaty R. (1993): Nuclear Phys. B **33A, B**, 141

Mandzhavidze N., Ramaty R., Bertsch D.L., Schneid E.J. (1996): in High Energy Solar Physics, ed. by R. Ramaty, N. Mandzhavidze, X.-M. Hua, AIP Conf. Proc.**374**, 172

Marschhäuser H., Rieger E., Kanbach G. (1994): in High Energy Solar Phenomena - A new Era of Space Measurements, ed. by J.M. Ryan, W.T Vestrand, AIP Conf. Proc. **294**, 171

McKenzie D.L., Feldman U. (1992): ApJ **389**, 764

Meyer J.-P. (1993): in Origin and Evolution of the Elements, ed. by N. Prantzos et al. (Cambridge), p.26

Miller J.A., Vinas A.F. (1993): ApJ **412**, 386

Miller J.A., Reames D.V. (1996): in High Energy Solar Physics, ed. by R. Ramaty, N. Mandzhavidze, X.-M. Hua, AIP Conf. Proc.**374**, 450

Murphy R.J., Share G.H., Letaw JR, Forrest D.J. (1990): ApJ **358**, 298

Murphy R.J., Ramaty R., Kozlovsky B., Reames D.V. (1991): ApJ **371**, 793

Murphy R.J., Share G.H., Grove J.E., et al. (1996): in High Energy Solar Physics, ed. by R. Ramaty, N. Mandzhavidze, X.-M. Hua, AIP Conf. Proc.**374**, 184

Ramaty R., Kozlovsky B., Lingenfelter R.E. (1979): ApJS **40**, 487

Ramaty R. (1986): in Physics of the Sun (Dordrecht: Reidel), p.291

Ramaty R., Murphy R.J. (1987): Space Sci. Rev. **45**, 213

Ramaty R., Mandzhavidze N., Kozlovsky B., Skibo J.G. (1993): Adv. Space Res. **13**, No 9, 275

Ramaty R., Mandzhavidze N., Koslovsky B., Murphy R.J. (1995): ApJ **455**, L193

Ramaty R., Mandzhavidze N., Kozlovsky B. (1996): in High Energy Solar Physics, ed. by R. Ramaty, N. Mandzhavidze, X.-M. Hua, AIP Conf. Proc.**374**, 172

Reames D.V., Meyer J.-P., von Rosenvinge T.T. (1994): ApJS **90**, 649

Rieger E. (1994): ApJS **90**, 645

Rieger E, Marschhäuser, H. (1990): in Max91/SMM Solar Flares: Max91 Workshop 3, ed. by R.M. Winglee, A.L. Kiplinger, p. 68

Rosner R., Vaiana G. (1978): ApJ **222**, 1104

Saba J.L.R., Strong K.T. (1993): Adv.Space Res. **15**, No 7, 13

Share G.H., Murphy R.J. (1995): ApJ **452**, 933

Share G.H., Murphy R.J., Skibo J.G. (1996): in High Energy Solar Physics, ed. by R. Ramaty, N. Mandzhavidze, X.-M. Hua, AIP Conf. Proc.**374**, 162

Shimizu T. (1995): PASJ **47**, 251

Simnett G.M. (1986): Solar Phys. **106**, 165

Talon R., Trottet G., Vilmer N. et al. (1993): Solar Phys. **147**, 137

Trottet G. (1986): Solar Phys. **104**, 145

Trottet G., Vilmer N., Barat C. et al (1993): Adv. Space Res. **13**, No 9, 285

Trottet G. (1994): Space Sci. Rev. **68**, 159

Trottet G., Chupp E.L., Marschhäuser et al. (1994): A&A **288**, 647

Trottet G., Barat C., Ramaty R. (1996): in High Energy Solar Physics, ed. by R. Ramaty, N. Mandzhavidze, X.-M. Hua, AIP Conf. Proc.**374**, 153

Vestrand W.T. (1988): Solar Phys. **118**, 85

Vestrand W.T. (1991): Phil. Trans. R. Soc.London, A, **336**, 349

Vilmer N., Trottet G. (1996) in Coronal Physics from Radio and Space Observations, ed. by G. Trottet, Lecture Notes in Physics, submitted

Vilmer N., Trottet G. Verhagen H. et al. (1996): in High Energy Solar Physics, ed. by R. Ramaty, N. Mandzhavidze, X.-M. Hua, AIP Conf. Proc.**374**, 311

Vlahos L., Georgoulis M., Kluving R. et al. (1995): A&A **299**, 897

Vogt E., Hénoux, J. C. (1006): Solar Phys. **164**, 345

Widing K.G., Feldman U. (1995): ApJ **442**, 446

Wild J.P., Smerd S.F., Weiss A.A. (1963): ARA&A **1**, 291

New Ground-Based Solar Instrumentation

Pierre Mein

Observatoire de Paris, Section de Meudon, F-92195 Meudon Cedex

Abstract. Solar physics requires more and more multiwavelength observations, not only with high spatial and time resolution, but also with wide coverage in space and time. We review briefly instruments dedicated to the solar interior, visible layers and the corona. In addition to the accuracy of spectroscopy and polarization measurements, we emphasize the coverage of data sets in the $k - \omega$ diagram. New image restoration methods are reviewed, in the context of the best compromise between spatial resolution and isoplanetic field of view. Ambitious projects do exist, as well as new generation telescopes under construction. Progress in the establishment of data bases and easier data exchange between observatories using complementary facilities look very promising for the future.

1 What are the new requirements of Solar Physics?

Solar phenomena are very complex. They involve mechanisms taking place simultaneously at different levels in the Sun, and connecting many different physical quantities. Multiwavelength observations, from ground and space, from the radio range to gamma rays, become more and more essential. Many instruments are coordinated in the framework of international campaigns; in which case, complementary performances are expected. In section 2, we propose a way to compare the capabilities in terms of resolution and field of view, in space and time. The $k - \omega$ diagram is well adapted to this analysis.

In addition to instrument resolution and field-of-view, specific requirements are needed for specific purposes. Low scattered light levels are necessary for coronal observations. Low instrumental polarizations are expected from polarimeters and magnetographs. Different techniques are needed for radio, infrared and visible light detectors. In each case, imaging devices and spectrometers must be considered.

This paper cannot be exhaustive. We shall concentrate on some improvements made during the past few years, and shall refer to specialized papers for more extensive analysis. We shall review briefly the instruments dedicated to the main problems of the Solar Interior (oblateness and heliosismology), the Photosphere and the Chromosphere (especially instruments devoted to high spatial resolution, imaging spectroscopy and polarimetry, magnetographs), and the Corona (radio and optical instruments). Special sections are devoted to infrared solar astronomy, and to large telescopes presently under construction. We conclude with some comments about the future of solar observations, multiwavelength campaigns and data bases.

2   Resolution and field of data sets

Solar phenomena, and especially waves, can be plotted in the well-known $k - \omega$ diagram. Figure (1) shows schematically the limitations of data sets produced by a given instrument in a given observing mode.

We assume that the data sets are made of successive cycles, with the time period δt, during the total time interval Δt. We assume that the equivalent pixel on the Sun is δl, and that the solar area covered during the cycle δt (with or without scanning) is the square $\Delta l \times \Delta l\ arcsec^2$. So, the highest detected frequency AB is typically $\pi/\delta t$, and the smallest one CD is $2 \times \pi/\Delta t$. The spatial resolution BD is $\pi/\delta l$, and the smallest detected spatial frequency is $2 \times \pi/\Delta l$. The resolution and field in space and time are represented by the rectangle ABCD.

The spatial resolution BD is generally limited by the size of the instrument (diameter of telescope, length of interferometer) and by the seeing conditions (which will eventually be improved by adaptive optics or *a posteriori* speckle techniques). At low frequencies (boundary CD), the largest period cannot exceed 12 hours, except in the case where several compatible instruments are at different longitudes, or for polar observations (helioseismology).

The limitations AB and AC must be considered simultaneously. Of course, AB is limited by the integration time. But this is valid only in the case of small fields of view (AC close to BD). Generally, the main limitation is due to the time necessary to store the numerical data, at least in the case of optical instruments and CCD detectors. If Δl increases by a factor 2 (AC moves to the left) the frequency resolution decreases by a factor 4 (AB moves downwards), for a given speed of data storage.

Since computing facilities are becomimg faster and faster, we can expect that this limitation will not be the most important one in the future. In that case, the time necessary to scan the observed area should become the new limitation (the step of the scan should be smaller for slit spectroscopy than for imaging devices). We see again that increasing Δl by a factor 2 implies a decrease in the resolution frequency by a factor 4 (AC to the left, AB downwards). This stresses the great interest of 2D observations providing high spatial resolution over wide areas, or in other words, large isoplanetism domains. They are necessary to reduce the number of steps for a given scan. We shall mention in section 4.2 the new techniques which are used for image restoration. They are very promising because they do not limit the isoplanetism domains as drastically as do adaptive optics when correcting the wavefronts at only one level of the earth atmosphere.

Figure (2) shows a $k - \omega$ diagram including plots which characterize very roughly some of the instruments mentioned in this paper, the performance of which can be found in the literature. They are given only as examples. We shall discuss them in the following sections of this paper.

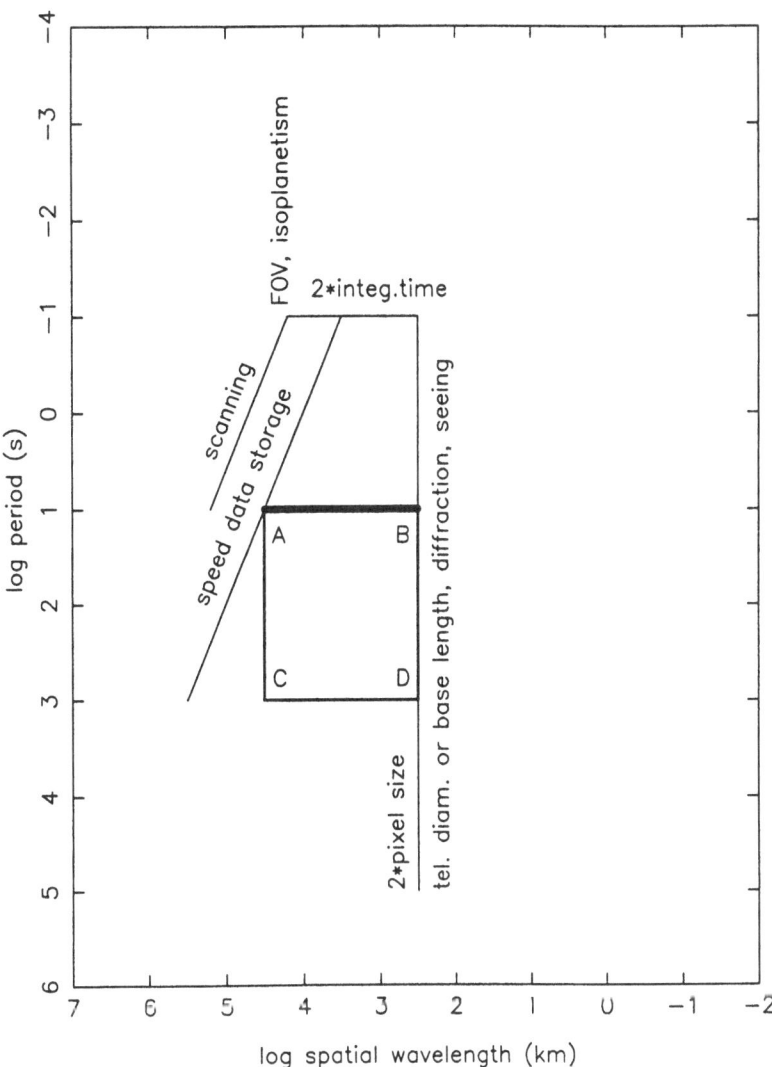

Fig. 1. $k - \omega$ diagram in log-scale. The rectangle ABCD represents the range of spatial and time frequencies covered by a given data set. The same instrument is able to provide different fields-of-view according to the required time resolution: the point A moves along the line labelled "speed of data storage". In any case, the shorter observable period is roughly twice the integration time. If the computing facilities are fast enough, the limitations are the instantaneous field of view and the integration time, or, in the case of scanning mode, the fields and times corresponding to the "scanning" line. The isoplanetism domain becomes the ultimate limitation if it is smaller than the field-of-view of the instrument.

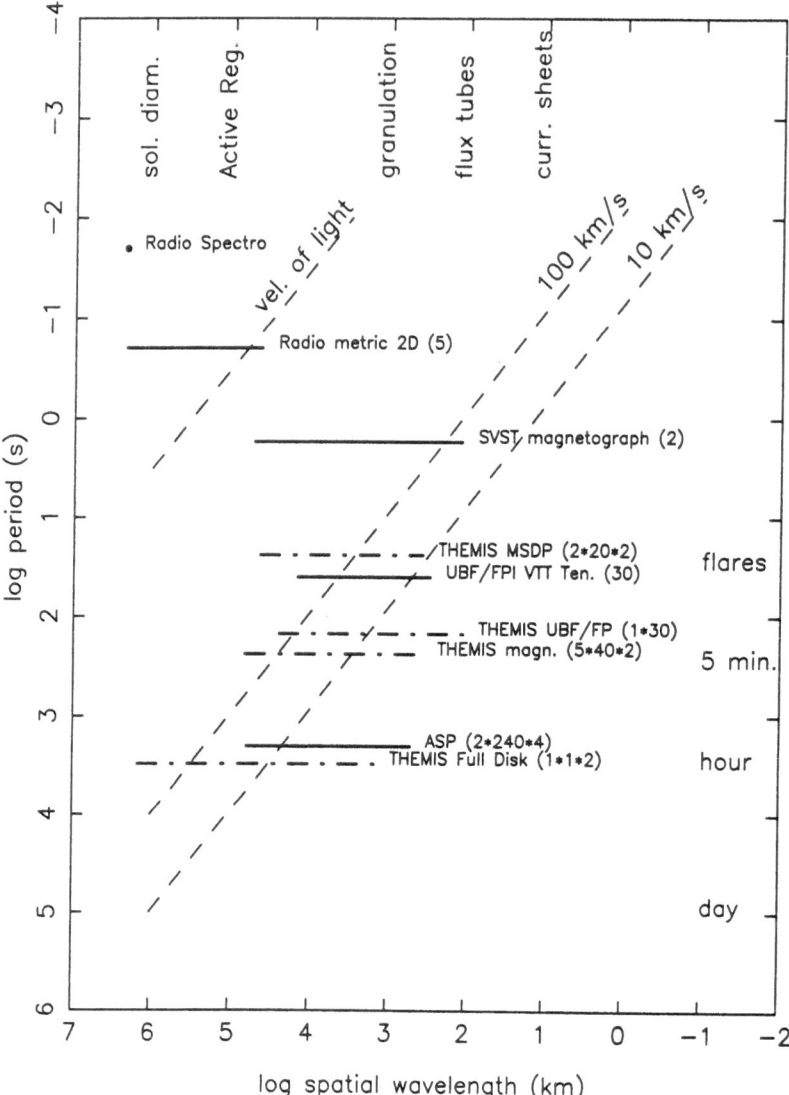

Fig. 2. Resolution and field of some solar instruments: rough estimates of capabilities in the $k - \omega$ diagram, as indicated by lines **AB** in Figure 1. Full lines refer to existing instruments, and broken lines to facilities under construction (expected values). Seeing limitations are not taken into account. The periods correspond to the case of square scanned areas (see the text). The numbers in brackets characterize the observations with respect to the number of frequencies (or the number of solar lines, times the number of wavelengths inside each line profile), and possibly the number of polarization states.

Only the upper lines AB of rectangles ABCD have been plotted in Figure 2, because the CD limit depends very much on seeing conditions (except for helioseismology). Typical scales of solar structures and events are indicated. The oblique lines correspond to the velocity of light, and to velocities which can be used as typical speeds of Alfvén and pressure waves in the visible layers of the Sun. For a given instrument, if a point A is located to the left of such a line, the propagation of particles or waves travelling with the corresponding velocity is effectively observed by the data set.

3   Heliosismology; solar diameter and oblateness

Currently the solar interior is studied by means of neutrino detection, diameter and oblateness measurements, and heliosismology. We shall not review the first topic, which deals with very specific technologies of particle detection.

Oblateness measurements are difficult because of seeing effects disturbing the profile of the solar limb. Measurements of the quadrupole moment of the Sun are in progress at the heliometer of Pic du Midi (Rösch et al., 1996). Fluctuations of the solar diameter versus the solar cycle are observed at Grasse with the Astrolabe (Irbah et al, 1994). Since 1992-93, similar measurements are available at Santiago (Noël, 1995) and San Fernando (Sanchez et al., 1995).

Heliosismology requires networks of identical instruments observing the intensity and/or the line-of-sight velocity of the Sun continuously. We know that the successful launch of SOHO is providing unique data from space. In addition to their intrinsic value, these data will help us to understand the limitations of ground-based observations. From this we will be able to define the necessary improvements that the ground-based networks will have to make in order to be effective during the solar cycles following SOHO.

From the ground, the Sun is generally observed as a whole, and the main improvements concern increasing the number of stations around the earth. IRIS (International Research on the Interior of the Sun) now observes velocity oscillations of Na D1 with 6 instruments. The last one was started at Culgoora in 1994 (Fossat, 1996). A wide collaboration is now organized between IRIS and most of the other networks, and a data base is being established. BISON (Birmingham Solar Osc. Network) observes the potassium line with 6 stations (Underhill and Speake, 1996). The Magneto Optical Filter, using Cacciani cells (Sodium lines), is now operated at JPL and Tashkent.

Other instruments also analyse fine structures of intensity and/or velocity and magnetic field. This helps especially to analyse high degree modes, and to eliminate the so-called "solar noise" introduced by solar activity. In addition to the number of stations, the spatial resolution is an important parameter. GONG (Global Osc. Network Group) uses now 6 stations, and the pixel size of 0.8" should be replaced by 0.2" (Leibacher et al., 1996). TON (Taiwan Oscillation Network) starts a fourth station at Tashkent, following Tenerife,

Huairou and Big Bear (Chou et al., 1995). The LOWL instrument, with a 25" pixel, devoted to low degree oscillations, is operated at Mauna Loa and is described by Tomczyk et al.(1995). The LOI instrument (Luminosity Oscillations Imager) similar to the corresponding SOHO experiment, using 12 pixels across the solar disk, is located at Tenerife (Appourchaux et al., 1995).

Let us mention also the MR5 instrument (Magneto. Res. analyser, 5 channels) located at Bordeaux, which observes Na D1 velocities with a 1' resolution. It is well adapted to the study of active region effects, especially in support of the GOLF experiment of SOHO (Robillot et al.,1993).

Although it is not a project really dedicated to systematic observations of helioseismology, the Precision Solar Photometric Telescopes (RISE/PSPT), operated at two sites with large CCDs, should provide accurate observations in both solar lines and the continuum (Kuhn et al., 1993).

4   The Photosphere and Chromosphere

4.1   Full disk and flare patrols

The full disk and flare patrols produce the key data to analyse the activity cycle and the solar dynamo. Systematic observations of the chromosphere at all longitudes are necessary to get continuuous data. We cannot review the instruments devoted to this important task. We mention only two examples of facilities for flare observations. At the Locarno station, CCD cameras are used in a blue light imaging spectrograph, an H_α imaging spectrograph, and an H_α filter (Rolli et al., 1995). Concerning the linear polarization in H_α flares, the PARIS instrument of Meudon (Polarimètre pour l'Activité des Régions d'Instabilités Solaires) should be operational in the near future (Vogt and Hénoux, 1996).

4.2   High spatial resolution

The physics of very fine solar structures like flux tubes or reconnection regions, is still a quite unresolved question, although it concerns the important problems of energy balance and heating mechanisms. Several solutions are proposed to increase the spatial resolution of ground-based instruments: interferometry and adaptive optics on the one hand, speckle and phase diversity methods on the other hand. Further details concerning high resolution techniques can be found in a review by C.U. Keller to be published in the IAU reports.

Active methods: interferometry and adaptive optics

Trials have been made in the visible range to perform solar interferometric imaging. The instrument proposed by Damé et al. (1996) is derived from space mission projects in the UV range.

In the case of classical solar telescopes, substantial improvements in resolution can be achieved by adaptive optics. Correlation trackers and associated tilting mirrors are now more and more common. They are especially efficient for spectroscopy and spectropolarimetry, which require integration times exceeding more than a few milliseconds. The Sacramento Peak VTT telescope records the image motions every 0.008s. A new device built by KIS and IAC is operated at the VTT telescope of Canary Islands (Collados et al., 1995; Schmidt et al., 1995).

Adaptive optics correcting higher orders of the disturbances due to the telluric turbulence have also been investigated, in particular at the VTT telescope of Canary Islands, but no device is working regularly so far.

Passive methods: speckle imaging and phase diversity technique

Unlike the adaptive optics which can be used even in telescopes feeding slit-spectrographs, the passive methods can be applied only to imaging devices. The simplest one consists of destretching individual short exposure images, and then adding them to reduce the data noise.

Speckle imaging is an efficient technique for restoring high resolution images from a great number of disturbed ones (von der Lühe, 1994). It can be applied directly to broad band images obtained by short exposures. Figure (3) shows an example obtained by the Göttingen group at the VTT telescope, with a $\pm 10nm$ band. It can also be applied to narrow-band imaging when simultaneous wide-band pictures are available. Such speckle spectroscopy can be made by a tunable filter (Keller et al., 1995).

The Phase Diversity technique uses simultaneously focused and defocused images to determine instantaneous wavefronts. The image restoration is very promising, especially because the number of necessary images is reduced, and because large fields of view can be restored. This is not the case with adaptive optics (using only one correcting mirror). The method was tested with data from the SVST (Löfdahl and Scharmer, 1994, Paxman et al., 1996). An example is given in Figure (4). The speckle imaging and the phase diversity method seem to be very promising to reconcile a large isoplanetic field of view with high spatial resolution.

4.3 1D - Spectroscopy and Spectro-polarimetry

Several spectrographs have been upgraded to observe many lines, and to include CCD detectors. They are often dedicated to flare observations. Five or six lines can be observed simultaneously by the Ondrejov Multichannel Flare Spectrograph (Kotrč, 1993), the imaging spectrograph of the Nanjing

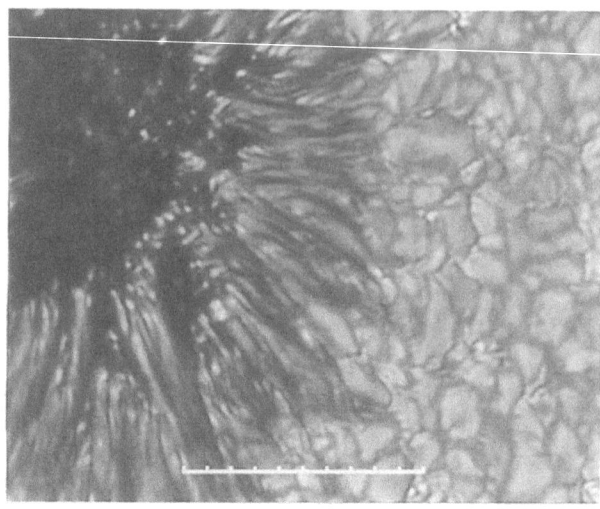

Fig. 3. Speckle reconstruction of a sunspot: spectral band $550 \pm 10nm$, tick marks at 1" distance; data from the German VTT telescope, Tenerife (*Courtesy C.R. de Boer and C. Denker*)

University (Huang et al., 1995) and the spectrograph of the Kiev Horizontal Solar Telescope (Burlov-Vasiljev, 1996). Polarization measurements are also available at the Yunnan SST.

Spectro-polarimetry using spectrographs is characterized by high spectral resolution and accurate photometry, due in particular to a low level of scattered light. Some instruments use simultaneous observations of two Stokes parameters (Mouradian et al., 1994). More usually, rotators or modulators provide four Stokes parameters, with rates as fast as possible to eliminate seeing effects. The Potsdam vector-magnetograph uses two lines simultaneously (Hofmann, 1994). The Advanced Stokes Polarimeter, with two beams, two lines, 0.25s period, and least square processing techniques (taking filling factors into account), is a very accurate magnetograph (Lites et al.,1993). We have plotted in Figure (2) the corresponding resolution in time and space. The Zurich Imaging Stokes Polarimeter ZIMPOL uses 100Hz piezoelastic modulators, and CCDs as synchronous detectors. The efficiency was increased by the use of micro-lenses in the second version (Povel, 1995). Averaging a number of exposures can lead to the observation of very low polarization rates (less than 10^{-4}).

A very promising instrument is under construction at Irkutsk. The POlarization Free SOlar Telescope uses a large modulator in front of the objective lens (95mm). Moderate spatial resolution, but high sensitivity for low polarization rates are expected (Grigoryev et al., 1996, this meeting).

Fig. 4. Restoration of solar granulation images in the G spectral band; *top right:* reconstruction by 5 × 5 subfields; *top left and bottom:* typical individual focused images; data from the SVST telescope in La Palma (*Courtesy M.G. Löfdahl and G.B. Scharmer*)

4.4   2D - Spectroscopy and Spectro-polarimetry

2D-spectroscopy is able to produce data with the highest spatial resolution. The images are not degraded by any slit-width and the 2D character allows the use of speckle or phase-diversity techniques. Moreover, the large field-of-view permits observation of large areas of the solar disk within short time scales.

Filters

Filters offer the fastest way to observe 2D areas in one wavelength, but the restoration of full line profiles requires wavelength scanning. Consequently, image destretching is necessary to achieve 2D-spectroscopy with high spatial resolution. A very interesting comparison between filters and spectrographs, used as magnetographs, can be found in papers by Lites et al. (1994), Lites (1996) and Zirin (1995).

Many instruments are now operational. The 2D spectrometer of the VTT (Tenerife) covers 20" × 20" with a pixel size of 0.2" and a 2.2pm passband; the time resolution is typically 20s for a full line profile (Bendlin, 1995; Bendlin et al., 1996; Balthasar et al., 1996). This filter is used also for spectro-polarimetry. Improvements are under construction to replace the UBF part of the filter by a second FPI, in order to increase the transmission. The SVST magnetograph (La Palma) is operated in the blue wing of the 525.0nm line, with a bandwidth of 15pm. By adding 4 exposures, high spatial reolution magnetograms are obtained. By adding 44 exposures, a sensitivity thresh-old reduced to 20 Gauss was obtained by Scharmer and Löfdahl. Typical performances of both instruments are plotted in Figure 2.

New magnetographs are now operational in Haleakala (Mickey et al., 1996) and Taiwan (Wang et al., 1995). Let us mention also the Pic-du-Midi tunable filter (12pm, 20 frames/s with destretching) operated by A. Dollfus and R. Muller.

A new filter, called the Panoramic Monochromator and built by Italy for THEMIS, will be mentioned in section 7.

Multichannel Subtractive Double Pass

The MSDP spectrographs produce simultaneously a number of 2D im-ages in all wavelengths covering some line profiles. The spectral band can be adjusted for each line to get the best compromise between exposure time and spectral resolution. A review of MSDP capabilities can be found in Mein (1995). CCD detectors are now available at the MSDP of the German Vac-uum Tower Telescope; recent observations are described in Mein et al., 1996 (this meeting). A new MSDP was installed at the large coronagraph of the Wroclaw University (Rompolt et al., 1994). We shall discuss in section 7 the high spectral resolution MSDP under construction for THEMIS.

5   The Corona

5.1   Radio wavelengths

Spectrography

Radio-spectrography is characterized by high time and frequency resolu-tion, but normally makes global measurements of the disk emission. Many instruments have been improved during the last few years, and extensive reviews can be found in the literature (Pick, 1994; Krüger and Voigt, 1995).

Following the order of decreasing frequency, we can mention the RATAN-600 spectrograph, working up to 18GHz, with time resolution of a few sec-onds (Bogod et al., 1993), the Bern instrument (3-8 GHz, 0.8s) and the new radiospectrograph of Ondrejov (100-4500MHz) described by Jiricka et al. (1993). The new radiopolarimeter in Trieste is being completed (1-3GHz, expected time resolution 0.1ms), while the old one (200-800MHz) has also been renewed. New software and image analysis techniques are now available at the Phoenix instrument operated by A. Benz et al. in Zurich (100-4000 MHz, 500 channels and 0.1s resolution) . The Tremsdorf instrument (Potsdam) now uses 4 sweep spectrographs, 14 frequencies and 2 multi-channels between 14 and 800MHz (Mann et al., 1992). In Porto, a new spectrograph (150-650 MHz) is under construction.

The ARTEMIS spectrograph has been moved from Nançay to Thermopyles. The size of the dish has been increased; acousto-optical and sweep-frequency will provide 128 channels within the range 110-600 MHz (sampling 0.1 and 0.01s); real-time compression will allow a higher data rate (Maroulis et al. 1993). The Nançay Decameter Array is presently working between 10 and 80 MHz (250 to 1000 frequencies). The four Stokes parameters are available, with time resolution of around 1s. Numerical analysis of the signals is in preparation.

Imaging radioastronomy

The spatial resolution of radio interferometers cannot be extended without extending the baseline. However many improvements can be made in 1D and 2D observations, with the increasing number of available frequency channels and the increasing time resolution. The Figure (5), proposed by L. Klein, shows the frequency ranges covered by the existing imaging facilities, and also the observing times which are quite important for full coverage of solar events. VLA, BIMA and OVRO are located at the right side of Figure 5. On the opposite side, the Nobeyama Radio-Heliograph (17 and 34 MHz, 10" and 5" resolution, 2D synthesis by software, Nakajima et al., 1994) observes before 7 UT. The GMRT and the Irkutsk SSRT (5.7 GHz, 15" resolution, 0.014s time resolution with new fast data acquisition system, Altyntsev et al., 1995) cover later time intervals up to 12 UT. The Nançay Radioheliograph (2D, 5 channels, 150-450 MHz, time resolution 0.1 and 0.5s; RH group, 1993, Kerdraon et al.,1996) is the only one observing around 12 - 13 UT. Its performance is plotted schematically in Figure (2).

A Solar Sub-mm Telescope project (Bresil-Argentina) must be mentioned in this section. It should be in operation in 1997, at 405GHz, and will be especially dedicated to flare observations (Kaufmann et al., 1994).

5.2   Optical Coronagraphs

Although well-known Lyot coronagraphs are still extensively used, solar astronomers try to build achromatic instruments allowing multi-line observations, as well as photon collectors larger than the usual lenses (Kim et al., 1995).

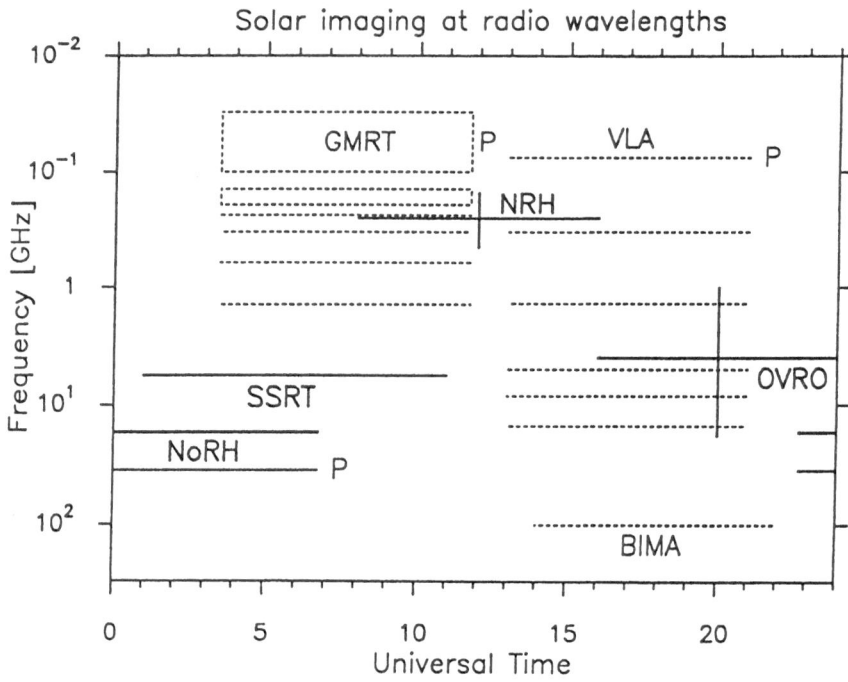

Fig. 5. Solar imaging radio facilities: frequency ranges versus observing times: full lines = dedicated to solar observations; broken lines = general radio astronomy (*Courtesy L. Klein*).

The most promising advance concerns the new technology of super-polished objective mirrors. Similar to the space-borne LASCO-C1, the mirror corona-graph of the Pic du Midi is presently working in the visible and near infrared ranges up to 2 solar radii from Sun center (Epple and Schwenn, 1995). The aperture is 6cm. After the MAC I instrument (Mirror Advanced Coronagraph, collaboration SPO, IAP) which employs off-axis reflection, the MAC II was made of a 15-cm aperture mirror and an annular field mirror, working as an inverse occulting disk. The MAC III project, with a 60 cm mirror, should provide high angular resolution in any wavelength range. The high thermal conductivity of the cooled mirror, as well as an electrostatic dust control, should ensure high quality results (Smartt and Koutchmy, 1995, Smartt et al., 1996).

The CLEAR project (Coronagraph and Low Emissivity Astronomical Re-flector), with 2 to 4 meter aperture, should be used for solar and nighttime as-tronomy, from 370nm to 35 microns. The Gregorian secondary mirror should be an adaptive optic (Beckers, 1995). This project must be considered in the framework of very large telescopes, with additional capabilities in solar and nighttime coronagraphy.

6   Infrared and Submm Solar Astronomy

In addition to the higher quality expected in the seeing conditions, the Zeeman effect is larger for IR lines, and magnetographs are increasingly used in that range. The use of large telescopes is necessary to improve both the photon flux and the diffraction limit. The 1.5 μm lines can provide weak field measurements and full Stokes polarimetry. The NIM-2 imaging polarimeter attached to the McMath-Pierce Telescope is based on a rapidly-tunable Fabry-Perot etalon (Rabin, 1995).

The level of scattered light due to the earth's atmosphere is especially low at long wavelengths, and infrared coronagraphy is a very attractive domain for solar research. Unfortunately, the available flux from the Sun is also low. Efforts are being made to solve this difficulty by building new generation large coronagraphs, as we have seen before (Koutchmy, 1995). The SiX line at 1430nm was detected at the Evans coronagraph (Penn et al., 1994). In the case of magnetic field measuements, another difficulty is the low level of the magnetic field. Attempts have been made in particular with the FeXIII line (Kuhn, 1995).

Advanced detectors are now becoming available, especially in the near infrared domain, up to 5 microns (Fowler, 1995; Vural et al. , 1995). Large arrays will soon allow polarimetry by beamsplitting techniques. A Fourier Transform Spectrometer instrument, working at 1.5μ, with 25" resolution, is described by Chang (1996). An interference spectrometer using a grating and a Fabry-Perot is proposed to provide a very small instrument which can be easily cooled (Kulagin, 1995). In the submillimeter range, a polarizing FTS was used to detect the 29 cm^{-1} HI emission-line at the limb (Clark et al., 1995).

Many details about solar infrared astronomy can be found in the proceedings of the Fifteenth NSO/Sacramento Peak Summer Workshop (1995).

7   Large Optical Telescopes

It is not possible to mention all the improvements being made to the existing large optical telescopes. They generally concern high resolution and the development of granulation trackers (VTT Sacramento Peak, VTT Tenerife). Polarimetry is also now such an important part of solar physics that fine analysis of the instrumental polarization is a major part of a typical large facility (e.g. Sacramento Peak). Some improvements concern the mechanics of the instrument, as in the Baikal telescope, which now uses a pneumo-mechanical axial support of the siderostat (Skomorovsky et al., 1996).

A few large telescopes are either in the design stage or are under construction. LEST is not financed so far, but many technical advances are indebted to the study of this project. The Dutch Open Telescope, mainly devoted to solar imaging, will be soon operational on La Palma (Rutten, 1996, this meeting).

The French-Italian THEMIS telescope will be soon in operation on Tenerife. Several descriptions can be found in the literature (Rayrole et al., 1994; Mein et al., 1989, 1994, 1995; Molodij et al., 1994; Schmieder et al.,1994). It is designed to achieve polarization measurements with high spatial resolution and with large spectroscopic facilities. Figure (6, *left*) shows the general layout of the azimuthal mount, with the spectrographs hanging below the rotating base, and the Italian narrow-band filter attached to the frame of the spectrographs. Figure (6, *right*) presents the main parts of the optics.

High resolution should be obtained thanks to the quality of the site, the aerodynamic cap-dome, the evacuated telescope, with efficient cooling of the secondary mirror, and the tilting mirror controlled by the granulation tracker (Molodij et al., 1996). The field of view is extended up to $4' \times 4'$ by the Ritchey-Chrétien optical design. The polarization analysis is achieved in the prime focus F1 to reduce instrumental effects; the temperature of the entrance window is strictly controlled to avoid any asymmetry and residual polarization of the beam.

THEMIS will be characterized by a wide set of possible observing modes, controlled by computers, which are rapidly exchangeable in many cases. In addition to usual slit-jaw images and possible wide-band pictures at the F2 focus, the Italian filter will produce high resolution monochromatic images. The "predisperser" can be used with 3 exchangeable gratings. One of them is an echelle grating providing the same dispersion as the main "echelle spectrograph". In additive mode, THEMIS will function as a vector magnetograph, observing simultaneously high resolution spectra in 10 wavelength ranges; 20 CCD cameras are put at the exit focus, to record simultaneously I+V and I-V, or I+U and I-U, I+Q and I-Q. In 2D subtractive mode, the MSDP optics put in between the predisperser and the main spectrograph, can select typically 30 channels in the profiles of two lines. This MSDP is designed to allow spectral resolutions up to 2.3pm (Mein and Rayrole, 1994). In both cases, the pixel sizes can be adjusted to achieve the best compromise with the seeing conditions.

Another subtractive mode is possible in 1D spectroscopy; it is devoted to fast scans of the full disk, to deduce magnetograms and dopplergrams from a one line-profile by the barycenter method. The principle is described in Mein and Rayrole (1989). Simultaneous H_α spectroheliograms will be obtained. The main purpose is to provide daily support for SOHO observations whenever possible.

We have plotted on Figure (2) the expected typical performance of the THEMIS modes, for an average value of the pixel size. The most accurate

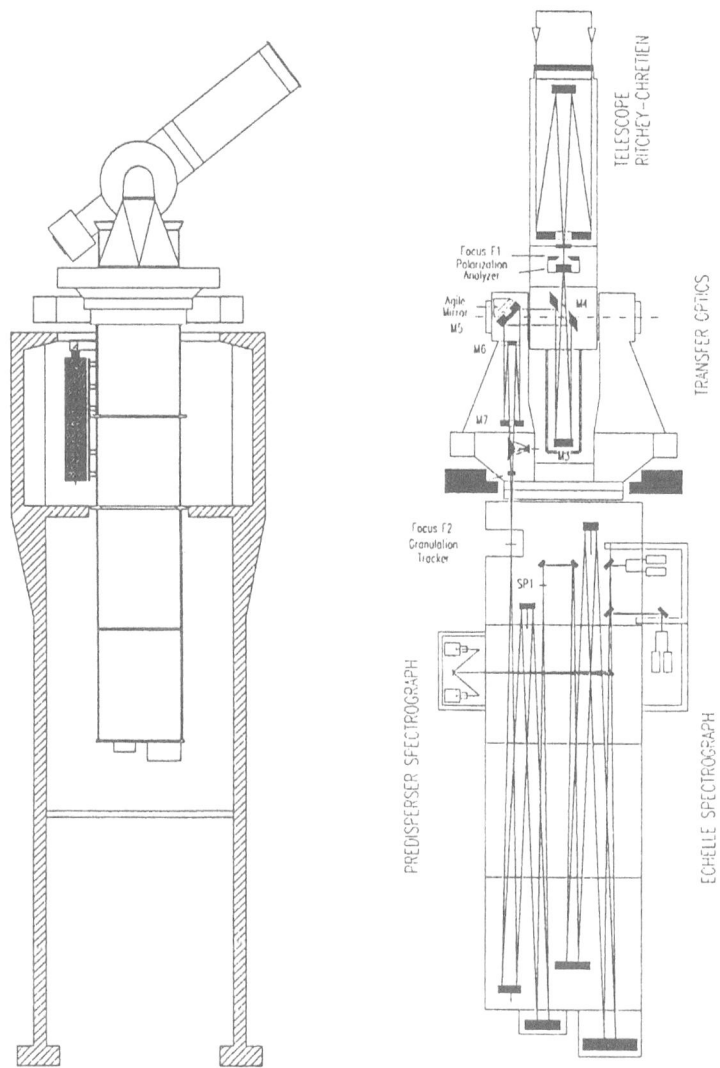

Fig. 6. THEMIS telescope: *left:* the general layout of the azimuthal mount, the vertical spectrographs and the Italian Panoramic Monochromator; *right:* the optical scheme: polarization analysis in the prime focus, tilting mirror and granulation tracker, two long spectrographs used in additive and/or subtractive mode, with a number of CCD cameras (*Courtesy J. Rayrole*).

spectro-polarimetry should be obtained by Multi-line Spectroscopy, while the highest spatial resolution should be possible with the Italian Filter and the MSDP; the fastest modes in spectroscopy should be the "Full Disk" and the MSDP.

8   Conclusion

We have noted many advances in the capabilities of new instruments, as well as in the improvements of existing facilities. Helioseismology is now available from many stations all over the world. Accurate polarization measurements, which are increasingly prevalent, are certainly a key performance for the understanding of the solar atmosphere, where most of the mechanisms are driven by the magnetic field. Super polished mirrors are very promising for multi-line investigations of the optical corona. The 2D observations are becoming more and more frequent for radio and optical instruments. Adaptive optics and efficient passive methods of *a psteriori* image restoration are very promising.

The speed of instruments dedicated to the observation of active phenomena is also a major performance. It is necessary not only for the analysis of fast events, but also for scanning large areas of the solar disk within short time scales, because magnetic loops may connect distant polarities across the corona. The schematic plot of Figure (2) shows that it is possible to find complementary performances, among the existing instruments, that fulfill the requirements of many programmes.

New facilities, such as the THEMIS telescope, are designed to run many modes with complementary capabilities, and to commute as quickly as possible from one mode into an other one. This is necessary in order to adapt the observing programmes to the kind of activity present at any time on the Sun, and also to support "on line" the programmes of several space probes.

Coordinated campaigns are often organized. They are very fruitful. The development of data bases is in progress in all fields of solar physics (helioseismology, solar activity,...), for ground-based data as well as space observations. A review of European data bases can be found in Roudier (1996). This is also a very important point.

We can say that the instrumental advances of the past few years are well oriented towards the needs of modelling and theory. They are very promising for the future.

References

Altyntsev, A.T., Grechnev, V.V., Zubkova, G.N., Kardapolova, N.N., Lesovoi, S.V., Rosenraukh, Y.M. and Treskov, T.A.: 1995, *Astron. Astrophys.*, **303**, 249.

Appourchaux, T., Toutain, T., Telljohann, U., Jimenez, A., Rabello-Soares, M.C., Andersen, B.N. and Jones, A.R.:1995, *Astron. Astrophys.*,**294**, L13.

Balthasar,H., Schleicher,H., Bendlin, C. and Volkmer,R.:1996, *Astron. Astrophys.*, in press.

Beckers, J.M.: 1995, in *Infrared Tools for Solar Astrophysics: What's next?*, 15th NSO/SPO meeting (ed. J.R. Kuhn and M.J. Penn), 145.

Bendlin, C.:1995, in *Tridimensional Optical Spectroscopic Methods in Astrophysics*, IAU Coll. 149, Marseille (ed. G. Comte and M. Marcelin), ASP Conf. Series **71**, 188.

Bendlin, C. and Volkmer, R.: 1995,*Astron. Astrophys. Suppl. Ser.*, **112**, 371.

Bogod, V.M., Vatrushin, S.M., Abramov-Maximov, V.E., Tsvetkov, S.V. and Dikij, V.N.: 1993: in *The Magnetic and Velocity Fields of Solar Active Regions*, ASP Conf. Series (eds. H. Zirin, G. Ai and H. Wang) Vol 46, 306.

Burlov-Vasiljev, K.A., Matvejev, Yu.B., and Vasiljeva, I.E.: 1996, JOSO Annual Report 1995 (ed. M. Saniga), 147.

Chang, E.S. and Drake Deming: 1996, *Sol. Phys.* **165**, 257.

Chou, D-Y. et al.: 1995, *Sol. Phys.* **160**, 237.

Clark, T.A., Naylor, D.A. and Davis., G.R.: 1995, in *Infrared Tools for Solar Astrophysics: What's next?*, 15th NSO/SPO meeting (ed. J.R. Kuhn and M.J. Penn), 139.

Collados, M. and Schmidt,W.: 1995, JOSO Annual Report 1994 (ed. M. Saniga), 100.

Damé, L., Derrien, M., Kozlovski, M., Ruillier, C.:1996, JOSO Annual Report 1995 (ed. M. Saniga), 52.

Epple, A. and Schwenn, R.: 1995, in *Infrared Tools for Solar Astrophysics: What's next?*, 15th NSO/SPO meeting (ed. J.R. Kuhn and M.J. Penn), 233.

Fossat, E., 1996: IRIS Newsletter No 9, 3

Fowler, A.M.: 1995, in *Infrared Tools for Solar Astrophysics: What's next?*, 15th NSO/SPO meeting (ed. J.R. Kuhn and M.J. Penn), 75.

Hofmann, A.: 1994, JOSO Annual Report 1993 (ed. A.v.Alvensleben), 112.

Huang, Y.R., Fang, C., Ding, M.D., Gao, X.F., Zhu, Z.G., Ying, S.Y., Hu, J. and Xue, Y.Z.: 1995, *Sol. Phys.* **159**, 127.

Irbah, A., Laclare, F., Borgnino, J. and Merlin, G.: 1994, *Sol. Phys.* **149**, 213.

Jiricka, K., Karlický, M., Kepka, O. and Tlamicha, A.: 1993, *Sol. Phys.* **147**, 303.

Kaufmann, P., Parada, N.J., Magun, A., Rovira, M., Ghielmetti, H.

and Levato,H.: 1994, Proceedings of Kofu Symposium, NRO report No. 360, 323.

Keller, C.U., Johanesson, A.: 1995, *Astron. Astrophys. Suppl.* **110**, 565.

Kerdraon, A. and Delouis, J.M.: 1996, in *Coronal Physics from Radio and Space Observations*, Lecture Notes in Physics (ed. G. Trottet), to be published.

Kim, I.S., Bougaenko, O.I., Brouevitch, V.V., Koutchmy, S., Neidig, D.F., Smartt, R.N. and Evseev, O.A.: 1995, in *Infrared Tools for Solar Astrophysics: What's next?*, 15th NSO/SPO meeting (ed. J.R. Kuhn and M.J. Penn), 239.

Kotrè, P., Heinzel, P. and Knižek, M., 1993: JOSO Annual Report 1992 (ed. A.v.Alvensleben), 114.

Koutchmy, S.: 1995, in *Infrared Tools for Solar Astrophysics: What's next?*, 15th NSO/SPO meeting (ed. J.R. Kuhn and M.J. Penn), 1.

Krüger, A.,and Voigt, W.: 1995, *Sol. Phys.*, **161**, 393.

Kuhn, J.R. and Foukal, P.V.: 1993, *Bull. Americ. Astron. Soc.*, **25**, 1184.

Kuhn, J.R.: 1995, in *Infrared Tools for Solar Astrophysics: What's next?*, 15th NSO/SPO meeting (ed. J.R. Kuhn and M.J. Penn), 89.

Kulagin, E.S.and Papushev, P.G.: 1995, in *Infrared Tools for Solar Astrophysics: What's next?*, 15th NSO/SPO meeting (ed. J.R. Kuhn and M.J. Penn), 243.

Leibacher, J. et al.: 1996, NOAO Newsletter No46, 57.

Lites, B.W., Elmore, D.F., Seagraves, P. and Skumanich, A.P.: 1993, *Ap.J.* **418**, 928.

Lites, B.W., Martinez Pillet, V. and Skumanich, A.: 1994, *Sol. Phys.* **155**, 1.

Lites, B.W.: 1996, *Sol. Phys.* **163**, 223.

Löfdahl and M.G., Scharmer, G.B., 1994: *Astron. Astrophys. Suppl.* **107**, 243.

von der Lühe, O.: 1994, *Astron. Astrophys.* **281**, 889.

Mann, G., Aurass, H., Voigt, W. and Paschke, J.: 1992, ESA-Journal, SP-348, 129.

Maroulis, D., Dumas, G., Bougeret, J-L., Caroubalos, C. and Poquerusse, M.: 1993, *Sol. Phys.* **147**, 359.

Mein, P. and Rayrole, J.: 1989, in *High Spatial Resolution Solar Observations*, Tenth Sacramento Peak Workshop (ed. O. von der Lühe), 12.

Mein, P. and Rayrole, J.: 1994, JOSO Annual Report 1993 (ed. A.v. Alvensleben), 116.

Mein P. and Rayrole, J.: 1995, JOSO Annual Report 1994 (Saniga ed.), 11.

Mein, P.:1995, in *Tridimensional Optical Spectroscopic Methods in Astrophysics,* IAU Coll. 149, Marseille (ed. G. Comte and M. Marcelin), ASP Conf. Series **71**, 350.

Mickey, D.L., Canfield, R.C., LaBonte, B.J., Leka, K.D., Waterson, M.F. and Weber, H.M.: 1996, *Sol. Phys.,* in press.

Molodij, G. and Rayrole, J.: 1994, JOSO Annual report 1993, 69.

Molodij, G., Rayrole, J., Madec, P.Y. and Colson, F.: 1996, *Astron. Astrophys. Suppl. Ser.,* **118**, 169.

Mouradian, Z., Scholiers, W. and Semel, M.: 1994, JOSO Annual Report 1993 (ed. A.v. Alvensleben), 110.

Nakajima,H., Nishio, M., Enome, S., Shibasaki, K., Takano, T., Hanaoka, Y., et al., 1994: "The Nobeyama Radioheliograph", Proceedings of the IEEE, **82**, No 5.

Noël, F., 1995: *Astron. Astrophys. Suppl. Ser.,* **113**, 131.

Penn, M.J. and Kuhn, J.R.: 1994, *Ap.J.* **434**, 807.

Paxman, R.G., Seldin, J.H., Löfdahl, M.G., Scharmer, G.B. and Keller, C.U.: 1996, *Ap.J.,* **466**, 1087.

Pick,M.:1994, JOSO Annual report 1993 (ed. A.v.Alvensleben), 104.

Povel, H.: 1995, *Optical Engineering 1995,* Vol.34, No 7.

Rabin, D., 1995: in *Infrared Tools for Solar Astrophysics: What's next?,* 15th NSO/SPO meeting (ed. J.R. Kuhn and M.J. Penn), 87.

The Radioheliograph Group: 1993, *Adv. Space Res.,* **13**, No9, 411.

Rayrole,J., Mein P., Cavallini, F.,:1994, in *Solar Surgace Magnetism,* Soesterberg NATO Advanced workshop, (ed. R.J. Rutten and C.J. Schrijver), 507.

Robillot, J-M., Bocchia,R. and Denis, N.: 6th IRIS workshop and GONG meeting, 1993.

Rolli, E. and Magun, A.: 1995 *Sol. Phys.* **160**, 29.

Rompolt, B., Mein, P., Mein, N., Rudawy, P. and Berlicki, A.: 1994, JOSO Annual Report 1993 (ed. A.v.Alvensleben), 87.

Rösch, J., Rozelot, J-P., Deslandes, H. and Desnoux, V.; 1996, *Sol Phys.* **165**, 1.

Roudier, T.:1996, JOSO Annual Report 1995 (ed. M. Saniga), 71.

Sanchez, M., Parra, F., Soler, M., Soto, R., 1995: *Astron. Astrophys. Suppl. Ser.,* **110**, 351.

Schmidt, W. and Kentischer, T.: 1995, *Astron. Astrophys. Suppl. Ser.* **113**, 363.

Schmieder, B. and Mein P.:1994, Third SOHO Workshop, Estes Park (ESA SP-373), 393.

Skomorovsky, V.I., and Firstova, N.M.: 1996, *Sol. Phys.* **163**, 209.

Smartt,R.N. and Koutchmy, S.: 1995, in *Infrared Tools for Solar Astrophysics: What's next?,* 15th NSO/SPO meeting (ed. J.R. Kuhn and M.J. Penn), 163.

Smartt, R.N., Koutchmy, S., Kim, I.S., Bougaenko, O.I., Carmichael, R.E., Hegwer, S.L. and Zimmermann, J.P., 1996: International Workshop on "Mirror Substrate Alternative", Grasse 1995 (Rozelot ed.), in press.

Tomczyk, S., Streander, K., Gard, G., Elmore, D., Hull, H. and Cacciani, A.: 1995, *Sol. Phys.* **159**, 1.

Underhill, C.J., and Speake, C.C.: 1996, *Sol. Phys.* **163**, 231.

Vogt, E. and Hénoux, J-C.: 1996, *Sol. Phys.* **164**, 345.

Vural, K. et al., 1995: in *Infrared Tools for Solar Astrophysics: What's next?*, 15th NSO/SPO meeting (ed. J.R. Kuhn and M.J. Penn), 85.

Wang, J., Ai, G., Song, G., Zhang, B., Ye, X., Nie, Y., Chiveh, T., Tsay, W. and Li, H.: 1995, *Sol. Phys.* **161**, 229.

Zirin, H.: 1995, *Sol. Phys.* **159**, 203.

Future Space Instrumentation
for Solar Physics

Ester Antonucci and George M. Simnett

[1] Osservatorio Astronomico di Torino, 10025, Pino Torinese, Italy
[2] Department of Physics and Space Research, University of Birmingham, Birmingham, B15 2TT, UK

Abstract. We review the space instrumentation that is currently being developed for studies of the Sun. Currently the main solar physics mission is SOHO, which has support from Yohkoh, Coronas I and a variety of "particles and fields" spacecraft such as Polar, Wind, Geotail and Interball. The principal new facility will be the TRACE mission, which is scheduled for launch in 1997/1998. For the medium term future, missions such as the Solar Probe, Coronas-F and Foton, plus the successor to Yohkoh are likely to be realised, at least in part. Other missions are in a definition phase, such as HESSI, SIMURIS and a STEREO mission of some form. New particle instruments which can detect solar emissions, such as ACE, will be launched before the year 2000. The ESA Horizon 2000+ program has some medium missions, yet to be defined, which should be devoted to solar studies.

1 Introduction

Before addressing the facilities which will be available in the future, it is worth reviewing what is available in the present, and what the basic limitations of the current instrumentation are. We need to try to identify what observations we would like to have, which are currently lacking, and what solar phenomena are likely to be understood better once we have them. The resources available to the various international funding agencies are necessarily restricted and therefore we recognise that even in the forseeable future some very desirable instruments, which currently are technically-feasible, may not in practice be realised.

The main solar missions of the 1990s have been Yohkoh (The Yohkoh (Solar-A) Mission, 1991) and SOHO (The SOHO Mission, 1995). This is clear from the variety of papers submitted to the recent international conferences. These missions are complemented by Coronas I, plus a number of missions in the solar-terrestrial physics area, which include Wind and the GOES satellites. A list of acronyms used in this paper is given in the Appendix. The main goal of the Yohkoh satellite is to observe energetic solar phenomena related to solar flares in the X- and γ-ray spectral regions. Thus it is important to have observations with good time resolution, which for the X-ray imaging telescopes is around 0.5 s. The best resolution of the wide-band spectrometer is $0.125 \, s$.

SOHO is not a solar flare mission, and has therefore concentrated on high spatial and spectral resolution at the expense of time resolution. SUMER has a resolution of around $1''$; CDS is $<2''$; and EIT is $2.6''$. The spectral resolution of CDS is ~ 0.03 nm, or $\lambda/\delta\lambda$ of 500 at the shortest wavelengths. For SUMER the spectral resolution is typically an order of magnitude better than CDS. The time resolution of the imaging instruments is limited by the spacecraft telemetry. Through UVCS and LASCO the corona is imaged out to 30 solar radii in the UV and visible regions, and for the first time UV spectroscopy of the extended corona is possible. The interplanetary radio instruments can, in many cases, take over from the coronagraphs and study phenomena out to 1 AU.

It is worth mentioning the capability of a spacecraft devoted primarily to extra-solar astrophysics. The Compton/Gamma Ray Observatory responds to solar X- and γ-rays with a (best) time resolution of around 1ms. However, for solar studies there is no spatial resolution and only crude spectral resolution. Nevertheless BATSE is currently the best instrument available for studying the intensity-time history of solar hard X-ray bursts.

The next question is what can we hope to achieve realistically in the foreseeable future, and does this match what we might ultimately need? We currently think that high time resolution is only needed for flare studies. This may, of course, ultimately prove incorrect, for the quiet Sun is far from quiet, and the study of spicules is one example where high time resolution could be useful. However, one might loosely argue that the time resolution needed is the spatial resolution of the instrument divided by the velocity of the phenomenon being addressed. Thus for $1''$ and 10^3 km s^{-1} there is little need to have a time resolution much better than 1s. Thus at the moment this is a reasonable goal for the imaging instruments which are looking at thermal radiation. If we turn our attention, however, to what we might ultimately need in terms of resolution, then we would realise that most plasma processes occur over length scales of metres, so that spatial resolutions of $10^{-5}{}''$ or even $10^{-6}{}''$ are desirable. This then would need to be fed back to the preceding discussion. In terms of spectral resolution, probably the most stringent requirement is to try to measure motions via Doppler shifts. To measure velocities to within 1 km s^{-1} (at about 0.1 of the resolving power), we require instrumentation with $\lambda/\delta\lambda$ of around 3×10^4. At the moment all these seem to be reasonable targets to aim at.

Hard X-ray and γ-ray detectors are different, as they are responding to non-thermal radiation which may well be coming from relativistic particles. Here 1 ms may not be enough. However, the other parameters, *i.e.* spectral and spatial resolution currently lag far behind the capabilities of the UV and EUV instruments, so there appears to be a need to improve these before worrying too much about increased time resolution. However, the ground-based studies of decimetric radio bursts have indicated that some solar bursts are composed of tens of thousands of individual bursts (Benz, 1985). This suggests that in the context of energetic particles providing a significant fraction of the energy for flares, then substantial fragmentation of the particles emitted from the primary acceleration process must be occurring. Benz put a typical upper limit of the dimension of the individual radio sources of 200 km (0.25''), and suggested that a typical event duration was 50 ms. Thus until we have hard X-ray detectors with the capability of taking an image with 0.1'' resolution every 10 ms we may not make significant progress. Although this is no reason for not trying to make advances over what is currently available. In terms of spectral resolution, much of the emission is a broad-band bremsstrahlung continuum, so that resolutions of $\lambda/\delta\lambda$ of 10 or 100 may be adequate to address the physical processes we suspect are responsible for the emission. This resolution is probably more than adequate at the moment for gamma ray line studies, as sensitivity will start to become an issue long before resolution is the limiting factor.

To complement these studies it will be important to have spacecraft which can measure the fields and particles which are coming from the Sun. Monitoring the particles is important over all energies, ranging from the solar wind to the highest energy electrons and ions emitted at the time of large solar flares. Spacecraft currently operational in these areas include Wind, Polar, Geotail and Interball. SOHO also has some capability in this area, and in the future ACE will be launched.

Having set the scene, we now discuss how the future missions currently under development or study will meet these goals.

2   ESA Horizon 2000 Plus Programme

The ESA space science programme is developing harmonically according to a long-term programme, which is articulated in two phases: the Horizon 2000 and the Horizon 2000 Plus programmes. The first phase started in 1984 with the approval of the first cornerstone constituted by the ensemble of the solar mission SOHO (Solar and Heliospheric Observatory) and the Cluster mission devoted to space plasma studies. The Horizon 2000 Plus extends the planning of ESA scientific activities initiated with Horizon 2000 to the year 2016. The two phases are conceived as an integrated programme ensuring balance among the different disciplines, continuity and coherence. The foreseen mission frequency is of about 6 missions per decade. It is very probable and desirable that some of these missions will be carried out in collaboration with NASA, with whom ESA has had very successful past joint missions.

The programme Horizon 2000 Plus was initiated with a call for mission concepts issued in June 1993 and has been formulated on the basis of the recommendations reached by a survey committee on October 1994. The programme consists of 2 or 3 cornerstones which are in principle self–standing ESA missions and 4 medium–sized missions. The cost envelopes considered for Horizon 2000 Plus are 625 MAU per cornerstone and 345 MAU per medium mission (1993 economic conditions). However, ESA recognises the budgetary constraints and has stressed that these are maximum values and that lower cost missions "should be regarded with particular favour". Thus there will be pressure on mission definition teams to avoid making huge advances in technical capability.

The proposals of the scientific community in response to the call for mission concepts were by far more numerous in the area of solar system studies than in any of the other disciplines. A breakdown of the responses by discipline is given in Table 1. Twelve out of the 41 proposals related to the investigation of the solar system were dedicated exclusively to solar research.

Table 1

RESPONSES TO CALLS FOR MISSION CONCEPTS

	Horizon 2000	Horizon 2000 Plus
Astrophysics	30	32
Solar System	**34**	**41***
Fundamental	–	29
Interdisciplinary	–	4
Miscellaneous	4	2

Topical teams of scientists were entrusted by ESA with the task of analysing the concepts presented by the scientific community. The general lines of the programme Horizon 2000 Plus were then established following the recommendations of the survey committee reached on the basis of the work of the topical teams.

The topical team working on the solar heliospheric and space plasma physics proposals considered two potential cornerstone missions. The first was a multi–spacecraft mission for stereo viewing of the Sun, where one spacecraft would be near the Earth and others might be some 60° away, possibly near the L4 and L5 Lagrange points. The second would be a spacecraft orbiting Mercury, which might also include observations of the Sun. A further mission, the Solar Probe, which is currently the 'green dream' of Horizon 2000, is conceived as a multi-agency project. Furthermore, several proposals were recognized to be suited as medium missions through competitive selection. Small monitoring missions were to be developed as joint projects by various space agencies.

3   ESA Solar-System Science

Solar system space research within ESA has traditionally proceeded in two main areas, that is, the investigation of the plasma in the Sun and in the solar system and the investigation of the solid bodies of the solar system. Missions developed in the first area were Ulysses and within the frame of the first Cornerstone: Cluster and SOHO. The unfortunate loss of Cluster in June 1996 has certainly reduced the scientific return from the first Cornerstone. Meanwhile the Giotto mission, and the future Rosetta mission, (the third Cornerstone), and Huygens, the Titan atmosphere probe, which is part of the ambitious NASA-led Cassini mission, have all been conceived to study the solid bodies of the solar system.

The Survey Committee recommendations for the future in these two main areas are the following. For the plasma in the Sun and solar system:

- participation in an international solar mission, or
- take advantage of International Space Station Alpha opportunities, or
- medium class missions.

In the area of the solid bodies studies:

- cornerstone-level mission to Mercury, and
- medium–class mission for Mars exploration

Therefore none of the proposals in solar physics research was identified as a possible mission at cornerstone level. There are two areas where ESA may collaborate, according to the Horizon–2000 Plus programme, with other agencies in the foreseeable future. These are solar tomography and the solar probe, and it it worth outlining the objectives and current status of missions in these areas.

3.1   Solar Tomography

Tomography of the solar corona is a technique necessary to remove the ambiguity inherent in the two-dimensional coronal imaging, because each pixel of an image is representing the line-of-sight integration of the volumetric emissivity through the coronal source. Tomography can be performed in two ways. The first is with a single spacecraft, and structure may be inferred by allowing the Sun's natural rotation to change the viewing angle, as attempted with Skylab and OSO data. However in these first attempts the time evolution of solar features could not be deconvolved from spatial variations. The second approach is to use multiple spacecraft to observe the Sun from different points at the same time.

The scientific research targets of a solar stereoscopy–tomography mission are essentially related to the study of the three–dimensional structure of the corona, including the formation and evolution of large–scale coronal structures (streamers, coronal holes, active regions), the formation of solar wind, the density distribution parallel and transverse with respect to the magnetic field.

In a study of three-dimensional image reconstruction by using an algebric reconstruction technique, Davila (1994) has demonstrated that under the hypothesis that four spacecraft will be available, the optimum fidelity of reconstruction can be obtained with an angular range of observing spacecraft of 135° and an angular spacing of 45° apart (evenly spaced, otherwise the fidelity of reconstruction is reduced, although still possible). The presence of a polar orbiter is not necessary for three-dimensional reconstruction. Furthermore the optimum number of observing spacecraft is four. The fidelity of reconstruction increases with spacecraft number but most of the improvement is obtained by addding the fourth spacecraft.

A possible tomography mission employing this concept has been studied by Dr J.M. Davila. The mission consists of 4 identical spacecraft observing the corona from different angular positions in the ecliptic. Of them 3 spacecraft are in an orbits with 0.5 AU perihelion and 1 AU aphelion (40° are reached in 3 months after launch) and 1 spacecraft is geosynchronous. The pointing stability is $1''$.

In this scenario the spacecraft weight is only 230 kg. Each spacecraft carries a basic lightweight instrument package ($\sim$ 50 kg):

- White Light Coronagraph
  FOV 1.3–10 $R_\odot$
  resolution $20''$/pixel
  envelope $10 \times 10 \times 150$ cm
  1K x 1K CCD
  spectral bandpass 4000-9000 Å

- Soft X–ray/UV Coronal Imager
  FOV 2 $R_\odot$
  resolution $2''$/pixel
  envelope $30 \times 30 \times 170$ cm
  $1K \times 1K$ CCD
  multilayer optics (this has a potential problem: degradation in space) or grazing incidence optics.

The Solar Stereoscopic and Heliospheric Mission considered as a programme element for solar studies by the ESA Solar Heliospheric and Space Plasma Physics topical team was ambitious, but not to the level of four spacecraft. The typical minimal mission (no redundancy, limited three-dimensional imaging) consisted of two spacecraft with identical payloads, one at L1 (0.01 AU from the Earth) and one at L4 or L5 (60° and 1 AU from the Earth).

Both satellites were equipped for studies of helioseismology (the phase shift of the observed frequency is proportional to the angle between the two space-craft), the dynamics and evolution of magnetic fields from the photosphere to the corona, and irradiance variations. The spacecraft weight was 1500 kg.

A two-spacecraft STEREO mission was proposed in 1995 in response to the NASA-MIDEX Announcement of Opportunity by Dr G.E. Brueckner et al.. The fundamental science objectives were:

- 1. To determine the role of magnetic reconnection in the heating, evolution and ejection of coronal structures by imaging large scale coronal structures, whose underlying magnetic drivers will be simultaneously determined on the disk of the Sun.
- 2. To measure the geometry, magnetic topology and evolution of coronal mass ejections and their impact with the Earth by providing complete observational coverage for their transit through the heliosphere.
- 3. To determine the structure, evolution and dynamic interaction of heliospheric phenomena such as streamers, the heliospheric current sheet, corotating interaction regions and mass ejecta by means of unique wide field observations of the heliosphere.

The proposed payload consisted of a soft X-ray imaging telescope similar to that developed for the GOES programme; a coronagraph similar to LASCO/SOHO; a Sun-Earth imager similar to the SMEI being developed for the US Air Force; and a plasma and particle analyser.

So far no firm flight opportunity has been realised for such a mission. There is currently a NASA study team looking at the 2-spacecraft option with a view to concentrating on the Sun-Earth connection. The CMEs would be studied from their initiation on the Sun, then on their passage through the interplanetary medium across 1 AU, and finally to their interaction with the Earth's magnetosphere. Discussions with possible European collaborators are due to take place in March 1997.

The results obtained with a tomography mission could be enhanced when coordinated with simultaneous space solar missions (Solar-B, SOHO, TRACE, GOES soft X-ray imager (1997)) and ground-based instruments.

3.2 Concepts for Solar Probes

In the ESA concept the main scientific goals of the solar probe Vulcan are to sample plasma, electromagnetic field and particle populations, to study structures in the solar atmosphere and corona very close to the Sun and to test theories of gravitation and relativy. The solar probe closest approach to the Sun will be ~ 4 $R_\odot$. This allows remote sensing with unprecedented resolution and *in situ* measurements in regions possibly within the Alfvénic point. More importantly it will provide a direct measurement of the solar coronal magnetic field. The scientific payload consists of *in situ* field and particle experiments, imaging and spectroscopic instruments and gravitational experiments.

This mission is very demanding from the technological point of view. Its trajectory requires either a Jupiter gravity–assist manoeuvre, or a solar sail, or electric propulsion. Another issue is that of the energy generation system. The close approach to the Sun requires also a thermal shield of special properties to protect the instrumentation at 4 solar radii.

The concept of the NASA solar probe as that of Vulcan as been developing for many years. After 17 years of study in April 1995 a Minimum Solar Mission Concept has been reached that focuses on the heating and acceleration of the solar wind in the polar coronal holes (plasma dynamics and properties, waves and plasma turbulence, polar magnetic field, spectrum of energetic particles, high–resolution imaging (≤ 70 km of coronal structures)). A 4 $R_\odot$ perihelion encounter (year 2007) is foreseen at solar activity minimum. The instrument payload is 8 kg, the power 8 watts, and the telemetry 500 bits/s. The prime mission consists of the period from 1 day before perihelion to 2h after perihelion.

During 1995 there has been a further updated of the Minimum Mission with the introduction of the 'transient viewing' concept: that is, viewing the Sun using a retractable mirror instead of viewing through the heat shield. In this design the mirror would be extended and retracted in 0.1 s. The plasma instrument would gather its data from the end of a boom extendable to the tip of the heat shield. Furthermore the end of the primary mission has been prolonged to 12 h after perihelion.

The NASA solar probe should also be considered in the framework of the Fire Program, which currently is envisaged as a joint US – Russian venture. Plamya, the Russian solar probe, has a 10 $R_\odot$ perihelion encounter which would occur simultaneously with the NASA Solar Probe with 4 $R_\odot$ perihelion encounter. A potential Solar Probe instrument package is given in Table 2, together with the instrument mass.

4   International Space Station Alpha (ISSA)

The International Space Station Alpha (ISSA) offers several accommodation sites to host scientific instrumentation. Large external payloads can be directly accommodated on the truss. Smaller payloads can be installed externally, by using accommodation facilities such as the Express Pallets. Of these two are in the Earth direction and two are in the zenith direction and therefore they are compatible with solar and astronomy missions. Smaller payloads can be also accommodated on the Material Exposure Facility.

The space station attitude will be driven by the compensation of dynamic torques, which are mainly aerodynamic and gravity gradients. The equilibrium attitude of the station itself is not particularly good for high quality solar imaging. Deviations of 3.5° peak to peak per axis and rotation rates of 0.02°/s per axis are expected. Therefore the utilization of the space station for solar research would in general require the availability of stabilized

Table 2
Solar Probe Instruments

Instrument	Mass
– Plasma spectrometer	2 kg
– Energetic particles	0.5 kg
– Plasma waves	0.5 kg
– Magnetometer	1 kg
– Visible Imager (256 x 256 pxls)	2 kg
– EUV Imagers (171 Å, 304 Å) (256 × 256 pixels)	2 kg

platforms or instruments with their own internal attitude control system.

Two options have been considered up to now for the utilization of the Space Station for solar research:

- Solar Platform for the early flight opportunities
- SIMURIS.

4.1 The Solar Platform

In 1996 ESA has completed an accommodation study for a solar payload to be assembled on an express pallet of the space station, taking advantage of the early flight opportunities. One of the express pallets in the zenith direction can support a stabilized platform which carries instruments which do not have extremely high stabilization requirements. The instruments considered in the accommodation studies are those already flown or previously selected for the EURECA platform (Table 3). These payloads could carry out mainly studies of the global Sun and of its variability over several years, emphasizing continuous space-based measurements of the solar irradiance – both the total flux and the spectral properties – from the ultraviolet to the infrared.

The solar instruments that will be selected by ESA for the early Space Station utilization period will be operative in the time-frame 2001 - 2004 and they will be retrieved by the Space Shuttle after this period. The ISSA platform is also an interesting site for prototypes of solar instrumentation and studies of the degradation of optical components in space. The latter has restricted the capability of some past instruments, such as the UVSP on SMM.

Table 3
ISSA External Payload Accommodation Study
Solar Payload

		pointing stability	FOV
ARMS	Short/long term variations of solar irradiance	10″/5s	4° cone
EUDOSSO	Long term solar diameter variations	5″/1s	±2°
HREUV	High resolution spectroheliometer	1″/1s	40′ x 40′
MGS	Ultraviolet spectroscopy	2″/1s	2.5°
SOLSPEC	Solar spectral irradiance and Stability	±1°	6°
SOVIM	Solar constant helioseismic irradiance Radiometric measurements	10″/8s	10°

4.2 'SIMURIS' - Solar Interferometric Mission for Ultrahigh Resolution Imaging and Spectroscopy

SIMURIS is based on an innovative concept, that of solar inteferometry for ultrahigh resolution observations of the Sun. An international workshop was held in Paris in 1992 to present and discuss the objective of SIMURIS and to try to find a way to realise it (Solar Physics and Astrophysics at Interferometric Resolution, 1992).

It can obtain both high resolution imaging and spectroscopy. SIMURIS has also been proposed as a potential scientific payload of the Space Station, which offers large power, telemetry, mass capacity. SIMURIS would however require the use of the ESA Instrument Pointing System, a 2 m stabilized platform which reaches the arcsec stabilization, to accommodate the following ultra-high resolution instruments.

- **Solar Ultraviolet Network (SUN)**
  linear interferometer: 2 m baseline, four 20 cm telescopes
  - spatial resolution 0.01″ (10 km on the Sun) at 1216 Å
  - wavelength band access 1175–12000 Å (far–UV – near–IR)
  - 2D imaging 0.01″ × 0.1″ by rotation of the baseline (3 deg/s)

  FOV_{max} 30″× 30″
- **Imaging Fourier Transform Spectrometer (IFTS)**
  scanning Michelson interferometer
  - 2D spatially resolved spectrometry
    FOV 20″× 20″
  - independent spectrum for each spatial element
  - independent spatial and spectral resolution
  - visible/near-UV regions (1175–9000 Å)
  - magnetic and velocity maps at
    $\sim$ 0.1″ spatial resolution and 5 10^4 spectral resolution
- **Extreme Ultraviolet Telescopes**
  - 133–211 Å
  - FOV 5′× 5′
  - 0.1″ spatial resolution, 10 Å spectral resolution

5   Future Solar Missions from ISAS and IKI

5.1   ISAS Satellite Programme for Solar Studies

The ISAS solar satellite program has been developing with frequent launches of satellites devoted for solar research, with progressively more complex and sophisticated payloads, over the last two decades.

Solar missions in Japan started with the launch of Taijo in February 1975 with 86 kg of payload consisting of a Soft X-ray Flux Monitor and a UV Flux Monitor. The second mission was Tansei-4, launched in February 1980 with a 185 kg scientific payload, which included a Bragg Crystal Spectrometer. Hinotori was launched in February 1981 and contained a 188 kg payload with a Hard X-ray Telescope (bi-grid modulation collimator), a Bragg Crystal Spectrometer and a Wide-Band X-ray Spectrometer. Yohkoh, which is still in operation, was launched in August 1991 with a large payload of 420 kg containing a Hard X-ray Telescope (Fourier synthesis), a Soft X-ray Telescope (grazing incidence mirror), a Bragg Crystal Spectrometer and Wide-Band X-ray/γ-ray Spectrometers.

The scientific research targets of the more recent ISAS solar missions were the following:

- Hinotori studied high energy phenomena in solar flares and was able to provide images of hard X-ray flares.

Table 4
Possible Solar-B Instruments

– **Soft X–ray Telescope**

20 cm normal incidence telescope (multi–layer coatings)
2K × 2K pixels CCD (pixel size 1.25″)
FOV 42′
wavelength bands:

* Fe XIX 108.4 Å, 6.3 10^6K
* Fe XV 284.2 Å, 2.0 10^6K
* Fe XI 108.4 Å, 1.3 10^6K
* Fe IX 171.1 Å, 0.9 10^6K
* visible light

– **Optical Telescope/Magnetograph**

50 cm Gregorian Telescope (F/15)
observing wavelengths 3800–6600 Å
diffraction limit 0.2" at 5000 Å
2K × 2K pixels CCD (pixel size 0.1″–0.2″)
FOV 200″–400″
three optical paths:

* broad band filter
* universal birefringent filter with 0.1Å passband (monochromatic images, magnetograms, Dopplergrams; Fe I 6302, Fe I 5576)
* spectrograph for Stokes polarimetry (magnetic velocity fields)

– **XUV Spectrograph**

primary mirror multi–layer coated
concave grating multi–layer coated
wavelength range 250–290 Å
temperature range 10^5 – $2\ 10^7$ K
resolution 2″× 0.02 Å(20 km s^{-1})
FOV 2″× 500″

– Yohkoh also studied high energy phenomena in solar flares. One of the important new features has been the soft X–ray telescope which has revealed the surprisingly dynamic nature of the solar corona.

ISAS is contemplating a new solar mission which if approved will become Solar-B. This is conceived with the primary objective of investigating the magnetic coupling between the solar surface and the corona. The photosphere is viewed as the energy source (formation of flux tubes, dynamics of granular convection), which may be studied with an optical telescope and a magnetograph. The energy is then transported into the corona (by waves, magnetic reconnection, explosive events, EUV jets). Finally there may be energy release in the corona (waves, reconnection, interaction of coronal loops with emerging and/or ambient fields) which can only be studied effectively with X–ray imaging and XUV spectroscopy.

A possible instrument complement for Solar-B is summarized in Table 4. The most likely orbit would be a Sun-synchronous polar orbit at a height of 600 km. The scientific payload weight is envisaged at 300 kg.

5.2   IKI Satellite Programme for Solar Studies

The Russian solar programme is the series of Coronas satellites. The first Coronas satellite was launched in 1994. It includes amongst its instruments a soft X-ray imaging telescope, operating at wavelengths shorter than 300 Å; an optical photometer for helioseismology; a radiospectrometer; a hard X-ray spectrometer; and a γ-ray spectrometer. A full list of the Coronas instruments is given in Table 5. There are two more spacecraft in this series, namely Coronas-F and Foton. The satellites have a near polar orbit, with an inclination of $\sim83°$ which will enable the Sun to be viewed continuously for recurrent intervals of about 20 days.

Table 5
Coronas Instruments

Instrument	Acromym
Solar XUV Telescope	TEREK
Solar X-ray Spectrometer	RES-C
Diagnostics of Energy Sources and Sinks in Flares	DIOGENES
X-ray and Gamma-ray Spectrometer	HELIKON
Integral Radiation and Intensity Spectrometer	IRIS
Solar Ultraviolet Radiometer	SUVR
Vacuum Ultraviolet Solar Spectrometer	VUSS
Solar Optical Photometer	DIFOS
Solar Radio Spectrometer	SORS
Solar Cosmic Ray Spectrometer	SCR

The solar cosmic ray spectrometer (SCR in Table 5) is in fact a set of three instruments. The first detects gamma rays from 0.1 - 100 MeV, neutrons above 30 MeV and protons from 200-500 MeV. The second is a set of solid state detectors to study composition. The third is a particle detector sensitive to electrons from 0.5 - 12 MeV, protons from 1-200 MeV and alpha particles from 30-60 MeV/nucleon.

The Russian solar programme is headed by the Institute for Terrestrial magnetism, Ionosphere and Radio Wave Propagation (IZMIRAN) of the Russian Academy of Sciences.

6   Other Solar Missions: Developed or Being Built

6.1   TRACE - The Transition Region and Coronal Explorer

TRACE is a single-instrument telescope (Strong et al, 1994) which is scheduled to be taking observations by 1998. It will be launched into a Sun-synchronous polar orbit and its scientific objective is to study plasmas in the temperature range 10^4-10^7 K with $1''$ resolution. The principal characteristics of TRACE are given in Table 6.

Table 6
TRACE Characteristics

Primary Mirror Diameter	30cm
Effective focal length	8.66m
Pixel Size	0.5×0.5 acrsec
CCD size	1024×1024 pixels
Field-of-view	8.5×8.5 arcmin
Image stabilisation	± 0.1 arcsec
Instrument mass	44 kg
On-board memory	1 Gbit
Data rate	3.6 Gbits/day

It will aim at studying simultaneously all levels of the visible solar atmosphere, and to do this it must observe at a variety of different wavelengths. The optics therefore employs a combination of different UV and EUV multilayers and filters. The UV continuum is employed to study the photosphere and the temperature minimum region. C IV is used for the transition region. C I, H Lyα and Fe II are used to study the chromosphere, while the iron lines Fe XI, XII and XV are used to study the corona. The imaging system employs a CCD camera which allows TRACE to take images with a high cadence over an area of the Sun $8.5' \times 8.5'$. If subframes are used the cadence will go as high as every 0.2 s. Solar observations will be obtained for a period of around eight months/year.

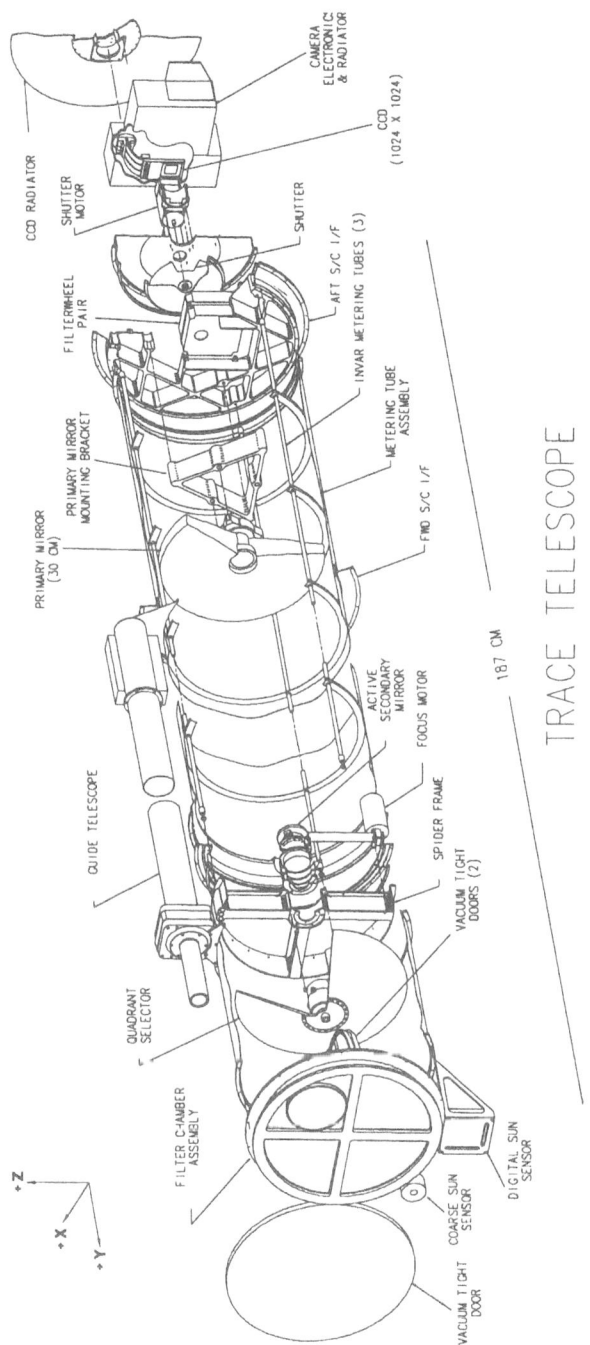

Fig. 1. A schematic view of the TRACE instrument

6.2   SPARTAN 201

This is a joint scientific payload developed by Goddard Space Flight Center, the High Altitude Observatory and the Smithsonian Astronomical Observatory. The scientific objectives are to investigate the physical properties of the corona and the base of the heliosphere. The payload consists of a white light coronagraph (Principal Investigator Dr R. Fisher), which gives information of the free electron concentration in the corona, and the Ultraviolet Coronal Spectrometer (Principal Investigator Dr J.L. Kohl), which gives information on the protons and some heavier ions such as O VI and can perform UV spectroscopy of the outer corona. It is deployed and recovered from the Space Shuttle. SPARTAN has been successfully flying in 1993, 1994 and 1995, and the next flight is scheduled for 1997. The September 1994 and 1995 flights were timed to coincide with the passage of the Ulysses spacecraft over the Sun's south and north poles, respectively. The 1997 flight is a collaborative one with the SOHO mission.

6.3   ACE – The Advanced Composition Explorer

The ACE spacecraft is a NASA mission designed to study solar energetic particles and galactic cosmic rays. It is scheduled for launch from Cape Canaveral, Florida in August, 1997 and it will be placed in a small-radius (158,000 km) "halo" orbit about the L1 Lagrangian point. It will be 10 to 1000 times more sensitive than past instrumentation to study energetic charged particles. The mission duration is planned to be 2-5 years, so that it should provide coverage over most of the forthcoming solar maximum.

Specific ACE science goals include a detailed study of the elemental and isotopic composition of solar particle emissions ranging from the solar wind to the output from major flares. Topics that should be addressed include the origin of the solar wind, the acceleration of solar energetic particles and the formation of the solar corona. It will provide some advance warning of geomagnetic storms which might be about to hit the Earth.

6.4   SMEI – The Solar Mass Ejection Imager

This mission is an all-sky photometer (Jackson et al, 1996) being developed by the University of Birmingham and the University of California, San Diego for the United States Air Force. The Principal Investigator is Dr S.L. Keil. The instrument consists of three identical fan beam photometers each $3° \times 60°$ mounted on a 3-axis stabilised earth-orbiting spacecraft such that they point away from the Earth and cover the sky from horizon to horizon in a 3-degree swath. Thus once an orbit the full 4π sr field will be sampled. The stray light is reduced to less than 10^{-13} of the direct solar signal and the photometric images are recorded by CCD cameras. As the full sky includes the Sun, when the viewing direction is within $\sim 18°$ of the Sun a shutter is used to shield the CCD cameras. The desired signal is that arising from the detection of Thomson-scattered sunlight off the electrons within a CME.

The primary objective of SMEI is to provide an accurate early warning that a CME will hit the Earth. It should be able to make this announcement when the CME is at around 0.5 AU, thus providing over 1 day's notice. This will be a significant improvement on data gathered on CMEs from spacecraft like SOHO and ACE which are at the L1 point. It is planned to fly on a spacecraft with the USAF Space Test Program in 2000.

7   Other Missions Which Are Still in Definition

7.1   High Energy Solar Spectroscopic Imager (HESSI)

The HESSI investigations during next solar maximum ($\sim$2000) are focused on the study of the energy release in solar flares, and the physics of particle acceleration during flares. It aims to achieve this by taking high resolution images of flaring regions from 2 keV up to gamma-ray energies. A consortium led by Dr R.P. Lin has been trying for some years to obtain funding for this project, and consequently the instrument capabilities keep being adjusted to match the mission constraints. Currently there is some hope that it could start at the MIDEX Minimum Science Level in the spring of 1997. If that doesn't happen, a descoped version will be submitted in response to the 1997 SMEX Announcement of Opportunity.

HESSI uses Fourier-transform imaging with 7 to 9 bi-grid modulation collimators and cooled germanium detectors. This is a reduction from the 12 collimators baselined in the original 1996 MIDEX proposal. The original imaging energy response out to $\sim$400 keV requires the production of thick fine grids for the collimator. The production of the thick fine grids is no longer practical, and so the 2-arcsecond imaging will only be possible from 2 keV up to about 50 keV. Coarser imaging (20 arcseconds) up into the gamma-ray line region should still be possible. Because of the thinner grids, the high-resolution gamma-ray spectroscopy will have similar sensitivity to that of the original 12 detectors. The time resolution will be in the tens of milliseconds region, and HESSI will have $\leq$ 1 keV energy resolution.

7.2   The Chinese Space Solar Telescope

The Space Solar Telescope was proposed in 1992. After a proving flight of a prototype balloon payload it is planned to launch a 2.5 ton spacecraft into a polar, Sun-synchronous orbit in the timeframe 2001-2002. This will therefore encompass the approaching solar maximum. The anticipated lifetime will be 3-5 years. The primary instrument will be a 1-m aperture optical telescope, complete with a spectrometer. This will be able to make two-dimensional spectra and will also determine magnetic and velocity fields. In addition the payload will include imaging telescopes in Hα, and the UV, soft X-ray and hard-X-ray regions of the spectrum.

References

J.M. Davila, 1994, Ap. J., 423, 871.

Jackson, B.V., A. Buffington, P.L. Hick, S.W. Kahler, R.C. Altrock, R.E. Gold, and D.F. Webb, "The Solar Mass Ejection Imager", in Solar Wind Eight, AIP Conf. Proc. 382 (eds. D. Winterhalter, J.T. Gosling, S.R. Habbal, W.S. Kurth, and M. Neugebauer), pp. 536-9, AIP Press, NY, 1996.

K.T. Strong, M. Bruner, T. Tarbell, A. Title and C.J. Wolfson, 1994, Space Sc. Rev., 70, 119

The SOHO Mission, (ed. B. Fleck, V. Domingo and A.I. Poland), 1995, Solar Phys., 162.

The Yohkoh (Solar-A) Mission, (ed. Z. Svestka and Y. Uchida), 1991, Solar Phys., 136.

Solar Physics and Astrophysics at Interferometric Resolution, (ed. L. Dame and T-D Guyenne), 1992, ESA SP-344

Appendix

The following lists the acromyms used in this article.

ACE	Advanced Composition Explorer
BATSE	Burst And Transient Source Experiment
CDS	Coronal Diagnostic Spectrometer
CME	Coronal mass ejection
EIT	Extreme-Ultraviolet Imaging Telescope
ESA	European Space Agency
EURECA	EUropean REtrievable CArrier
EUV	Extreme Ultra-Violet
FOV	Field-Of-View
GOES	Geostationary Operational Environmental Satellite
HESSI	High Energy Spectroscopic Imager
ISSA	International Space Station Alpha
LASCO	Large Angle Spectroscopic COronagraph
MAU	Million Accounting Units
MIDEX	Middle-sized Explorer
NASA	National Aeronautics and Space Administration
OSO	Orbiting Solar Observatory
SIMURIS	Solar Interferometric Mission for Ultrahigh Resolution Imaging and Spectroscopy
SMEI	Solar Mass Ejection Imager
SOHO	Solar and Heliospheric Observatory
STEREO	Solar-TERestrial Event Observer
SUMER	Solar Ultraviolet Measurements of Emitted Radiation
TRACE	Transition Region And Coronal Explorer
UVCS	Ultraviolet Coronagraph Spectrometer
UVSP	Ultraviolet Spectrometer and Polarimeter
XUV	X- and Ultraviolet

Lecture Notes in Physics

For information about Vols. 1–455
please contact your bookseller or Springer-Verlag

New Series m: Monographs